# Springer Proceedings in Complexity

Springer Proceedings in Complexity publishes proceedings from scholarly meetings on all topics relating to the interdisciplinary studies of complex systems science. Springer welcomes book ideas from authors. The series is indexed in Scopus. Proposals must include the following: - name, place and date of the scientific meeting - a link to the committees (local organization, international advisors etc.) - scientific description of the meeting - list of invited/plenary speakers - an estimate of the planned proceedings book parameters (number of pages/articles, requested number of bulk copies, submission deadline). Submit your proposals to: christoph.baumann@springer.com

More information about this series at http://www.springer.com/series/11637

Şefika Şule ERÇETİN • Şuay Nilhan AÇIKALIN
Editors

# Chaos, Complexity and Leadership 2018

## Explorations of Chaotic and Complexity Theory

Springer

*Editors*
Şefika Şule ERÇETİN
Hacettepe University
Ankara, Turkey

Şuay Nilhan AÇIKALIN
Middle East Technical University
Ankara, Turkey

ISSN 2213-8684 ISSN 2213-8692 (electronic)
Springer Proceedings in Complexity
ISBN 978-3-030-27671-3 ISBN 978-3-030-27672-0 (eBook)
https://doi.org/10.1007/978-3-030-27672-0



This Springer imprint is published by the registered company Springer Nature Switzerland AG.
The registered company address is: Gewerbestrasse 11, 6330 Cham, Switzerland

*You are like the fire of the volcanoes flowing into my heart*
*You are the abundance of clouds raining into my eyes*
*You are one of the greatest miracles I have ever witnessed*
*Always choose the beauties like your heart and yourself*
*Let them through into your lives*
*With everlasting love*
*To my daughters and grandchildren*

**Prof. Dr. Şefika Şule ERÇETİN**

*To my mom who is my hero, best friend, and light of my life*

**Şuay Nilhan AÇIKALIN**

# Preface

By 2018, we finalized the sixth series of the International Symposium on Chaos, Complexity, and Leadership. There is no doubt six years is a very short period of time in science. Just a few days ago, before those sentences were written, scientists revealed the first image of the black hole in the world history that has fascinated us with its beauty and uniqueness. Science and scientific developments are like whirlwind regarding its density and consequences. In such an inspirational scientific environment, studies of chaos and complexity cannot be exception. Although chaos and complexity studies have had intense literature in physics, biology, and somehow mathematics, through the years, it became one of the focal areas in social sciences as well. In 2012, when we first organized this conference and published book, we had the aim to open new discussions in different fields with chaos and complexity studies because we believed that our world needs more than linear and traditional approach. Today, we feel honored to see how our discussions have brought many senior and young scholars together from various fields all over the world.

As well as scientific developments, the changing nature of our world in social and political realm exposes new questions. The 2000's motto of "globalized world" is replaced by the notion of "nonlinear world" or "networking world." The reader will find the unique and interesting chapters that lighten up how social phenomenon could be understood from the perspective of chaos and complexity studies. Also, this book includes exploration of leadership from a nonlinearity perspective and contains a practical application of theoretical concepts to management and leadership and an analysis of chaotic and complex systems from scientific, quasi-scientific, and social science fields. In addition to this, the book will illuminate current research results and academic work from all branches of science, mathematics, physics, education, economics, political science, statistics, and management sciences. The glaring gist thereof shall be an exploration of chaotic and complex systems, as well as chaos and complexity theory in lieu of their applicability.

As Francis Crick says, "In nature, hybrid species are usually sterile, but in science the reverse is often true. Hybrid subjects are often astonishingly fertile, whereas if a scientific discipline remains too pure it usually wilts." That's why, this book presents itself as a serious and excellent discussion pertaining new approaches to chaos, complexity, and leadership from the interdisciplinary point of view. It is extremely important that this book will provide an essential reference source to improve current literature in related fields and provide newer ideas for future developments.

In our last words, we would like to thank our contributors for their vital contributions and our publishing house Springer for their efforts. We hope you will enjoy the book.

Ankara, Turkey

Şefika Şule ERÇETİN
Şuay Nilhan AÇIKALIN

# Contents

**Foreign Policy in a 'Networked World': Exploring Britain's Response to the Arab Uprisings** .......... 1
Mark Garnett and Simon Mabon

**A New Method in the Analysis of Chaotic Systems: Scale Index** .......... 21
Nazmi Yılmaz, Mahmut Akıllı, and K. Gediz Akdeniz

**Reminiscence of Alija Izetbegovic and His Leadership** .......... 27
Bakir Sadović

**Some Conceptual and Measurement Aspects of Complexity, Chaos, and Randomness from Mathematical Point of View** .......... 33
Fikri Öztürk

**Relationships Between Stock Markets: Causality Between G8 Countries and Turkey** .......... 67
Kamil Demirberk Ünlü, Nihan Potas, and Mehmet Yılmaz

**The Color Revolutions in the Former USSR Countries, Viewed in the Light of Chaos Theory** .......... 77
Özlem Demirkıran

**Brexit in the Light of Chaos Theory and Some Assumptions About the Future of the European Union** .......... 87
Erjada Progonati

**Intra-Specific Competition in Prey Can Control Chaos in a Prey-Predator Model** .......... 97
Md Saifuddin and Santanu Biswas

**Angela Merkel's Chancellor Democracy and Leadership in Times of Crisis** .......... 107
Başar Şirin

**The Achilles' Heel of Strategic Management: Strategic Leadership in a Chaotic Environment** ........ 123
Halil İbrahim Özmen

**Behaviours of Error-Prone Variables on Low-Chaotic Autoregressive Models** ........ 137
Şahika Gökmen and Rukiye Dağalp

**The Energy Policy of the European Union and Its Implications for Turkey** ........ 145
Ahmet Atilla

**Customer Perception of the Quality of Online Banking Services (with Special Reference to SBI and ICICI): A Study on Chaos and a Complexity Perspective** ........ 155
Niyasha Patra and Nilanjan Ray

**Insightful Leadership** ........ 175
Halime Güngör

**The Role of Trust in Principals in Readiness for Change Within Schools** ........ 181
Hanifi Parlar, Mahmut Polatcan, and Ramazan Cansoy

**A Chaos Theory Perspective on Migration of Sub-Saharan Africans to Europe** ........ 193
Mahamadou Yahaya

**Chaos in the Future: Artificial Intelligence as a Strange Attractor of the Future** ........ 203
M. Şahin Bülbül

**Financing Higher Education in Sub-Saharan Africa: A Proposed Model Based on the Experiences of Ugandan Higher Education Institutions and Exemplary Practices from the Asian Tigers** ........ 211
Miiro Farooq and Ssekamanya Siraje Abdallah

**Autophagic Leadership: Promoting Self-Renewal of Organizations** ........ 227
Mustafa Özmusul

**A Qualitative Research Analysis of Chaotic Circumstances Affecting the Happiness of Teachers** ........ 233
Deniz Görgülü

**Analysis of Organizational Memory in the Context of School Administrators** ........ 249
Abdullah Balıkçı

**A Way for Organizations to Cope with Uncertainty: Mimetic Isomorphism** ........ 271
İnci Öztürk

**Index** ........ 285

# Foreign Policy in a 'Networked World': Exploring Britain's Response to the Arab Uprisings

**Mark Garnett and Simon Mabon**

**Abstract** This chapter uses the concept of the networked world to engage with British foreign policy after the Arab Uprisings. Many have accused London of conducting a contradictory foreign policy, underpinned by hypocritical claims about supporting democracy and the rule of law, yet when the opportunity to do this emerged, it appeared that Britain was much more concerned with maintaining the stability of allies, seemingly whatever the political – and human – cost. In Bahrain, Britain maintained support for its long-standing ally, the Al Khalifa, despite egregious human rights violations, whilst in Libya, support was provided to opposition groups, seemingly without consideration of the long-term implications. Despite this apparent contradiction and suggestion of double-standards, we argue that there is a degree of coherence within British strategy. In this chapter, we argue that to understand the trajectory of British foreign policy at this time, we need to consider the concept of the networked world, which featured prominently within the foreign policy agenda of the coalition government. We begin by considering this concept of the networked world, placing it within the context of a Conservative-led coalition, which shaped the character of the network. We then turn to a consideration of the cases of Bahrain and Libya, which provide rich scope for analysis of contrasting responses to the popular protests of the Arab Uprisings.

**Keywords** Foreign policy · Networked world · Britain's response · Arab uprisings

M. Garnett (✉) · S. Mabon
Lancaster University, Lancaster, UK
e-mail: m.garnett@lancaster.ac.uk; s.mabon@lancaster.ac.uk



Ş. Ş. ERÇETİN, Ş. N. AÇIKALIN (eds.), *Chaos, Complexity and Leadership 2018*, Springer Proceedings in Complexity, https://doi.org/10.1007/978-3-030-27672-0_1

## Introduction

In his first major speech after taking office as Foreign Secretary in Britain's 2010–15 coalition government, William Hague spoke of a changing world in which the concept of 'networks' was increasingly important:

> Today, influence increasingly lies with networks of states with fluid and dynamic patterns of allegiance, alliance and connections, including the informal, which act as vital channels of influence and decision-making and require new forms of engagement from Britain.[1]

Elaborating on the theme, Hague referred to the new ease of communication between foreign ministers but also explained what he meant by 'informal' networks:

> There is now a mass of connections between individuals, civil society, business, pressure groups and charitable organizations which are also part of the relations between nations and which are being rapidly accelerated by the internet.... If the increasingly multipolar world already means that we have more governments to influence and that we must become more active, the ever accelerating development of human networks means that we have to use many more channels to do so, seeking to carry our arguments in courts of public opinion around the world as well as around international negotiating tables.[2]

Hague clearly hoped that his speech, delivered in the grand surroundings of the Locarno Room at the Foreign and Commonwealth Office, would be widely reported as the coalition's foreign policy 'mission statement' and that the concept of a 'networked world' would receive considerable attention from media commentators. This proved to be the case; and although the speech was not given a unanimous welcome, the idea of a 'networked world' was at least taken seriously.[3] Far from being a soundbite crafted for a specific occasion, the theme of a 'networked world' could be detected in many of Hague's speeches prior to 2010 and was highlighted in a pamphlet written by the former Shadow Foreign Secretary Michael Ancram and published shortly after the coalition came to office.[4]

After the Conservatives and the Liberal Democrats had formed a government, they were faced with the legacy of previous decisions, notably relating to Afghanistan. In unveiling a new approach to international politics, Hague was, at least in part, hoping to show that the appropriate lessons had been learned from these perceived mistakes. In practice, however, the coalition government can be accused of making mistakes of its own, at least some of which seemed to emanate from the kind of thinking which inspired the most egregious errors of the Blair

---

[1] Foreign & Commonwealth Office and William Hague, *Britain's Foreign Policy in a Networked World*, (01.07.10) Available at: https://www.gov.uk/government/speeches/britain-s-foreign-policy-in-a-networked-world%2D%2D2.

[2] Ibid.

[3] See for example, Tim Dunne, *A foreign policy for the 17th century,* (The Guardian, 06.07.10) Available at: https://www.theguardian.com/commentisfree/2010/jul/06/foreign-policy-william-hague-bilateralism.

[4] Michael Ancram, *Farewell to Drift: A New Foreign Policy For A Networked World,* (2010) Available at: http://www.globalstrategyforum.org/wp-content/uploads/Farewell-to-Drift-.pdf.

Government. On the basis of a recent article, this outcome could be predicted, since the 'networked world' was an unsatisfactory basis for a new approach to foreign policy.[5] Indeed, Philip Leech and Jamie Gaskarth conclude their analysis with the comment that 'networked foreign policy, arguably, is no policy at all'.

This chapter uses the concept of the networked world to engage with British foreign policy after the Arab Uprisings. Many have accused London of conducting a contradictory foreign policy, underpinned by hypocritical claims about supporting democracy and the rule of law, yet when the opportunity to do this emerged, it appeared that Britain was much more concerned with maintaining the stability of allies, seemingly whatever the political – and human – cost. In Bahrain, Britain maintained support for its long-standing ally, the Al Khalifa, despite serious human rights violations, whilst in Libya, support was provided to opposition groups, seemingly without consideration of the long-term implications. Despite this apparent contradiction and hypocrisy, we argue that there is a degree in coherence within British strategy. In this chapter, we argue that to understand the trajectory of British foreign policy at this time, we need to consider the concept of the networked world, which featured prominently within the foreign policy agenda of the coalition government. We begin by considering this concept of the networked world, placing it within the context of a Conservative-led coalition, which shaped the character of the network. We then turn to a consideration of the cases of Bahrain and Libya, which provide rich scope for analysis of contrasting responses to the popular protests of the Arab Uprisings.

In focusing on Libya and Bahrain, this chapter follows the approach of Leech and Gaskarth, whose analysis devotes considerable attention to these instructive examples. It is not part of our purpose to defend the coalition's foreign policy record in practice, either in relation to the 'Arab Uprisings' or to its responses to other key issues, such as Russia's aggressive behaviour towards Ukraine. British involvement in Libya was a miscalculation which bears comparison with Tony Blair's Iraq adventure; by contrast, the policy adopted towards Bahrain was defensible, but its presentation in the region and beyond was maladroit at best.

Our argument, rather, is that Britain's mistakes between 2010 and 2015 cannot be attributed to William Hague's notion of a 'networked' foreign policy, which in reality was a plausible approach, which furnishes a satisfactory explanatory framework for Britain's responses to the various manifestations of the Arab Uprisings. This conclusion is suggested when one moves from the essentially *quantitative* analysis presented by Leech and Gaskarth to a more holistic understanding of networks which incorporates *qualitative* elements – i.e. networks as they present themselves in the minds of foreign policy decision-makers, including the assumption of shared understandings which arise from historical interactions. The reasons for British foreign policy failures in relation to the Arab Uprisings lie elsewhere, most notably

[5] Philip Leech and Jamie Gaskarth, 'British Foreign Policy and the Arab Uprisings', *Diplomacy & Statecraft*, 26:1 (2015) pp139–160.

in the coalition's desire to bring a mixture of 'Realism' and 'liberalism' to bear on its foreign policy decisions.

## A 'Networked World'?

The argument developed by Leech and Gaskarth rests heavily on detailed statistical evidence relating to Britain's bilateral relations with specific states, comparing these data to Britain's varied responses to dramatic events beginning in Tunisia in December 2010 – the episode which is now known as the 'Arab Uprisings'.[6] On this basis, they argue that 'the nature of the security network in each case correlates more closely with the British government's response to each situation than the economic or societal networks, implying that security priorities more than economic or social pressures shaped British policy towards the region'.

In itself, this conclusion hardly suggests that Hague's 'networked world' was an inadequate basis for the making of British foreign policy. At most, Leech and Gaskarth seem to be saying that, in practice, the Coalition's approach towards those states affected by the Arab Uprisings was mainly shaped by security considerations. In the post-9/11 context, this outlook could be criticized as unimaginative, but it could scarcely be classed as irrational. In any case, it can be argued that Leech and Gaskarth arrive at their conclusions on the basis of quantifiable factors – security and trade – and that these are characteristic of what might constitute a Realist approach to international relations.

The premise of the present chapter is that bald statistics can only form part of the policy-making considerations arising from interactions within a 'networked world' and that we must also take this broader understanding of networks into account when engaging with questions about foreign policy. Connections exist within a network between different nodes – individuals, civil society, government, and businesses – and it is the strength of these connections that determines the vitality of networks. From the perspective of a state (like Britain) seeking to operate within a networked world, connections can rarely be established at will. Crucially, they depend upon a sense of familiarity and empathy which ordinarily arise from prolonged interactions. In short, in all diplomatic activity 'a bird in the hand is worth two in the bush' – when policy actors are fairly confident that agreements of various kinds will be honoured, such exchanges will assume a degree of importance which cannot be conveyed by statistics alone.

Although the word 'networks' is relatively new in International Relations, the importance of new information technologies was appreciated very quickly. In referring to transgovernmental co-operation, Joseph Nye and Robert Keohane alluded to 'sets of direct interactions among sub-units of different governments that are not controlled or closely guided by the policies of cabinets or chief executives of

[6]Ibid.

those governments'.[7] From this perspective, what came to be called 'networks' took on a subversive character. It was entirely possible, however, that governments could participate in these informal networks, and manipulate them to their own advantage. Informal networks are impervious to quantifiable analysis and have become a more potent factor within international relations, particularly so since the technological advancements of the internet age.

As Manuel Castells stresses, the networked society transcends historical social connections as a consequence of technology, creating the scope for the development of new types of networked interactions.[8] The same issues apply within the diplomatic sphere, as the nature of diplomatic relations takes on a new technological – and public facing – dimension to complement existing formal networks. The informal nature of international exchanges allows states to project a positive image of themselves by being *seen* to be acting either rationally or righteously, in a nod towards soft power.[9] Networks, then, occur formally and informally, shaped by governments as well as people across the world, offering states the potential to influence domestic opinion as well as affecting the views of an audience which extends far beyond national boundaries.

In order to understand the type of network in operation, one must consider a range of different factors. To begin, it is necessary to consider the domestic context within which actors participate, which includes their ideological positions, facilitating the development of pre-existing networks and the creation of new linkages. One must then explore the varying fortunes of networks over time, which allows one to ascertain the strength of the linkages. This approach also allows us to trace change within the context of a particular network and the broader international system. Thus, to isolate individual 'nodes' within the network is problematic.[10] Ultimately, one must consider the temporal dimension to a network, which should not be seen to be *a-historic*; rather, networks develop – and evolve – over time and a static understanding of networks – a snapshot, particularly one focused upon statistics – is of limited use in the making of foreign policy. One should not forget that attempting to shape networks can have repercussions beyond causal relationships and indeed, beyond state borders, and in an increasingly networked world, change in one network may result in change within another.

[7]Robert Keohane and Joseph Nye, 'Transgovernmental Relations and International Organizations', *World Politics* 27:1 p43.

[8]Manuel Castells, *The Rise of the Networked Society* (Oxford: Blackwell, 1996).

[9]Joseph Nye, *Soft Power: A Means to Success* (Cambridge, MA: Peresus, 2004).

[10]Castells, Op. Cit.

## 'Liberal Conservatism' and the Arab Uprisings

Despite the suspicions of Gaskarth and Leech, Hague was not merely using the concept of a networked world as an eye-catching theme for his 2010 keynote address. Rather, he was responding to a phenomenon which had been recognized by many scholars of international theory. Foreign policy success depends, to an increasing extent, on the ability of a state to utilize networks across a range of institutions and actors. Part of Hague's purpose was thus to argue that Britain must maintain its existing networks – both formal and informal – and develop new ones when appropriate. Such a conclusion meant that the Foreign and Commonwealth Office (FCO) should be protected from the full impact of government spending cuts in an era of economic austerity.

However, an awareness of the importance of networks is not a *sufficient* condition of foreign policy success. An extensive range of connections with governments and non-state actors across the world should ensure better-informed decisions along with increased capacity to achieve aims, but these outcomes are also shaped by normative considerations in a range of locations. In other words, governments are not merely passive recipients of information gleaned from networks; they filter and prioritize that information in keeping with pre-existing interpretive frameworks, so that (for example) they will tend to pay more attention to networks which convey messages that coincide with their stated values.

Before taking office in May 2010 as part of a coalition with the Liberal Democrats, the Conservatives had espoused a 'liberal Conservative' approach to foreign policy. It would be liberal because Britain should support human rights and champion 'the cause of democracy and the rule of law at every opportunity'; but it would be Conservative because it would be 'hard headed and practical'. The party's manifesto pledged that 'We will work patiently with the grain of other societies, but we will always support liberal values because they provide the foundations for stability and prosperity'. Such a position reflected the prominence of normative issues within the global arena whilst accepting the limitations of one's ability to achieve them. In essence, it was difficult to distinguish from the Blair Government's much-vaunted 'ethical foreign policy', decontaminated by the media 'spin' which depicted the latter approach as an aspiration to make morality the *primary* factor in decision-making

This approach clearly reflected William Hague's own views, which he outlined in several speeches as Shadow Foreign Secretary (2005–2010).[11] In July 2009, for example, he delivered a major speech which made explicit reference to the Middle East and the Gulf. Britain, he argued, 'should embark on the elevation of its links' in these areas, for political and economic reasons. He went on to address likely criticisms from those who would point out that some of these states 'do not conform

[11] See, for instance 'The future of British foreign policy', speech of 21 July, 2009, Available at: http://conservative-speeches.sayit.mysociety.org/speech/601323.

to our own democratic and liberal values'.[12] Clearly taking aim at the foreign policy of the Blair Government, he argued that Britain should not try 'to prescribe the form of government in all the countries with whom we need friendly relations':

> British leaders will rightly always argue that democracy and freedom are the soundest basis for national security and international peace....Yet in foreign policy idealism must always be tempered with realism: even those countries like many of the Gulf States, which are making democratic reforms, will do so at varying paces and sometimes over an extended period.

Despite Conservative reliance on Liberal Democrat support in order to form a government in 2010, the general policy approach signalled by Hague remained unaffected and was restated in his 2010 'networked world' speech. Indeed, the 'liberal Conservative' formula seemed almost to have been designed in the expectation of a coalition. However, it remained to be seen whether it would provide a satisfactory basis for decision-making. The tensions between 'idealism' and 'realism' – the source of so much contention among scholars of International Relations – could not be transcended by a verbal formula. In practice, foreign policy actors in the UK would be presented with dilemmas which would force them to choose between the need to promote 'democratic and liberal values' and the preservation of stability. Such tensions were identified by Hague before taking office. By using the Gulf States as an example in his speech of July 2009, Hague identified this as one area in which the blend of approaches could be sustained – by urging the case for democracy, but not in a way which might destabilise existing regimes. As we shall see, however, in this case, it proved impossible for British policy to be 'liberal' and 'Conservative' at the same time, and here as elsewhere, the government's decisions were prompted by the strength of its networks.

## Bahrain and the 'Networked World'

On 17 December 2010, Mohammad Bouazzizi, a Tunisian street vendor, self-immolated and, in doing so, triggered an outbreak of unrest across the region.[13] Bouazzizi took the action out of desperation at economic conditions, a dearth of political accountability and ultimately, a lack of hope. Inside a month, the authoritarian Tunisian President Ben Ali had been toppled. Ideas of revolution spread quickly, across North Africa to the Arabian Penninsula. Ideas of *isqaat* and *malakiyyah destouriyyah* featured prominently across protest movements, as people took to the streets, challenging what appeared to be antiquated ideas of legitimacy and monarchical political organization. Shaped by differing contexts, the demands of opposition groups included the abdication of monarchs and the removal of corrupt ministers.

---

[12] Ibid.

[13] See Katerina Dalacoura, 'The 2011 Uprisings in the Arab Middle East: political change and geopolitical implications' *International Affairs*, 88:1 (2012) p72.

In the Gulf states, there were symptoms of instability in Bahrain and Kuwait and in the Eastern Province of Saudi Arabia. In Oman, the Sultan reshuffled his cabinet in response to protests. However, the most serious challenge to established governments in the Gulf occurred in Bahrain, where the House of Khalifa ruled a population with a Shi'a majority.[14] On 14 February 2011 – 3 days after the Egyptian President Hosni Mubarak was forced from office – protestors were attacked by security forces using tear gas and rubber bullets. In response, on the following day thousands of protestors, mainly but not exclusively Shi'a, marched to the Pearl Roundabout in the centre of the capital, Manama, and occupied it.

Asked to make a statement on these events in the House of Commons, the Foreign Secretary William Hague said that he had spoken to his Bahraini counterpart, Khalid bin Ahmed Al Khalifa, stressing the need for restraint and respect for human rights. In particular, Hague told Al Khalifa that Britain strongly opposed 'any interference in the affairs of Bahrain by other nations' – a clear reference to neighbouring states Saudi Arabia and Iran[15] – 'or any action to inflame sectarian tensions between Bahrain's Sunni and Shi'a communities'. The penetration of regional politics and security by both Riyadh and Tehran had destabilised the Middle East and, given Bahrain's location, the spectre of external involvement loomed large.

The Bahraini uprising would have been a serious issue for any British Foreign Secretary, threatening the stability of a long-standing Gulf ally. As a Foreign Affairs Committee report into the UK's relations with Saudi Arabia and Bahrain (henceforth FAC report) noted, the 'UK-Bahrain relationship is one of the UK's oldest and closest bilateral relationships in the Gulf'.[16] The first treaty between Britain and Bahrain was signed in 1820, and in 1868, Bahrain became a British Protectorate, a status which it retained until Britain's withdrawal from East of Suez in 1971.[17] In 1935, following the discovery of oil reserves, the British navy's Middle Eastern Command was relocated from Iran to Bahrain and remained there until 1971.

Britain's ties to Bahrain were a potential source of political embarrassment in February 2011, since the government had recently agreed to supply the Al Khalifa regime with riot control equipment, including tear gas, amidst allegations that such weapons were used in crushing protestors. Britain therefore stood accused of supporting an unsavoury regime which was about to be swept aside by the forces of democratic change.

The situation in Bahrain was especially problematic for William Hague since he had just returned from a visit to the Kingdom. Furthermore, the Gulf had featured heavily in his 'networked world' speech of July 2010. Indeed, he had used Bahrain itself to illustrate the importance of networks, revealing that he and

---

[14]See: Simon Mabon, 'The Battle For Bahrain', *Middle East Policy* 29:2 (2012).

[15]Ibid.

[16]House of Commons Foreign Affairs Committee, *The UK's relations with Saudi Arabia and Bahrain,* (2013) p15.

[17]See: Jeffrey Pickering, *Britain's Withdrawal from East of Suez* (Basingstoke: Macmillan, 1998).

the Bahraini foreign minister 'follow each other avidly on Twitter'. But Hague's vision of a networked world was one in which 'relations between states are no longer monopolised by Foreign Secretaries or Prime Ministers'.[18] Labour's Denis McShane asked Hague if he had made contact with any representatives of the opposition on his recent visit and jeered that before taking office, current British ministers had benefited from 'a regular gravy train' of visits to Bahrain, 'paid for by the rulers of the statelet'.[19]

Hague was able to reply that he had met 'a variety of opposition human rights organisations, including the Bahrain Human Rights Society, the Migrant Workers Protection Society and the Bahrain Women's Union'.[20] As for visits to Bahrain, he contended that it was entirely appropriate to engage in dialogue with senior figures within the regime: 'In fact, on every occasion when I went there in the last five years, Ministers of the previous Government were there at the same time'.

Such retorts, although effective debating points making it more difficult for the Labour Opposition to derive political capital from the crisis, did little to shore up the case for Hague's 'networked world'. Neither the regular contact between the governments, nor the dialogue with opposition groups, had alerted Hague and his colleagues to the imminent prospect of serious unrest. In a memorable passage of his July 2010 speech, Hague had declared that 'The country that is purely reactive in foreign affairs is in decline'.[21] Even in the case of Bahrain – where Britain's networks were deep and broad yet upon reflection were largely in the realm of high politics – developments had forced the country to be largely (if not 'purely') reactive. Nevertheless, when the British government was forced to react, its long-established networks ensured that it could convey its views directly to senior Bahraini officials but not across the influential Shi'a villagers. This reflects the fact that, unlike the static, statistical view of networks presented by Gaskarth and Leech, Britain's intimate ties with Bahrain were reinforced by longevity. Over the decades, the two states had developed a high degree of *empathy* which could certainly be strained at a time of crisis but was unlikely to dissipate overnight.

Thus the concept of 'networks', when properly understood, does help to explain Britain's attitude at the onset of the crisis in Bahrain. It can be acknowledged, however, that an appreciation of the importance of effective communication with elements of civil society rather than governments does not in itself constitute a foreign policy: rather, it refers to a *method* by which foreign policy should be conducted. William Hague's speech gave a clear signal of the principles which would inform Britain's dealings with the 'networked world', and arguably this is the respect in which the government is open to serious criticism.

Addressing the Kuwait National Assembly a week after the violence in Bahrain, Prime Minister David Cameron articulated this new approach:

---

[18] FCO and Hague, Op. Cit.

[19] Denis McShane, *House of Commons Debates*, Vol. 523, col. 1136, 11 February 2011.

[20] William Hague, *House of Commons Debates*, Vol. 523, col. 1137, 11 February 2011.

[21] FCO and Hague, Op. Cit.

> For decades, some have argued that stability required highly controlling regimes and that reform and openness would put that stability at risk. So, the argument went, countries like Britain faced a choice between our interests and our values. And to be honest, we should acknowledge that sometimes we have made such calculations in the past. But I say that is a false choice.
>
> As recent events have confirmed, denying people their basic rights does not preserve stability, rather the reverse. Our interest lies in upholding our values in insisting on the right to peaceful protest, in freedom of speech and the internet, in freedom of assembly and the rule of law. But these are not just our values, but the entitlement of people everywhere; of people in Tahrir Square as much as Trafalgar Square.
>
> So whenever and wherever violence is used against peaceful demonstrators, we must not hesitate to condemn it... *political and economic reform in the Arab world is not just good in its own right* but it's also a key part of the antidote to the extremism that threatens the security of us all.[22]

The difference was subtle but significant. Morality and 'hard headed' calculations of national interest still shared a direction of travel, but morality was clearly ahead.

From this perspective, one can readily perceive why the Gulf – and Bahrain in particular – was a favourite example for Hague. The empathic relationship with Bahrain, sustained by regular contacts with the ruling family and even with selected dissident groups, fostered confidence that this state would move towards democracy in graduated steps 'over an extended period'. The potential payoffs were economic as well as political/strategic; Bahrain is the smallest yet fastest growing export market for Britain in the Gulf, reflected in a 39% increase in trade between 2009 and 2012,[23] with further scope for improvement. Defence equipment represented a significant proportion of British exports to Bahrain, and this would be difficult to sustain unless the regime was seen to be engaged in a process of democratic reform, however 'gradual'. Ironically, of course, a successful programme of reform would reduce the need for the tools of repression which Bahrain habitually acquired from Britain; thus a gain for democracy was likely to come at a cost for Britain's crucial armaments industry.

Cameron praised the government of Bahrain for withdrawing its security forces and promising to embark on 'a broad national dialogue'. Indeed, the Bahraini regime had launched an independent inquiry into the events of 14–15 February. However, the unrest and the repression were not over. Days after Cameron's Kuwait speech, more than 100,000 demonstrators marched to the Pearl Roundabout; 25 February was designated a 'national day of mourning' and was marked by further demonstrations. By 14 March, the Gulf Co-operation Council had decided to intervene, and a force including 1000 Saudi troops was despatched to Bahrain. This was precisely the kind of 'external interference' which Hague had warned against in February. On 16 March, when the government declared a three-month state of emergency, there was a new flurry of phone calls, with Cameron telling the British-educated King Hamad of his 'serious concern' and urging 'restraint on all sides'.

[22]Foreign Affairs Committee, Op. Cit., p22.

[23]Ibid., p81.

William Hague resorted to the same old-fashioned mode of communication with his Twitter follower Sheikh Khalid.

If the preferred balance between 'idealism' and 'realism' had been broken, with the former now taking priority, there was good reason to expect a swing in the opposite direction. Significantly, exchanges after his House of Commons speech of February, William Hague had stressed that 'Our relationship[s are] with nations rather than individuals', suggesting that Britain would be willing to deal with a new Bahraini regime provided that it aspired to democratic legitimacy. However, he hastened to add that it would be 'the height of folly' to discontinue dialogue with the Al Khalid. At no point did the British government even hint that it would welcome 'regime change' in Bahrain.

Yet if the British government wanted the regime to survive – even in a less repressive form – its public and private criticisms opened the possibility that relations would never be restored to their erstwhile amity. There were strong signs of dissatisfaction with Britain's position among influential supporters of the regime. Despite clear evidence of continuing brutality, including mass detention of dissidents and well-founded allegations of torture, Formula One motor racing – controlled by the British-born businessman Bernie Ecclestone – returned to Bahrain in April 2012. In July, the FCO minister Lord Howell of Guildford restated the familiar position in terms which suggested that, in respect of Bahrain, 'realism' now held sway over 'idealism', even if 'liberals' found this objectionable:

> The Government there must go further and implement meaningful political reforms as well. That message is not only for our own consumption; it is one that I and my fellow Ministers in the FCO have delivered to the Bahraini Foreign Minister, the Minister of the Interior and the Minister of Justice, all of whom have visited the UK over the past month. Some have criticised us for this engagement and we may hear more criticism, but we believe that dialogue is essential if the reforms that we all want are to take place.[24]

By the autumn of 2012, the governments were ready for full-hearted *rapprochement*. A defence agreement was signed, covering a range of matters including education and technical co-operation as well as the obvious fields of internal security and counter-terrorism. Celebrating the repaired relationship, Crown Prince Salman told British ministers that

> You have stood head and shoulders above others. You have engaged all stakeholders. You have kept the door open to all sides in what was a very difficult and sometimes unclear situation. Your engagement and your help in police reform and judicial reform, and your direct engagement with the leadership of the Kingdom of Bahrain and with members of the opposition, has saved lives, and for that I will be personally eternally grateful. Thank you.[25]

In December 2014, it was announced that the Royal Navy would be based once again in Bahrain, in a facility which was granted on very favourable terms. It was

---

[24]Lord Howell of Guildford, *House of Lords Debates: Middle East: Recent Developments* Vol. 738 col. 1400 13 July 2012 https://hansard.parliament.uk/Lords/2012-07-13/debates/12071335000198/MiddleEastRecentDevelopments#contribution-12071340000131.

[25]Speech by HRH Crown Prince Salman, 07.12.12, at the Manama Dialogue, 2012.

difficult to identify a tangible benefit for Bahrain, since it was already host to a major US naval base – the one the British had abandoned in 1971, but which the Americans had greatly expanded since 2010. It was thus not unduly cynical to interpret the new facility as a thank-offering to Britain for its refusal to disown the regime during the Arab Uprisings – or, even as a means of ensuring that it would think twice before reprising its role of ‘critical friend’ if the Al Khalifa were ever to face a similar crisis. Whatever the motive, the gesture helped to ensure that in May 2016, King Hamad – a long-standing family friend – was seated beside Queen Elizabeth II during the celebrations of her 90th birthday at Windsor Castle. This prestige part of the network between Britain and Bahrain was clearly as resilient as ever, and in a speech in Manama in December 2016, it was announced that Britain would formally return to ‘East of Suez’.[26]

Superficially, Britain’s engagement in the Bahrain crisis could be summarised along the lines of ‘All’s Well that Ends Well’. However, it had exposed the difficulty of implementing a ‘liberal Conservative’ foreign policy, particularly in a time of chaos. In a crisis like the one which faced Bahrain in February–March 2011 – where religious divisions and regional rivalries had to be taken into account, as well as the question of human rights and long-term commitments – it was difficult to see how Britain could maintain a balance between ‘idealism’ and ‘realism’. In practice, it shifted from one perspective to the other in a fashion which threatened to generate mistrust among all interested observers outside the regime itself, along with accusations that Britain was simultaneously riding two horses, often in different directions.

Those who wanted radical reform had their hopes raised then dashed; for their part, devoted supporters of the Al Khalifa could no longer view the British as unquestioning allies. Insofar as a ‘networked world’ is one in which countries must reach beyond governing circles, and strive to earn and retain respect in a newly globalised ‘civil society’, Britain can only regard its approach to Bahrain during the Arab Uprisings as a setback. The fact that the two countries were able to re-establish good relations so quickly is a testament to the *depth* and *duration* of previous contacts – factors which are difficult to capture in any quantitative assessment. Far from being ‘no foreign policy at all’, the importance of the phenomenon which William Hague tried to encapsulate in the phrase ‘networked world’ was proven during the Bahrain crisis. Arguably, indeed, this was an instance when networks saved Britain from a full-scale diplomatic humiliation, rather than the undoubted soft-power setback inflicted by a well-intentioned but impracticable approach to foreign policy. In other manifestations of the ‘Arab Uprisings’ Hague and his ministerial colleagues would not be so lucky.

[26] Jeevan Vasagar, *Britain revives military engagement east of Suez* (Financial Times, 23.12.16) Available at: https://www.ft.com/content/3477fe5a-c809-11e6-8f29-9445cac8966f.

## Libya and the 'Networked World'

One possible explanation for the 'pivot' towards a liberal outlook with regard to Bahrain in March 2011 is the fact that the British government had adopted a strongly moral approach in response to a simultaneous crisis in Libya. Major protests took place in Beghazi on 15 February – the same day as the mass march on the Pearl Roundabout in Manama. However, whereas the serious unrest in Bahrain was confined to the capital, in Libya it quickly spread and by the end of February the regime of Muammar Gaddafi had lost control of key cities, including Benghazi and Misrata. In the space of a few weeks, the melange of actors involved in Libyan politics had erupted along tribal, ethnic and religious lines, setting in motion a series of events that would result in the long-term fragmentation of the state and further regional destabilisation.

The recent history of British relations with Libya was ambiguous to say the least. After Italian forces were driven out in 1943, Britain and France became occupying powers but withdrew in 1951, when a united Kingdom of Libya was established. This regime enjoyed friendly relations with Britain, but after Gaddafi seized power in a 1969 *coup*, the atmosphere deteriorated. In 1984 gunfire from inside London's Libyan embassy killed policewoman Yvonne Fletcher. Two years later, the UK gave permission for the USA to launch bombing raids against Gaddafi's regime from its territory, and in December 1988, the destruction of Pan Am Flight 103 over the Anglo-Scottish border was widely attributed to Libyan operatives. Whilst Libyan involvement in the Lockerbie bombing is disputed, the country was certainly a key source of supplies for the Provisional Irish Republican Army (PIRA) in the latter stages of its campaign to force British withdrawal from Ireland.

Against this background, there was considerable surprise in March 2004 when Prime Minister Tony Blair travelled to meet Gaddafi and engaged in a public handshake. In the British tabloid press, Gaddafi had been reviled as a murderous dictator at the time that Saddam Hussein was gassing his own citizens without incurring undue opprobrium. His rehabilitation followed his decision to relinquish weapons of mass destruction (WMD) in the wake of the terrorist attack on New York in September 2001. This was a pragmatic decision which could not be regarded as a sign of personal repentance; but it was presented as a considerable triumph for British diplomacy, and the prospect of Libya losing its 'pariah' status in the West was augmented by lucrative agreements relating to the extraction of Libyan oil.

Britain had actually resumed diplomatic relations with Libya (broken off after the murder of Yvonne Fletcher) in 1999. However, the improved climate amounted to something short of full-hearted 'friendship', and Gaddafi himself remained an object of suspicion. When his forces attempted to regain control of Benghazi, the Arab League appealed to the United Nations Security Council to authorise a 'no fly zone' over Libya. The French President Nicholas Sarkozy was quick to respond and was joined by Lebanon and the UK in proposing a UN resolution to that effect, which also authorised member states to 'take all necessary measures...to protect civilians and civilian populated areas under threat of attack'. Russia and China (along with Brazil, India and – significantly – Germany) abstained, so UNSCR 1973 was adopted.

Britain had already sponsored UNSCR 1970, which imposed sanctions against the Libyan regime, and recalled its Ambassador before the end of February. Recommending the new resolution to MPs, Cameron anticipated potential criticisms:

> to those who say it is nothing to do with us, I would simply respond: Do we want a situation where a failed pariah state festers on Europe's southern border, potentially threatening our security, pushing people across the Mediterranean and creating a more dangerous and uncertain world for Britain and for all our allies as well as for the people of Libya?[27]

To those who argued that Cameron was on the verge on repeating the Iraqi misadventures of his political idol, Tony Blair, the Prime Minister could respond that his policy had been authorised by the UN; that regional powers were demanding urgent action; and that the likelihood of a major refugee crisis, affecting Europe as a whole, made intervention a matter of national interest.[28] However, the obvious differences between the two situations could not obscure Blairite assumptions about the nature of the people he was hoping to relieve. In Cameron's view, 'The Libyan population wants the same rights and freedoms that people across the Middle East and North Africa are demanding, and that are enshrined in the values of the United Nations Charter'.[29]

The obvious response, from those who were watching Britain's policy towards other countries involved in the 'Arab Uprisings', was that protestors in Bahrain were asking for similar rights; and yet David Cameron had not rushed to the United Nations on their behalf. In turn, Cameron could argue that while the Bahraini regime might be imperfect, it was never likely (even in extremis) to commit the kind of atrocities of which Gaddafi was deemed capable. However, his remarks reinforced the view advanced above, that (despite all his attempts to differentiate his policy from the Blairite approach to Iraq) he was now actuated by 'liberalism' rather than the 'idealism tempered by realism' to which William Hague had referred in a less agitated context. Furthermore, his confidence in the Bahraini regime was based on a well-established relationship between the two countries, whereas his assumptions about the Libyan people (and about Gaddafi's murderous intentions) were the product of guesswork.

In its 2016 report into British policy towards Libya, the House of Commons' Foreign Affairs Committee quoted the former Ambassador to Libya, Sir Dominic Asquith, who had admitted that 'the database in terms of people, actors and the tribal structure – the modern database, not the inherited historical knowledge – might well

[27]David Cameron, "*Britain will remain at the forefront of Europe in leading the response to this crisis*" (14.03.11) Available at: https://www.gov.uk/government/news/britain-will-remain-at-the-forefront-of-europe-in-leading-the-response-to-this-crisis.

[28]Sam Goodman, *The Imperial Premiership: The role of the modern Prime Minister in foreign policy making, 1964–2015* (Manchester: Manchester University Press, 2016), 270.

[29]Cameron, 2011, Op. Cit.

have been less than ideal'.[30] In plain language, this meant, 'In 2011 we didn't have a clue about Libya'.[31] The Committee's report identified the source of the problem: 'the UK's understanding of Libya before February 2011 was constrained by both resources and the lack of in-country networks for UK diplomats and others to draw on'.[32]

The lack of a 'fit for purpose' network in Libya had been illustrated in spectacular fashion even before the passage of UNSCR 1973. Britain despatched a diplomatic team, accompanied by members of the Special Air Service (SAS) to Eastern Libya, in order to establish contact with Gaddafi's opponents.[33] This belated attempt to establish 'networks' turned into a fiasco, precisely because of a lack of even basic knowledge of the situation in Libya. The team arrived by helicopter without previous warning, and was promptly arrested. The British government found itself having to use its diplomatic skills to secure the team's release from detention by the very opposition forces it was supposed to be helping. This incident followed an announcement by William Hague that, amidst fighting in the capital, Tripoli, Colonel Gaddafi had fled Libya, seeking refuge not in Saudi Arabia (the usual destination for fugitive dictators like Ben Ali) but in distant Venezuela. Although Hague was referring to reports, rather than confirming media speculation about Gaddafi's movements, it was assumed that he would not have spoken without feeling confident about his sources. Unfortunately, he had been misled: Gaddafi never sought sanctuary overseas, although Tony Blair advised him to do so.[34]

In retrospect, Hague claimed that Gaddafi's own regime had little understanding of the various Libyan groups, 'so there is not much hope that a foreign intelligence service would have a more profound understanding'. Whilst this was a reasonable explanation for Britain's failure to establish reliable networks, it was hardly a defence of the coalition government's decision to take an active role in Libya, with the declared aim of facilitating regime change. President Sarkozy at least had clear objectives relating to France's perceived national interest and his own political ambitions. Cameron, by contrast, seems only to have thought of protecting civilians and toppling Gaddafi – laudable aims, but difficult to realise without some knowledge of the people he was dealing with. In place of detailed intelligence,

---

[30] House of Commons Foreign Affairs Committee, *Libya: Examination of intervention and collapse and the UK's future policy options* (14.09.16) Available at: https://www.publications.parliament.uk/pa/cm201617/cmselect/cmfaff/119/119.pdf.

[31] Ibid.

[32] Ibid.

[33] Lizzie Dearden, *Tony Blair urged Gaddafi to stand down and find 'a safe place to go' during Libyan uprising*, (The Independent, 02.10.15) Available at: http://www.independent.co.uk/news/uk/home-news/tony-blair-urged-gaddafi-to-stand-down-and-find-a-safe-place-to-go-during-libyan-uprising-a6676356.html.

[34] *Libya: Colonel Gaddafi 'flees' to Venezuela as cities fall to protesters* (The Telegraph, 21.02.2011) Available at: http://www.telegraph.co.uk/news/worldnews/africaandindianocean/libya/8338948/Libya-Colonel-Gaddafi-flees-to-Venezuela-as-cities-fall-to-protesters.html.

Cameron's assumptions clearly derived from wishful thinking. As he told the House of Commons on 28 February 2011:

> In many parts of the Arab world, hopes and aspirations that have been smothered for decades are stirring. People—especially young people—are seeking their rights, and in the vast majority of cases they are doing so peacefully and bravely. The parallels with what happened in Europe in 1989 are not, of course, precise, and there have been many disappointments in the past, but those of us who believe in democracy and open society should be clear that this is a precious moment of opportunity.
>
> While it is not for us to dictate how each country should meet the aspirations of its people, we must not remain silent in our belief that freedom and the rule of law are what best guarantee human progress and economic success. Freedom of expression, a free press, freedom of assembly and the right to demonstrate peacefully are basic rights—they are as much the rights of people in Tahrir square as they are of people in Trafalgar square. They are not British or western values, but the values of human beings everywhere.[35]

Cameron's remarks contained clear echoes of his speech to Kuwait's National Assembly; the main difference lay in the enhanced emotional appeal, which he evidently felt was more suited to this domestic audience. After his statement, he fielded questions from 53 backbenchers, not one of whom expressed concern about the impact of intervention on an unfamiliar and potentially unstable country. However, one Conservative MP, John Baron, whilst approving Cameron's support for democratic reform, wondered if the new evangelistic tone would 'have any effect on future relationships with our other autocratic friends in the region?'. Cameron replied that on his own recent trip to the Gulf

> I was quite struck by the fact that a number of our very strong and old allies, such as Kuwait, Qatar and Oman, are in favour of taking further steps towards democracy and more open societies. Far from being dismayed by our very clear reaction that democracy, freedom and that sort of progress are good things, they were fully in support of them.[36]

Once again, Cameron's assessment was obviously a product of his wishes rather than reflecting facts on the ground.

Published accounts of the policy-making process during the Libyan crisis have revealed considerable tension between experienced officials and the politicians. The latter seemed to be preoccupied with the gruesome precedents of Srebrenica and Rwanda during the 1990s. For Cameron and his allies, the fact that the moral issues – and the likely political price of inaction – seemed similar was more important than any differences of detail between these cases. Britain's new National Security Council – whose establishment had been trumpeted in Hague's 'networked world' speech as a safeguard against the kind of mistakes which tarnished Tony Blair's record – had clearly not lived up to its billing during its first serious test.

The ultimate effects of British policy towards Libya are still unclear, although the Commons' Foreign Affairs Committee report published in 2016 has been highly critical of David Cameron's role. Back in September 2011, however, it seemed that

[35] David Cameron, Statement on Libya and the Middle East, *House of Commons Debates*, Vol. 524, col. 25, 28 February 2011.

[36] Ibid., col. 31.

the military intervention authorized by UNSCR 1973 had been an overwhelming success. Although Gaddafi was still at large, Sarkozy and Cameron were greeted in Tripoli like liberators. Britain would eventually be thanked by the Bahraini regime for its approach during the Arab Uprisings, but the cheers which greeted the two European leaders provided something more gratifying – the sense that gratitude was felt by the Libyan people as a whole.

## Conclusions

While Gaskarth and Leech have argued that the British response to the Arab Uprisings demonstrates the limitations (even the vacuity) of a 'networked' foreign policy, the present chapter prompts a very different conclusion. A 'network', we argue, is not something static and one-dimensional, capable of being extrapolated from statistical evidence even in important areas like trade. Rather, the term denotes deep-rooted and long-established ties between governments, but also between governments and influential elements within civic society, including those who dissent from the existing regime. Such networks are usually developed over decades, rather than years, and result in *empathic* understanding between countries. Of course, when such relationships are underdeveloped – or lack construction – networks are much weaker.

In a mature network (like the Britain-Bahrain relationship), contacts with a government's critics can serve a dual purpose. Apart from giving the friendly state advance warning of trouble ahead, they can also provide dissidents with an indirect channel of communication with their domestic political opponents, possibly promoting timely political reforms. The problem in respect of Bahrain, arguably, was that the British government's relations with the Al Khalifa regime had led to an *excess* of 'empathy', so that although the FCO was in limited contact with dissidents, it was so close to the regime that its warnings were resented rather than heeded. In terms of Britain's 'soft power' in the region, the temporary disruption of amicable relations was all too reminiscent of a lovers' tiff, and the inevitable reconciliation was too effusive.

In Libya, as we have seen, Britain had failed to follow up Blair's initiative by establishing networks worthy of the name. There were trade relations – fostered in particular by the ubiquitous British arms manufacturers, as well as oil companies – and indeed there were numerous British nationals working within Libya at the time of the Arab Uprisings, whose delayed repatriation gave rise to criticism of the FCO. Such aspects are integral to the development of networks, but do not constitute networks in this context alone. Moreover, the British had limited contacts even with those dissident groups which could claim democratic credentials; and, equally important, they provided an inadequate grasp of the complexity of the country whose future the British Prime Minister hoped to affect.

William Hague's vision of a 'networked world' suggested different responses to the unfolding situations in Bahrain and Libya. In the first case, existing networks were sufficiently strong to permit a forceful, but primarily private intervention, designed to promote dialogue between the government and opposing forces which the British knew reasonably well. Instead, the British made no secret of an impatience with the Bahraini regime – a desire for rapid and radical reform which contrasted sharply with the previous attitude of benign encouragement for a glacial process of change. In regard to Libya, the 'networked world' approach should have dictated a secondary role for Britain, based on a recognition that a more interventionist approach would take it beyond the depth of its contextual knowledge. Important British interests, as well as moral principles, were undoubtedly involved; but in the absence of reliable networks, the chances of serious and costly mistakes were too obvious to authorise a more dynamic policy response. Instead, after the short-lived outburst of acclaim, David Cameron ended up with something akin to the worst of both worlds; France, rather than Britain, was regarded as the major source of any fleeting benefits which accrued to the Libyan people, but the British involvement had been sufficiently deep (and publicized) to ensure that the country would take its full share of any subsequent blame when the Libyan situation deteriorated.

If Britain's response to the Arab Uprisings should be regarded as a failure, it seems eccentric to identify the 'networked world' concept as the main culprit. It would be a mistake to conclude that a country's foreign policy can only be as effective as the networks it enjoys; as the Bahrain example shows, there are occasions when some aspects of a network can be *too* strong. However, the 'networked' approach denotes a *method* of conducting foreign policy, rather than describing the *content* of that policy. In other words, it is one thing to establish satisfactory networks, and quite another to use them effectively. The foreign policy of the coalition government was not predestined to fail because the incoming Foreign Secretary identified 'networks' as the key to future success. Rather, the coalition espoused an approach to foreign relations which attempted to blend liberal ideas with a consideration of the national interest. In the abstract this seemed to be a plausible compromise – especially at a time of economic stringency, when Britain could not afford to reject opportunities by striking Blairite moralistic poses. But during the Arab Uprisings, it was exposed as what it always had been – i.e. not a short-hand description for a coherent approach but rather an attempt to disguise a stark choice. In certain instances, the British government would always favour a 'hard-headed' 'realist' line, even if it felt constrained to make gestures on behalf of democratic principle. In others, it would adopt a full-hearted (even messianic) liberal policy, and then scramble in vain to find 'hard-headed' justifications for its decisions.

Reviewing the British responses to Bahrain and Libya, one might conclude that the government was 'hard-headed' when it had a vague idea of what it was doing, and indulged in an emotional 'spasm' when it was singularly ill-informed. From this perspective, it could be argued that when even supposedly liberal-democratic governments enjoy adequate 'networks', they tend to take a 'realist' approach; 'idealism' only takes control when they are unaware of the complexity of the situations they are hoping to influence. In short, on this view the establishment of an adequate 'network' generates a bias towards the *status quo*, as certainly happened in the case of Britain's relationship with Bahrain.

However, this is a travesty of the argument presented by William Hague in his speech of 1 July 2010. Ultimately, Hague was arguing that a well-informed foreign policy is likely to be more effective than one based on guesswork and wishful thinking; and this seems a defensible proposition. Beyond this, in asserting that Britain had to operate within a networked world, Hague was arguing that at a time of government spending cuts, the FCO should be protected, in order to ensure that Britain could deepen its existing networks and develop new ones. His own record during the Arab Uprisings is open to question, but the message of the 'networked world' speech was reinforced, rather than discredited, by those tumultuous events. If Britain had enjoyed better contacts with the Bahraini opposition, or meaningful links of any kind in Libya, it would assuredly have played a more constructive part in what was a period of dangerous upheaval as much as a 'precious moment of opportunity'.

# A New Method in the Analysis of Chaotic Systems: Scale Index

**Nazmi Yılmaz, Mahmut Akıllı, and K. Gediz Akdeniz**

**Abstract** In recent years, there has been great interest in the application of wavelet analysis in a variety of disciplines to investigate the characteristics of chaotic systems. The scale index is a wavelet-based method introduced in 2010 that has been used effectively in determining the degree of aperiodicity, hence chaotic characteristics of a signal. In this chapter, we will discuss previous works involving Scale index method including our works. We will also mention some social and economic systems that this method can potentially be applied.

**Keywords** Scale index · Wavelet analysis · Chaotic systems

## Introduction

Modern science is built upon the combinations of linear models that are based on the linear pieces of seemingly the non-linear nature. Science should progress and move towards new ways of thinking. Chaos theory, stated as the new science or the new paradigm, has been developed for this aim. It is known that chaos theory deals with non-linear systems that are long-term aperiodic and sensitive to the initial conditions (Strogatz 1994). The analysis of the systems which are non-linear and have chaotic characteristics can be made possible by only non-linear methods such as Fourier analysis, wavelet transform, Lyapunov exponents, Poincaré map, recurrence quantification analysis and Kolmogorov-Sinai entropy. Wavelet transform is a widely used method in analysing aperiodicity of non-stationary

N. Yılmaz (✉)
Koç University, İstanbul, Turkey
e-mail: nayilmaz@ku.edu.tr

M. Akıllı
Wone Lighting, R&D Department, İstanbul, Turkey

K. G. Akdeniz
Disordered Systems Working Group, İstanbul, Turkey



Ş. Ş. ERÇETİN, Ş. N. AÇIKALIN (eds.), *Chaos, Complexity and Leadership 2018*, Springer Proceedings in Complexity, https://doi.org/10.1007/978-3-030-27672-0_2

chaotic systems. The scale index parameter which is based on wavelet transform can be used as a measure of aperiodicity and chaos in a dynamical system and complement the other chaos analysis methods.

On the other hand, notable works have been performed very recently using chaotic theory metaphors in the studies of social movements too, particularly in the critics of the political behaviours (Açıkalın and Artun 2019; Akdeniz 2014, 2019; Akdeniz and Anastasopoulos 2016). In this work, we aim to shortly review various applications of scale index method and demonstrate and discuss its potential use in the fields of social and political sciences.

## The Scale Index Method

The scale index method is based on wavelet transform and used in quantitatively measuring aperiodicity and chaotic behaviour of a dynamical system. In practical application, for the calculation of the scale index parameters of a signal *f*, a finite time interval must be considered. Firstly, the inner scalogram of the signal $f$ at a scale $s$ must be determined which is derived from the continuous wavelet transform (Mallat 1999; Benitez et al. 2010)

$$S^{inner}(s) = \left\| Wf\left(u, sI\right\|_{J(s)} = \left(\int_{c(s)}^{d(s)} |Wf(u, s)|^2 du\right)^{1/2} \tag{1}$$

The inner scalogram is normalised so that the result will be independent from the scale.

$$\overline{S}^{inner}(s) = \frac{S^{inner}(s)}{(d(s) - c(s))^{\frac{1}{2}}} \tag{2}$$

The inner scalogram $\overline{S}^{inner}(s)$ can be shown as a matrix with single column or a row for each energy scale level $s$ in a finite scale interval $[s_0, s_1]$. The scale that the inner scalogram $\overline{S}^{inner}(s)$ will reach its maximum value in the given scale range is called $s_{max}$. Between $s_{max}$ and $s_1$, if the inner scalogram $\overline{S}^{inner}(s)$ never has a value too small from $\overline{S}^{inner}(s_{max})$, the signal will than be defined aperiodic in the scales between $s_0$ and $s_1$. The scale index parameter derived from $\overline{S}^{inner}(s)$, is written by

$$i_{scale} = \frac{\overline{S}^{inner}(s_{min})}{S^{inner}(s_{max})} \tag{3}$$

Here $s_{max}$ is the smallest scale for $\overline{S}^{inner}(s) \leq \overline{S}^{inner}(s_{max})$ in the scale interval $[s_0, s_1]$. $s_{min}$ is the smallest that satisfies $\overline{S}^{inner}(s_{min}) \leq \overline{S}^{inner}(s)$ in the scale interval $s \in [s_{max}, 2s_1]$ (Bolós 2017).

## Applications of Scale Index Method

The scale index method has been used in variety of recent works in many different fields, such as classical dynamical systems, speech signals, image encryption, meteorology, engineering, seismic waves, EEG signals and pneumocardiogram signals. Our previous works using this method will be mentioned and application in seismic waves and in weak periodic signal detection will be outlined in this section.

In a recent work, the scale index parameters were calculated from time series of Pneumocardiogram (PNCG) signals of rats (Yılmaz et al. 2018a). The results showed the PNCG signals are aperiodic, hence chaotic, agreeing with the maximum Lyapunov exponents.

In another study, the scale index method was used in Thirring model which is an exactly solvable quantum field theory to demonstrate the periodic characteristics of the stable bifurcation points of the Akdeniz-Smailagic fermion-like instanton solutions (Yılmaz et al. 2018b).

### *Sale Index in Seismic Waves*

The work on seismic waves was conducted to distinguish independent earthquakes from the aftershocks of a main earthquake by analysing their chaotic characteristics using scale index method (Yılmaz 2016). Time series of seismic waves of four different earthquakes were used. The data was collected from 19 different seismic stations. It was observed with the calculated scale index parameters that the *Van-Edremit* (M: 5.6) earthquake (*Bayram* hotel earthquake) has considerable difference in chaotic characteristics compared to the *Van* earthquake; therefore it can be stated that it is not an aftershock of the *Van* earthquake (Fig. 1).

### *Scale Index in Weak Periodic Signal Detection*

Duffing oscillator is a clear example of weakly non-linear oscillators and is very sensitive to initial conditions in the edge of chaos. Duffing equation is

$$\frac{d^2x}{dt^2} + \delta\frac{dx}{dt} - x + x^3 = \gamma cos(t) \tag{4}$$

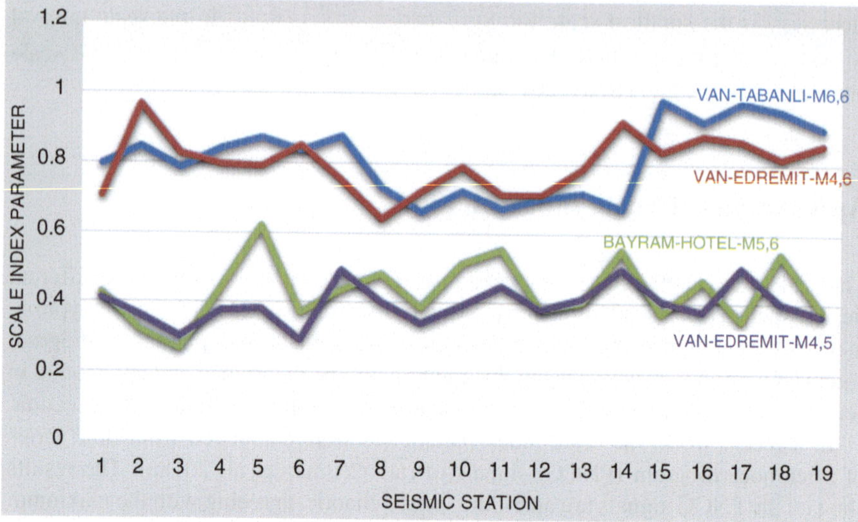


**Fig. 1** Scale index graph for the seismic waves of the four earthquakes recorded in 19 seismic station data. Blue line represents Van-Tabanlı_23/10/2011,M = 6.6, red line is Van-Edremit_23/10/2011,M = 4.6, green line is Van-Edremit_09/11/2011,M = 5.6 and purple line is Van-Edremit_09/11/2011,M = 4.5 (Yılmaz 2016)

where $\delta$ is the damping ratio, $\gamma$ cos $(t)$ is the driving force of the system which is periodic and $-x + x^3$ is the restoring force of the system which is non-linear. If $\delta$ is ncreased while $\gamma$ is is kept constant, the state of the Duffing oscillator at the edge of chaos will become stable for certain $\gamma$ parameters.

In one of our latest works (Akıllı and Yılmaz 2018), the scale index method and the Duffing oscillator were used together to detect weak periodic signals (WPS) in the Electroencephalogram (EEG) data. EEG signals were embedded into the Duffing equation, $\delta$ and $\gamma$ was fixed at critical chaotic state and the equation was solved for a range of frequency ($w^0$) values.

In this work, weak periodic signals in the Electroencephalogram (EEG) signals were detected by using Duffing oscillator system with scale index method (Yılmaz et al. 2018b). EEG signals were embedded into the Duffing equation, $\delta$ and $\gamma$ was fixed at critical chaotic state and the equation was solved for a range of frequency ($w_0$) values.

$$\frac{d^2x}{dt^2} + 0.5\frac{dx}{dt} - x + x^3 = 0.825\cos(\omega_0\tau) + 10^x\left(input_{EEG}\right) \tag{5}$$

It was shown that, when the reference signal frequency matched with a wps frequency in EEG signals, Duffing oscillator system in the the edge of chaos jumped to the large-scale periodic state, enabling us to identify the WPS. Then, the scale index parameters of the Duffing oscillator versus frequency of the reference signal

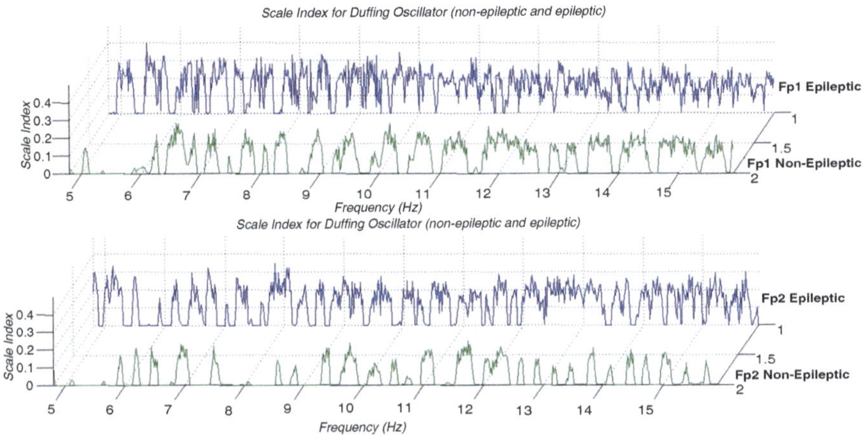


**Fig. 2** The scale index parameters of the Duffing oscillator with EEG signals (Fp1,Fp2): Differences between non-epileptic and epileptic EEG signals were observed. (Akıllı and Yılmaz 2018)

graphs were plotted and analyzed. Subsequently, different results for EEG signals of non-epileptic and epileptic persons were obtained, in terms of the occurrence of WPS and their frequency values.(Fig. 2).

## Discussion

By the above works, it can be clearly understood that the scale index method would have a great potential of application where the non-linear behaviour is present such as econometry, social sciences and political sciences as a complement to classical methods.

For example, this method can be used in determining the consensus among pre-election polls, in detecting and prediction of the trend changes and forecasting in the stock markets and currency rate fluctuations. Another use of the method can be to predict whether the outcome of the chaotic awareness realities in the simulation world will lead to an evolution in the society (Zuhur) (Akdeniz 2014).

## References

Açıkalın, Ş. N., & Artun, E. C. (2019). The concept of self-organized criticality: The case study of the Arab uprising. In Ş. Erçetin & N. Potas (Eds.), *Chaos, complexity and leadership 2017* (ICCLS 2017. Springer Proceedings in Complexity). Cham: Springer.

Akdeniz, K. G. (2014). Is Arab spring a complex utopia? In S. Banerjee & Ş. Erçetin (Eds.), *Chaos, complexity and leadership 2012* (Springer Proceedings in Complexity). Dordrecht: Springer.

Akdeniz, K. G. (2019). The chaotic awareness reality and the complexity in Turkish novels. In Ş. Erçetin & N. Potas (Eds.), *Chaos, complexity and leadership 2017* (ICCLS 2017. Springer Proceedings in Complexity). Cham: Springer.

Akdeniz, K. G., & Anastasopoulos, N. (2016). Chaotic awareness and simulacra in the recent emergence of the self-organized multitude. *Chaos, Complexity and Leadership, 2014*. https://doi.org/10.1007/978-3-319-18693-1_3.

Akıllı, M., & Yılmaz, N. (2018). Study of weak periodic signals in the EEG signals and their relationship with postsynaptic potentials. *IEEE Transactions on Neural Systems and Rehabilitation Engineering PP, 26*(10), 1918–1925. https://doi.org/10.1109/TNSRE.2018.2867515.

Benitez, R., Bolo's, V. J., & Ramırez, M. E. (2010). A wavelet-based tool for studying non-periodicity. *Computers & Mathematcs with Applications, 60*, 634–641.

Bolós, V. J. (2017). The windowed scalogram difference: A novel wavelet tool for comparing time series. *Applied Mathematics and Computation, 312*, 49–65.

Mallat, S. (1999). *A wavelet tour of signal processing*. London: Academic Press.

Strogatz, S. G. (1994). *Nonlinear Synamics and Chaos: With applications to physics, biology, chemistry and engineering*. Singapore: Addison-Wesley.

Yılmaz, N. (2016). *Analysis of regional earthquakes with wavelet scalogram, scalogram scale index and power spectrum methods* (PhD Thesis). Department of Physics, İstanbul University, İstanbul, Turkey.

Yılmaz, N., Akıllı, M., Özbek, M., Zeren, T., & Akdeniz, K. G. (2018a). *Analysis of pneumocardiogram signals of rats by the scale index method*. Turkish Physical Society 34th International Congress.

Yılmaz, N., Canbaz, B., Akıllı, M., & Önem, C. (2018b). Study of the stability of the fermionic instanton solutions by the scale index method. *Physics Letters A, 382*(32), 2118–2121.

# Reminiscence of Alija Izetbegovic and His Leadership

**Bakir Sadović**

**Abstract** I will write about one particular leader, the late President Alija Izetbegovic, who led Bosnia and Herzegovina at the time of chaos, namely, before, during, and after the war in my country. He was the leader who had authority, knowledge, and a vision, but he also showed compassion, offered solutions, and inspired hope when hope was almost lost. These, in my view, are the traits of a good leader, which is why I will dedicate this article to him. As an assistant to President Alija Izetbegović, I was a witness to our country's efforts to avoid the conflict, the war events, the political and diplomatic peace talks, and the reconstruction of the nation, as well as the reconciliation process that is still ongoing. This experience was both traumatic and priceless for me. That's why, although there are various articles and books on Alija Izetbegović and his leadership, this chapter would be unique regarding my personal experience with him.

**Keywords** Alija Izetbegovic · Leadership

## Introduction

At first, it appears that Quantum Mechanics – in which I have a keen interest – has little to do with human nature. But such conclusion could be deceiving. As you know, Quantum Mechanics is full of paradoxes that are sometimes difficult to comprehend. It teaches us that matter has two natures – a particle and a wave, simultaneously. However, it also touches on big questions such as "How much can we really know about the true nature of the world that surrounds us?" Quantum mechanics sets limits and introduces us to a realm of possibilities where some things can happen, but they do not have to. Just as the matter can have two natures simultaneously, so can a man. Alija Izetbegović wrote his whole life about the dualistic nature of man – as a physical and spiritual being. The synthesis of these

B. Sadović (✉)
Bosnia and Herzegovina, Ankara, Turkey



Ş. Ş. ERÇETİN, Ş. N. AÇIKALIN (eds.), *Chaos, Complexity and Leadership 2018*, Springer Proceedings in Complexity, https://doi.org/10.1007/978-3-030-27672-0_3

two human natures by Izetbegović was found in Islam, and he called it the "the third way." I will try to explain how I saw him as a man and a leader of my people.

## Alija as a Family Member

First of all, Alija Izetbegović was the brother of my grandmother, so I grew up with his presence in our family. Just as in Turkey, in my country, every family has some older and more respected member whom everyone asks for advice and invites for a talk when a problem arises in the family or when there are important decisions that need to be made. Our extended family had Alija. He was present in my life while I was growing up, although, as a teenager, I sometimes found that presence rather strange.

We would see each other on Bayrams or at birthday parties. He would always find time to talk to my sister and I about the books we were reading, or the movies we were watching. It was interesting and somewhat peculiar for us that someone older than our parents would show interest in the world of our secrets, a world which we thought belonged only to us, the youth. But Alija never stopped surprising us. He had read all the books we did, had seen all the movies that we saw, and was ready to "question" us about Hermann Hesse's books or to discuss the justness of the punishment in the novel *Crime and Punishment* by Dostoevsky.

He would ask us questions about why some hero in a particular book made one decision, and not the other, and would ask us to express our thoughts. In school at that time, we would only repeat what the textbooks said and no one ever asked us what we thought. This is why it was intellectually refreshing to talk to someone who valued our opinions. This is something I have always liked about Alija, even later in life, when I worked as his assistant. Alija was not educated to be a leader, but fate gave him this role. I had an opportunity, and a privilege, to witness how a diligent, wise, and humble man developed into a leader.

He was a passionate collector of pieces of information of every kind. He collected them while he was developing his "philosophy of the third way," which he shaped in his book *Islam Between East and West*. He started writing it as a young man and continued as a mature adult. He gathered pieces of information during his time in prison, but also as a President. He always had a thick folder on his desk in which he collected everything interesting about society, science, and art.

His book *My Escape to Freedom*, which he wrote in prison in the 1980s, and whose title symbolized the "escape" of the soul and mind into freedom, is actually a continuation of "the philosophy of the third way," to which he dedicated his entire philosophical thought. In this "escape," Alija analyzed the books he read and compared them to the multitude of information, which he collected from available sources, while also giving his comments on the current world events.

## Izetbegović's Spirit for Bosnians and Muslims

Izetbegović was released from prison in 1988, the time when the entire Eastern Bloc of socialist countries was politically shaking. Yugoslavia fell apart only 3 years later, and bloody wars in Croatia and Bosnia and Herzegovina ensued shortly after that. To understand the work and ideas of Alija Izetbegovic, one must know something about the position of Muslims in the former Yugoslavia.

As you know, when the Ottomans left the Balkans, the positions of Muslims in the whole region started to weaken. At the end of the nineteenth century, hundreds of thousands of Muslims left their homes and resettled in Turkey. Persecution and murders of entire villages where Muslims lived were left in their wake, and reforms were taking away their land, which resulted in a new exodus.

After World War I, the Kingdom of SHS was formed, where Muslims did not play a large role because of their small numbers. Following the end of the World War II, a new communist Yugoslavia was established with a strong military and political apparatus. Still, in comparison to the other countries of the Eastern Bloc, socialist Yugoslavia had different, more humane policies. Peace and the right to free education enabled the citizens of the former country to increase their quality of life.

The fall of the Berlin Wall marked the beginning of reforms in communist countries, and the same was true for Yugoslavia.

Uncertainty and insecurity about the future survival of Yugoslavia on the one hand and fear of a repetition of history on the other were the reality for Muslims in the former country at the end of the 1980s. That is when Izetbegović, with like-minded people, decided to begin articulating the political thinking of the Muslims in Yugoslavia. He understood that one era had come to an end, realizing that, over the next few years, people would need someone to gather and lead them.

Because of historical grievances and prejudices, the Muslims had the most complex task. They needed to find leaders who were like them, who were brave and wise enough to preserve their identity and tradition, and who were considerate and moderate enough to gather the largest number of followers around them.

Izetbegović and his political friends founded the Party of Democratic Action, SDA. I was invited to be one of the 40 founders, which I gladly accepted.

The results of the first elections in Bosnia and Herzegovina held in 1990 showed that SDA positioned itself as the strongest Muslim party in BiH. Alija Izetbegović was elected president of the Presidency of BiH, which was a proof that Muslim people in that former Yugoslav republic trusted him.

A few months after the elections, Alija called me and asked if I had time for a walk. This was the summer of 1991 when the war was already started in Croatia, and in our country armed attacks began. As we walked, Alija spoke to me about his predictions for what would happen in Bosnia and the region. He spoke to me about his political investigator from 1982, the year when Izetbegović was sentenced to long-term imprisonment. And I quote:

> "He was a real professional and did everything he could, in accordance with the laws of that time, to enable judges to give me the longest sentence possible, but he never hit me or raised

> his voice at me," Alija told me, and added: "He is still in the same position as he was in the Communist era, I don't plan to replace him. He loves Bosnia just as I do, and, regardless of our differences, he can help us save our people and our country. I would like you to be my link with the secret service headed by that man", Alija said. Naturally, I accepted.

That was Izetbegović as a leader – he was ready to forgive even those who had served an unjust regime that harmed him personally. He was ready to forgive for his own sake, and for the sake of his people and the country. Some viewed his decisions as controversial, but he insisted on them. He fought for every single Bosnian-Herzegovinian patriot to enlarge his following and increase the people's trust. He more than justified that trust in the years that followed, during the war in our country, because he inspired people to fight for their freedom, and not to take revenge on others.

Alija bravely stated his opinions to the East and to the West. He thought that Muslims in Yugoslavia and Muslims in the world needed their self-confidence back. So, in an interview with the German "STERN" in November 1994, he was asked to comment on reports in the Western media about the so-called Islamization of Bosnia and Herzegovina.

And this is what Alija said in response.

I quote:

> "I will be completely honest, these are not just rumours. The return to religion is almost a universal phenomenon in all the areas where communists brutally suppressed religion for 50 or 70 years. Islamization, as you call it, exists in Bosnia, but in the same fashion, Christianization exists, that is, a resurgence of interest in faith by Bosnian Catholics and Bosnian Orthodox believers. But Christian Europe does not recognize Christianization, and is not sensitive to that phenomenon, which I understand and do not frown upon.
>
> But I have to correct you in one thing – my tolerance is not of European, but of Muslim origin. Europe has some prejudices which it simply cannot get rid of, despite obvious facts. In Bosnia, in this war for example, hundreds of churches and mosques have been destroyed. They were all destroyed by "Europeans", not a single one was destroyed by Bosniaks... Fascism and Communism are not Asian, but European products. Even then, Europe did not show a great sensitivity to the appearance of Fascism in the Balkans. I value Europe, but I think that she has an overly flattering view of herself", Alija said.

In his Lecture to the German Association for Foreign Affairs in 1995, he said this (and I quote):

> "I came here in my official function, but also – why not say it – as a Muslim from Bosnia. I personally feel as both a Muslim and a European, and I do not think that one excludes the other. I do not accept that there are differences between people and civilizations that cannot be overcome. If each civilization is first of all a group of values – which are, ultimately, moral values that are believed in – then we can talk about a possible unity of civilizations.
>
> A long time ago, as a young man, I wrote an article on Kant's categorical imperative. The article's thesis was that the basic moral principle formulated by Kant... is practically outside of space and time. For me, this meant that there are no differences between civilisations that cannot be overcome and that all cultures are similar or even equal if one goes down to their roots. To me, this question was a matter of human equality. There is an exciting sentence in Qur'an that begins with the words, 'Come and gather around the word that is common to us...' The invitation is meant for Christians and Jews. So, I invite you to stop building artificial partitions between Islam and Christianity, between East and West. Rather, look into this to see if intolerance is caused by the selfishness and injustice of the

West. Furthermore, many differences that you can see and feel are not essential, they come from differences in the level of cultural and social development."

Izetbegović was a critic of the laggard Muslim countries and he was not afraid to clearly express himself at the Islamic Conference of 1997 in Tehran. He said the following (and I quote):

"Forgive me, for I will now be very open. Sweet lies do not help, but sour truths can be a cure. The West is not corrupted and degenerated... It is strong, educated and organized. Their schools are better than ours and their cities are cleaner than ours. The level of human rights in the West is higher and social care for the poor and less able, better organised. Westerners are, most often, responsible and hard-working people. Those are my experiences with them. I know, as well, the dark side of their progress, and I am not letting this out of my sight.

Islam is the best – that is the truth – but WE are not the best. Those two are different things and we always switch them. Instead of hating the West, we should compete with it. Did the Qur'an not order us to do just that: 'Strive to achieve the virtue of deeds . . . ' With the help of religion and science, we can create the power that we need. It is a long and hard road, it is an exhausting climb up the hill, the hill that the Qur'an talks about, but there is no other way. So, let us establish funds for education everywhere, Do not let any child of ours remain uneducated. Rich Muslim countries should help the poor ones in this regard."

Alija presented painful truths about weak educational institutions in Muslim countries, and especially the low status of women in the countries in which they are not allowed to be educated. In these facts, Alija saw the main reason why Muslims were lagging behind the rest of the world.

Here we can stop for a moment and ask ourselves: Why was Izetbegović saying this? A country destroyed in the war such as Bosnia needed friends in the West and the East alike; Bosnia needed help to fix the roads, hospitals, and schools, so why was Izetbegović criticizing both the East and West? He once told me that if you don't respect yourself, then no one else will respect you. He also said that if you betray your beliefs and just sit quietly in front of the richer and stronger because you need their help, real help will never come. Be upright and stay strong in your beliefs, and you will be ready to face every challenge . . . .

I remember one example that shows just how much Izetbegović held on to his principles. His party of SDA needed to elect members of the Executive Board. After the vote, among the 15 chosen members, there was not a single woman. Izetbegović was rarely angry with his party colleagues as he was then. He said:

"Shame on you! You are behaving like male chauvinists. You were not able to recognize our hard-working female colleagues after all this time and elect them too! Now I will use my powers as party President and appoint a few of our female colleagues into the Executive Board." And he did that.

When I remember this event now, I am more convinced than ever that the best leadership requires overcoming the bad sides of democratic decision-making. As many recent events have shown us, democratic decision-making and referendums do not always guarantee the best choice, nor the best outcome.

During our travels and conversations, Alija encouraged me to read analyses of artistic and scientific works, and I especially liked to talk with him about Bergson, Kant, and Spengler, whom he particularly appreciated among philosophers. He asked me to explain why the works of Shakespeare and Dostoevsky tell us more about human nature than the analyses of Freud and Jung. Why do we root for tragic heroes in books and movies? Why do we, deep down, refuse to accept the end of the world as science predicts it – with all the coldness, entropy, and meaningless?

He liked to listen to other's opinions about these dilemmas and he enjoyed questioning his own. To the very end, he defended freedom of will as one of the main principles in life, but insisted that with choice always came responsibility for one's decisions. He viewed this world as an endless drama each one of us has to go through – from the moment we were born, to the moment we die. He believed that our goodness is defined by how well we can fight the evil within us. The real war, he told me during the siege of Sarajevo, was not outside, but within us all.

One of my last conversations with Alija was during my Master's studies. I spoke to him about an essay that I was working on at that time, titled, "The Idea of Europe in the Works of Alija Izetbegović." Humbly, and with a gentle smile, he asked me: "So, have you found some ideas about Europe in my books, then?" and continued:

> "Many people credit me with founding the Bosnian army and enabling the defense of Bosnia and Herzegovina, but I share this credit with many other people all over Bosnia, and those abroad who helped us to defend ourselves. Still, there is one thing that I consider to be my greatest, personal contribution – I always strived to gather people who wanted to help Bosnia, putting aside selfishness – be it personal, ethnic, or party-related – in order to create a country that will have enough room for all its peoples and citizens," he said.

For his work, Alija Izetbegović received many international awards and recognitions, such as awards from Turkish state and universities, the award from the Center of Democracy in the USA, and awards from all over the Muslim world.

I hope that these personal memories that I have shared with you will help you understand why Alija was not just a Statesman and a visionary but a true leader of his people, both in war and postwar period.

## Conclusion

Leadership must always be viewed in the context of a certain time period and challenges a society is faced with. Difficult times, such as wars, large economic crises, and political turbulences, require leaders with authority, knowledge, and a vision. But they also call for leaders who have compassion and who offer solutions and inspire hope, especially in the worst of times.

# Some Conceptual and Measurement Aspects of Complexity, Chaos, and Randomness from Mathematical Point of View

**Fikri Öztürk**

**Abstract** One of the main purposes of the mankind is to understand and explain the dynamics of real-world phenomena, i.e., modeling them, and also to build predictive models for their behaviors. The most powerful tool in modeling a deterministic dynamic is mathematics, and the most powerful tool in modeling a stochastic dynamic is statistics. Some characteristics of deterministic chaotic systems are well known, as well as of stochastic systems. Distinguishing deterministic dynamical systems from stochastic ones, based on observed data, is a difficult and yet unsolved statistical problem. Natural phenomena and human behaviors dynamics are very complex. If the existing complexity and chaos in natural dynamical systems, also inherited in their mathematical models, is not well understood, then management and control processes in such systems may result in catastrophes. This study aims to reveal and emphasize the role of mathematics in formulating the conceptual and measurement stages of complexity, chaos, and randomness.

**Keywords** Chaos · Complexity · Randomness · Mathematics

## Introduction

Physicists, chemists, biologists, geologists, and astronomers are trying to understand the truth in the real world. Engineers, economists, sociologists, and all others have their own topic areas and field of interests. When we want to touch a real-world phenomenon with our mind, firstly we have to solve the conceptual stage, and after that we have to solve the measurement problem of qualitative or quantitative characterization of the concept. Conceptualizing stage of a particular feature of a phenomenon is more difficult than measurement stage, such as with the concept of mass, time, or randomness. Probability is a normalized measure of

F. Öztürk (✉)
Ankara University, Ankara, Turkey
e-mail: ozturkf@ankara.edu.tr



Ş. Ş. ERÇETİN, Ş. N. AÇIKALIN (eds.), *Chaos, Complexity and Leadership 2018*, Springer Proceedings in Complexity, https://doi.org/10.1007/978-3-030-27672-0_4

randomness. The problem of measuring randomness seems to have been solved by the development of probability theory, stochastic processes, and statistical theory. Within the development of dynamical systems theory, great progress has been made in conceptualizing and measuring the magnitude of chaos. But, for the concept of complexity the situation changes. For complexity, there is no accepted, single definition. This complicates the problem of grading and measuring complexity. At the modeling stage of real phenomena, in order to introduce mathematics into work, the measurement step should be solved. Keep in mind that mathematical thinking and views are also useful and can be used in the conceptualization of complexity, chaos, and chance.

Measuring is a real-world problem. Science of measurement is the fastest growing science in recent years. It should not be confused with measure theory in mathematics. The notion of measure in mathematics is an abstract idealization of measuring in reality. Analysis of measurements is a statistical topic, for example, the theory of errors. "Any measurement that you make without any knowledge of uncertainty is meaningless" says the well-known and famous Walter Lewin in his Classical Mechanics Lecture 1: Units, Dimensions, and Scaling Arguments, easily accessible on the Internet. The International System of Units developed in 1960, from the meter-kilogram-second (MKS), is the most widely used system of units. Making measurements has its own methods. Nowadays, the working principles of sensors and measuring devices are electronically based. Measuring devices will go toward quantum-based in the future.

In Newtonian mechanics, the concepts distance-mass-time are basic (primary) notions and their units are determined separately. The working principles of measurement instruments related to these concepts have developed in parallel with the progress of science, such as in mass spectroscopy. Each measuring instrument contains its own error. Indeed, every measurement is erroneous. The result of a measuring process is an estimated value (generally the mean of observations) and a confidence interval (including an unbiased estimated value for the variance of the measurement error, calculated by Gaussian error formula).

A measurement in quantum mechanics is an action that determines a particular property (position, momentum, energy, etc.), described by all possible values as a density (wave) function and resulting in single "collapsed" value, when a measurement is performed. John Wheeler's conjecture is that the universe is a self-synthesizing system, where everything is built on the unpredictable outcomes of billions upon billions of elementary quantum phenomena. The question is that how the universe should be so that the physical correlations between the outcomes of experiments will be exactly those predicted by quantum theory (Cabello et al. 2014). Even possible values are random, the average becomes a definite value, for large number of measurements. Indeed, the definite value on which we agree after measuring process is an estimation of the expected value, the value we want to specify. Here, the Central Limit Theorem, which is a theoretically proofed theorem in statistics, works like a natural low in the world of randomness in reality, as well as in measuring process, and ends with understanding the microcomplexity of natural phenomena by macrosimplified model. The measurement problem in quantum mechanics is a difficult problem. The situation is also the same in all other fields. Almost nothing can truly be measured. Any measurement is only an approximation, that is, an estimation, as saying statistically.

Complexity, chaos, and randomness (chance), actually, are words that describe real-world phenomena and behavior. They are often used in everyday conversations and in intellectual languages. Complexity, chaos, and randomness as concepts have a large and broad contents, in such a degree that they have gone beyond concepts and have become research fields. In the next sections of this study, we strive to make a mathematical look at these concepts without going into their depths.

In section "Mathematics and Modeling", the role of mathematics in understanding and modeling of real phenomena is emphasized. Section "Mathematics and Complexity" dwells on complexity, section "Integer Complexity" dwells on chaos, and section "Fractals" dwells on randomness. Some conceptual and measurement aspects of these notions are revived. The study ends with a conclusion section.

## Mathematics and Modeling

With regard to its methodology, it can be said that mathematics is an abstract discipline or science; more accurately, mathematics is an abstract formal science. Let us have a look at Euclidian geometry. The basic notions of geometry are: point, line, and plane. Point, line, and plane, also having some descriptions, are considered as fundamental objects in Euclidean geometry. Basic notions are notions without definitions, because every definition has to be based on another notion. Other notions, beside basic notions, have their own definitions using basic notions and already defined others. In his book *Elements*, Euclid gives five axioms (unproved statements, assumed as true) for plane geometry. The first four axioms are very simple statements that address our intuition as the obvious truths of the real world. The fifth axiom is: For a given line and a point out of the line in a plane, only one line can be drawn through that point, parallel to the given line. At the first hearing of the fifth axiom, every one wonders the situation on the parallel lines at "infinity." The fifth axiom is slightly longer and isn't a simple trusting proposition. For centuries, this axiom has been a research topic of mathematicians and has led to the emergence of non-Euclidean geometries. Let's recall that Euclidean geometry is constructed on five groups of axioms. Geometry, under the name planimetry, deals with lengths, areas, and volumes of real objects. In such applications, measurement is the ratio of the amount which has to be measured to the amount, chosen, and specified as a standard unit. The standard unit of length is meter, which represents a definite physical quantity. In mathematics a measure is a function from a sigma-algebra to nonnegative real numbers, giving zero value to the empty set and satisfying the additivity property. The concept of length in real-world measurements corresponds to the Lebesgue measure defined on the Borel sigma-algebra of real numbers, where, single points, as well as finite and countable infinite sets of points have zero lengths. In the physical world, the zero length is meaningless. Also, time of zero length is meaningless in reality. In the real world, we can never fall in situations, like in Zeno paradox. The distance, that is, the length in the physical world, is bounded from below. The minimum length is around

$$L_p = \sqrt{\frac{2\pi hG}{c^3}}$$

where under the sign of the square root, there are the three constants of nature: Newton's constant $G$, which sets the strength of gravity; the speed of light $c$, which opens up the extended present; and Planck's constant $h$, which determines the scale of the quantum granularity. The length $L_p$ is called the Planck length. Numerically, it is equivalent to approximately $10^{-33}$ cm (Rovelli 2018).

Mathematics is an abstract science with axiomatic methodology. To go from concrete physical samples to abstract mathematical thoughts is a nice teaching method, which is frequently using by teachers. But, abstract mathematical notions are not like the notions in physics, chemistry, biology, which are ontological. It is important to grasp the difference between reality and abstraction.

Natural sciences use mathematics as a tool, such as in Newtonian mechanics, where the objects are represented by points. The methodological basics and the abstractness of mathematics is different from any other branch of the science. Natural sciences seen from methodological point of view are experimental sciences. In their methodology, like in mathematics, there are basic notions and principles, instead of axioms. Theorems are named laws. There is no need to confirm the principles experimentally. They are some kind of laws, naturally accepted and assumed valid in advance. A new hypothesis in order to become as a law has to be verified experimentally. Physics, chemistry, biology, geology, and astronomy are natural sciences. They are using the methods of experimental sciences. Sometimes the data obtained as measurements are direct observations from the nature, as in astronomy, but not from a designed experiment.

Newtonian mechanics is deterministic with regard to its methodology of inference and uses mathematical language and Euclidian geometry as a modeling tool. In observational environment, there are some problems in predicting future events deterministically, by the principle of cause and effect. There are environments where Newtonian mechanics does not work. In general relativity theory, the space and time are not separate, they become as unique concept space-time. In Newtonian mechanics distance, mass and time are separate basic notions, measuring independently from each other. In general relativity theory, gravity is a curvature. In Newtonian mechanics, gravity is a force. The general relativity theory and the quantum mechanics use non-Euclidian geometry and more advanced mathematics such as partial differential equations and differential geometry. For example, mathematical notions like tensors have to be known in order to understand the Einstein's general relativity theory. General relativity theory describes the universe in a very short and simple but not easy equations called Einstein Field Equations:

$$R_{\mu v} - \frac{1}{2} R g_{\mu v} + \Lambda g_{\mu v} = \frac{8\pi G}{c^4} T_{\mu v}$$

where $R_{\mu v}$ is the Ricci curvature tensor, $R$ is the scalar curvature, $g_{\mu v}$ is the metric tensor, $\Lambda$ is the cosmological constant, $G$ is Newton's gravitational constant, $c$ is the speed of light in vacuum, and $T_{\mu v}$ is the stress-energy tensor (Wikipedia). The general theory of relativity is the most beautiful of theories, according to the great Russian physicist Lev Landau (Rovelli 2017).

Quantum mechanics is stochastic with regard to its methodology of inference and uses the statistical language predominantly, as a modeling tool. Quantum mechanics is based on four principles. In addition to Newton's three principles, the fourth principle is the Heisenberg's uncertainty principle which calls statistics in addition to the mathematical language. The aim of Quantum mechanics is to understand the subatomic real-world phenomena.

We can say that mathematics is the mostly used tool in understanding and describing some behaviors or phenomena in the real world; that is, mathematics is the most powerful modeling tool. Mathematical models are strict and abstract explanations of quantitative and qualitative changes of real-world behaviors or phenomena, not forgetting that they are obtained under some assumptions and simplifications of complex relations in reality. Mathematical functions relating variables are simple kind of mathematical models. It's a difficult task to discover the real functional interrelationship among variables directly, except linear, some polynomial and periodical ones. Linearity in relations is the most easily catched one. Sometimes it is more easily to capture linearity in relations among derivatives. Then, the mathematical model becomes a differential equation, a bit difficult model from mathematical point of view. Frequently mathematical models are nonlinear differential and integral equations, which can be solved only numerically.

The last year of the second millennium, the year 2000 was declared by UNESCO as "World Mathematical Year" (Fig. 1). Numerous mathematical activities were held in this year. "How should mathematics prepare itself for the twenty-first century?" was discussed, also. In the nineteenth century, physics became the locomotive of sciences. In order to understand the concept of speed, Newton needed the concept of derivative. And he had to discover this mathematical concept. Which branch of the science will become as a locomotive in the twenty-first century? Which mathematical concepts will be needed? Sociology was predicted to be the locomotive of science in the twenty-first century. Social phenomena, especially the human behaviors, are very complex, not as body robotics and cellular physiology, modeled like motions in Newtonian mechanics and chemical reactions in chemometric. The three-body problem, studied by Henri Poincaré, is the first example of complexity in physics, in the since of lack of adequacy of the classical mathematical models to deal with many body interactions. Real-world phenomena are complex. Mathematical models, like all other types of models, are always simplifications of real truths.

What is the essence of complexity? How can it be measured? Concepts like complexity, chaos, and randomness are used in describing some behavioral aspects of real phenomena. What is the role of mathematics in construction of conceptual

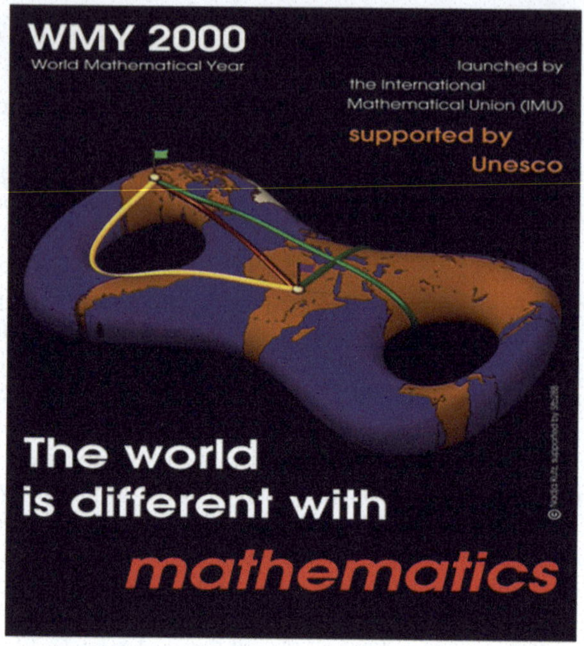


**Fig. 1** World Mathematical Year 2000. UNESCO's announcement

basics of complexity, chaos, and randomness? It can be said that randomness and chaos have satisfactory mathematical explanations and can be measured. Complexity is a bit vague and unsettled concept.

## Mathematics and Complexity

Complexity, as a word, is antonym of simplicity. But, what is simplicity? Complexity itself has to be conceptualized as a term. Complexity is an interdisciplinary term. There are numerous and different definitions of complexity, some of which are only intuitive descriptions. The feeling that the flow of simplicity to complexity increases complexity suggests the need for a measure of complexity.

### *Integer Complexity*

Mathematics as an abstract and axiomatic science has strictly defined concepts (perfect things). In mathematics, there is an easy understandable example for complexity, called integer complexity. Consider representing of natural numbers by arithmetical expressions using 1's, addition, multiplication, and parentheses.

The minimum number of 1's needed in such a representation is called the integer complexity of $n$ and is denoted by $\|n\|$. For example,

$$\begin{aligned}
2 &= 1 + 1\\
3 &= 1 + 1 + 1\\
4 &= 1 + 1 + 1 + 1 = 2 \cdot 2 = (1 + 1) \cdot (1 + 1)\\
5 &= 1 + 1 + 1 + 1 + 1 = 1 + 2 \cdot 2 = 1 + (1 + 1) \cdot (1 + 1)\\
6 &= 1 + 1 + 1 + 1 + 1 + 1 = 2 \cdot 3 = (1 + 1) \cdot (1 + 1 + 1)\\
7 &= 1 + 1 + 1 + 1 + 1 + 1 + 1 = 1 + 2 \cdot 3 = 1 + (1 + 1) \cdot (1 + 1 + 1).
\end{aligned}$$

and

$$\|1\| = 1,\ \|2\| = 2,\ \|3\| = 3,\ \|4\| = 4,\ \|5\| = 5,\ \|6\| = 5,\ \|7\| = 6,\ \|8\| = 6,$$
$$\|9\| = 6,\ \|10\| = 7,\ \|11\| = 8,\ \|12\| = 7,\ \|13\| = 8,\ \|14\| = 8,\ \|15\| = 8,\ \ldots$$

(Iraids et al. 2012). The logarithmic complexity $\|n\|_{\log}$ is defined as $\|n\|/\log_3 n$, and $3 \leq \|n\|_{\log} \leq 4.755$. How is the density spectrum of the logarithmic complexity values of natural numbers? Almost nothing is known with certainty about the structure of this distribution, that is, are the values dense somewhere in the segment [3, 4.755] (Cernenoks et al. 2014)?

Sometimes real-world phenomena look very simple for us and sometimes they look too complicated, in terms of our feelings and perceptions. But, it does not matter to mathematics. In mathematics, to some extent, complexity means complications in logical inferences and increasing the difficulties in proofs and problem-solving. The integer complexity is a well-defined purely mathematical notion, which is easily understandable. We can question what this concept does in reality. It seems that the notion of integer complexity tells us some things outside the intuitional simplicity. The algorithms for calculating integer complexity are itself very complicate (Cordwell et al. 2018).

## *Fractals*

It can be said, that, the concept of complexity entered in the world of mathematics through the gate of fractal geometry. Triangles, rectangles, parallelograms, circles, prisms, pyramids, spheres are simple figures, with regard to their shapes. These figures are useful in modeling the forms and structures of objects. How can we model the form of galaxies, clouds, coastlines, ferns, snowflakes, etc.? Benoit Mandelbrot (1982) invented fractal geometry, which is basically different from other geometries like Euclidian and projective geometries. For example, our understanding of length in the plane Euclidian geometry in essence is the metric in the two-dimensional Euclidian space, which is an inner product space. Rectangles and circles, being subsets of the two-dimensional Euclidian space, are also classified as two-dimensional figures. Figures in fractal geometry, called fractals, may have fractional dimensions, where the definition of dimension is different from that in classical geometry. Fractals can describe shapes, which are jagged, tangled, splintered, twisted, and fractured.

Let $(\mathbf{X}, d)$ be a complete metric space, that is, a Cauchy sequence in $\mathbf{X}$ has a limit $x$ in $\mathbf{X}$ and let $\mathbf{H}(\mathbf{X})$ denotes the space of compact subsets of $\mathbf{X}$, other than the empty set. Define:

$$d(x, B) = \min\{d(x, y) : y \in B\}$$
(the distance from the point $x$ to tne set $B \in \mathbf{H}(\mathbf{X})$)
$$d(A, B) = \max\{d(x, B) : x \in A\}$$
(the distance from the set $A \in \mathbf{H}(\mathbf{X})$ to the set $B \in \mathbf{H}(\mathbf{X})$)
$$h_d(A, B) = d(A, B) \vee d(B, A)$$
(the Hausdorff distance between points $A$ and $B$ in $\mathbf{H}(\mathbf{X})$)

where $a \vee b$ means the maximum of the two real numbers $a$ and $b$. $(\mathbf{H}(\mathbf{X}), h_d)$ is a complete metric space, referred as the "space of fractals" (Barnsley 1988). Fractal geometry studies complicated subsets of geometrically simple spaces. These subsets are generated by simple mappings of the space into itself and possess some invariance properties as in Fig. 2 (Bensoudane et al. 2008; Barnsley 1988).

Consider a triangle with vertices at (0,0), (1,0), (0,1) and apply the geometrical transformation, described in Fig. 2b three times, using Matlab statements below.

```
i=0;j=0;k=0;
for a=0:0.01:1
   i=i+1;
   for b=0:0.01:(1-a)
         j=j+1;k=k+i+j; x(k)=a;y(k)=b;
   end;end
X=[x' y'];plot(X(:,1),X(:,2),'.')
for f=1:3
   k=size(X,1);figure;
   W1=1/2*X;W2=W1+[1/2*ones(k,1) zeros(k,1)];W3=W1+[zeros(k,1)
1/2*ones(k,1)];
   WX=[W1
        W2
        W3];
   plot(WX(:,1),WX(:,2),'.')
end
```

The complicated geometrical figure, called fractal, which will emerges at the end is the well-known Sierpinski triangle (Fig. 3). What about the mathematics of the transformations, where such fractals emerge. An easy understandable one is the Iterated Function System.

Iterated Function System (IFS) is a finite set of contractive operators $\mathbf{T} = \{T_i\}_{i \in \Sigma}$, $\Sigma = 0,.., N - 1$, that operates on points of $\mathbf{X}$, where

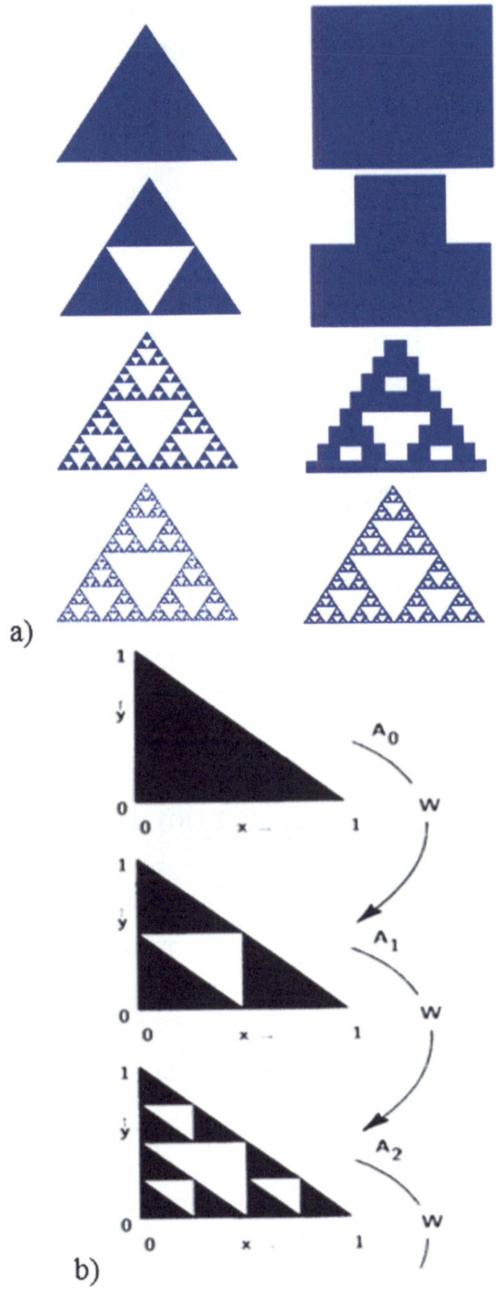


**Fig. 2** Simple spaces and easily conveyed mappings, resulting in fractal called Sierpinski triangle (**a**) Two different simple figures and mapping resulting as Sierpinski triangle (taken from Bensoudane et al. 2008), (**b**) The above transformation W is implemented in Fig. 3 (from Barnsley 1988)

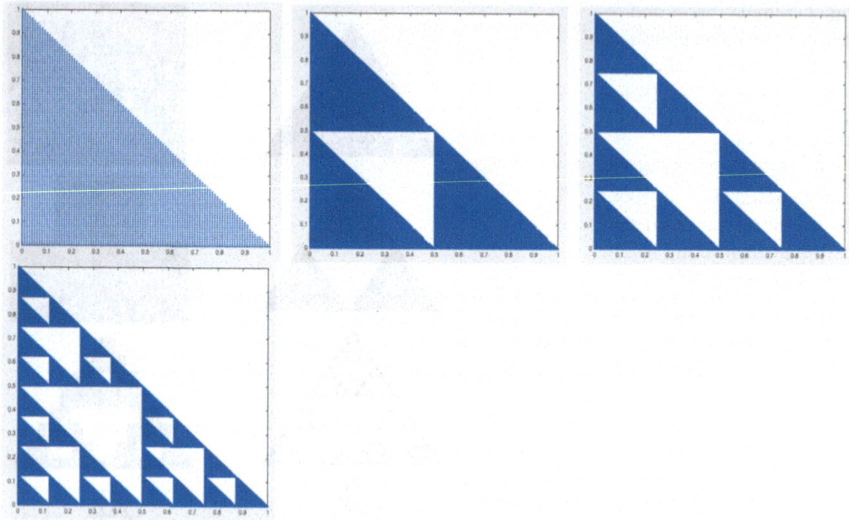

**Fig. 3** Sierpinski triangle. The first three mappings

$$T_i : \mathbf{H}(\mathbf{X}) \to \mathbf{H}(\mathbf{X})$$
$$A \to T_i(A)$$

and there is a constant $c$ $(0 \leq c < 1)$ such that

$$d_h\left(T_i(A), T_i(B)\right) \leq c.d\,(A, B) \quad (\forall A, B \in \mathbf{H}(\mathbf{X})).$$

Then the operator,

$$\mathbf{T} : \mathbf{H}(\mathbf{X}) \to \mathbf{H}(\mathbf{X})$$
$$A \to \mathbf{T}(A) = \bigcup_{i=1}^{N-1} T_i(A)$$

is contractive and admits a unique fixed point, called attractor, defined as

$$\mathbf{A} = \lim_{x \to \infty} \mathbf{T}^n(A), \qquad A \in \mathbf{H}(\mathbf{X}).$$

The attractor **A** verifies the following self-similar property (Barnsley 1988; Bensoudane et al. 2008). Barnsley uses the word "deterministic fractal" in place of "attractor," as a fixed point of a contractive transformation.

$$\mathbf{A} = T_0(\mathbf{A}) \cup T_1(\mathbf{A}) \cup T_2(\mathbf{A}) \cup \cdots \cup T_{N-1}(\mathbf{A})$$

## *Fractal Dimension*

Fractals are complex figures. There are different measures to compare fractal figures with regard to complexity. One of them is the fractal dimension, which is not necessary to be an integer; it may be a fraction, beyond to be a rational number. Let (**X**, *d*) be a complete metric space and let *A* be a compact subset of **X.** Consider a covering of *A* by closed balls of radius *r*, that is,

$$A \subset \bigcup_{i=1}^{n} B\left(x_n, r\right).$$

Let

$$N\left(A, r\right) = \min\left\{\#\left\{x_1, x_2, \ldots, x_n\right\} : x_1, x_2, \ldots, x_n \in A, A \subset \bigcup_{i=1}^{n} B\left(x_n, r\right)\right\}$$

denote the least number of balls of radius *r* needed in the covering. The fractal dimension of *A* is defined as

$$D(A) = \lim_{r \to 0} \frac{\ln\ N\left(A, r\right)}{\ln\left(1/r\right)}$$

when the limit exists. Let, for $0 < a < 1$, $r_n = ca^n$ ($n = 1,2,3, \ldots$ and $c$ is a constant). When the above limit exists, then (Barnsley 1988).

$$D(A) = \lim_{n \to \infty} \frac{\ln\ N\left(A, r_n\right)}{\ln\left(1/r_n\right)}$$

Instead of closed balls, closed squares can be used. Cover $\mathbf{R}^2$ by closed squares of side length 1/2*n* (see Fig. 4). According to the Box Counting Theorem (Barnsley 1988), the fractal dimension of the Sierpinski triangle is

$$D(A) = \lim_{n \to \infty} \frac{\ln\ N_n(A)}{\ln 2^n} = \lim_{n \to \infty} \frac{\ln 3^n}{\ln 2^n} = \frac{\ln 3}{\ln 2} \approx 1.58.$$

## *Hausdorff-Besicovitch Dimension*

Another measure for the complexity of fractals is the Hausdorff-Besicovitch dimension defined with the help of Hausdorff measure. The Hausdorff measure coincides with Lebesgue measure on Lebesgue measurable subsets in Euclidian spaces ($\mathbf{R}^n$,<.,.>). Euclidian spaces are inner product spaces with inner product defined as $<x, y> = \sum x_i y_i$. Euclidian spaces are complete metric spaces with metric

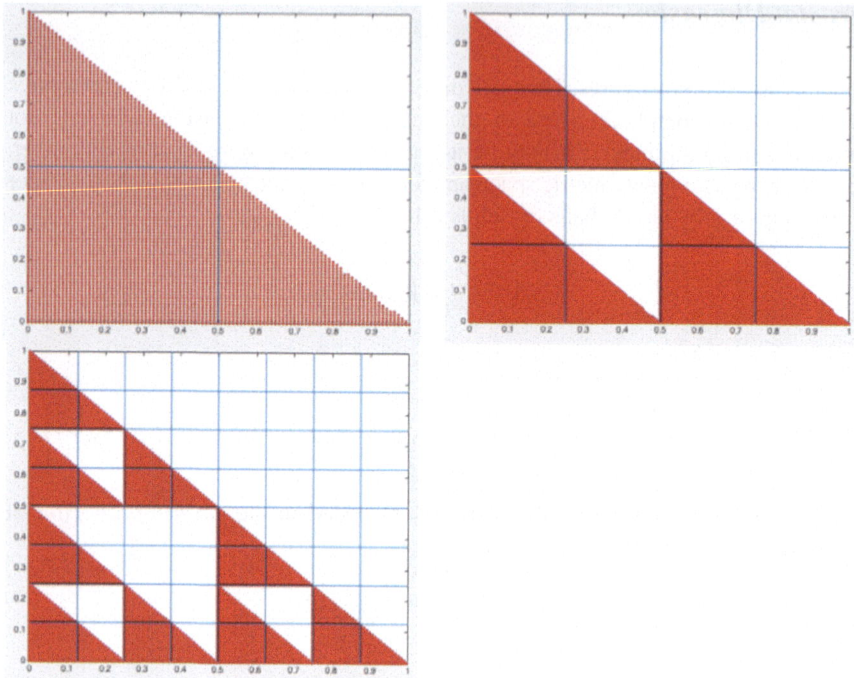

**Fig. 4** Box counting for Sierpinski triangle

$d(x, y) = ||x-y||_2 = (<x-y, x-y>)^{1/2}$, where $||.||_2$ stands for Euclidian norm. The diameter of a nonempty subset $U$ of $\mathbf{R}^n$ is defined as $|U| = \sup\{d(x,y): x,y \in U\}$. If $E \subset \bigcup_{i=1}^{\infty} U_i$ for $0 < |U_i| \leq \delta$, it is said that $\{U_i\}_{i \geq 1}$ is a $\delta$-cover of $E$. For $E \subset \mathbf{R}^n$, $s > 0$ and $\delta > 0$, let $H_\delta^s(E) = \inf \sum_{i=1}^{\infty} |U_i|^s$ where the infimum is over all (countable) $\delta$-covers $\{U_i\}_{i=1}^{\infty}$ of $E$. The $s$-dimensional Hausdorff outer measure of $E$ is defined as

$$H^s(E) = \lim_{\delta \to 0} H_\delta^s(E) = \sup_{\delta > 0} H_\delta^s(E).$$

For any $E$, $H^s(E)$ is nonincreasing as $s$ increases from 0 to $\infty$. For $s < t$,

$$\mathcal{H}_\delta^s(E) \geq \delta^{s-t} \mathcal{H}_\delta^t(E)$$

which implies that if $H^t(E)$ is positive, then $H^s(E)$ is infinite. Thus, there is a unique value, $\dim_{HB}(E)$, called the Hausdorff-Besicovitch dimension of $E$, such that $H^s(E) = \infty$ if $0 \leq s \leq \dim_{HB}(E)$, $H^s(E) = 0$ if $\dim_{HB}(E) < s < \infty$ (Falconer 1982, 2003). The s-dimensional Hausdorff outer measure of the Sierpinski triangle $A$ is

$$\mathrm{H}^s(A) = \lim_{\delta \to 0} \left( \inf \sum_{i=1}^{\infty} |U_i|^s \right)$$

$$= \lim_{n \to \infty} \left( 3^n \left( \frac{\sqrt{2}}{2^n} \right)^s \right) = \begin{cases} 0, & s < \ln 3 / \ln 2 \\ \left(\sqrt{2}\right)^{\ln 2/\ln 3}, & s = \ln 3 / \ln 2 \\ \infty, & s > \ln 3 / \ln 2 \end{cases}$$

where $U_i$ is a closed circle with radius $\sqrt{2}/2^i$, $i = 1,2,\ldots$ and $3^n$ is the minimum of the numbers of circles needed in the covering (Fig. 5). The Hausdorff measure of the Sierpinski triangle is $\mathrm{H}^s(A) = \left(\sqrt{2}\right)^{\ln 2/\ln 3} \approx 1.44$, which is a measure for its "field size." The Hausdorff-Besicovitch dimension of the Sierpinski triangle is $\dim_{HB}(A) = \ln 3/ \ln 2 \approx 1.58$, which is equal to its fractal dimension D($A$). Generally, $0 \le \dim_{HB}(A) \le D(A) \le m$ for $A \in H(\mathbf{R}^m)$. Hausdorff-Besicovitch dimensional measure is useful in comparing the "sizes" of fractals having equal fractional dimensions (Barnsley 1988).

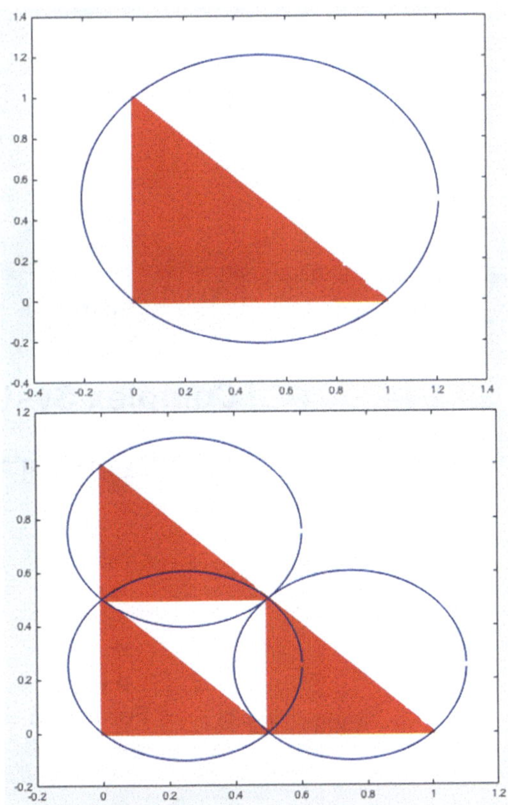

**Fig. 5** Covering stage

## *Complexity Theory*

Complexity theory is a scientific field that studies the common properties of systems considered "complex" in nature. Complex system is a broad term encompassing a research approach to problems in many diverse disciplines (Fig. 6). For example, Tomé and Açıkalın (2017) explain the contribution of complexity theory to the understanding of the international system and its change, emphasizing that the international reality is much more complex and unpredictable in its evolution than those aspects that are in keeping with cognitive structures and natural expectations of conventional international relations theories. They notify five fundamental notions, which compete for a better understanding of complex systems: system, pattern, network, scale, and nonlinearity, once again emphasizing that the most important feature of complex systems is the concept of emergence. Emergency is a function of the nature of recursive causation in complex systems.

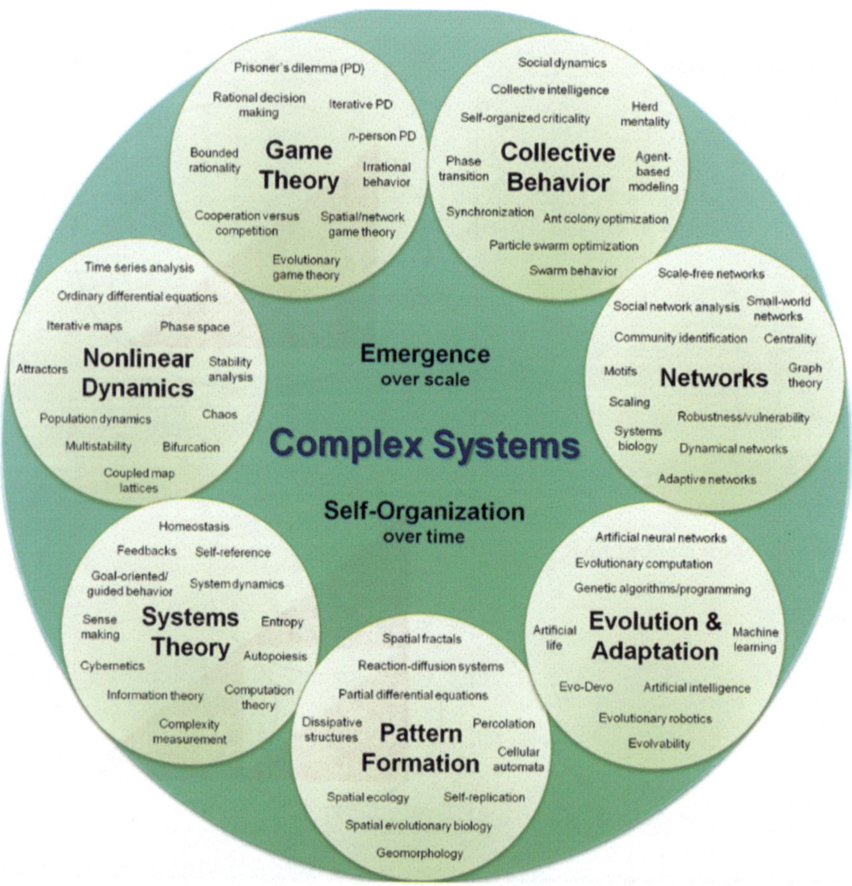


**Fig. 6** (Wikipedia)

## Mathematics and Chaos

As is well known, the stability theory has been extensively explored and developed for a long time. Also, the concept of chaos was not new when it was put in focus in the second half of the last century, after the invention of the computers. The development of chaotic theory has been intensified by enormous developments in computing and graphical capabilities. Chaos is a typical behavior of deterministic nonlinear dynamical systems.

### *Dynamical Systems*

Dynamical systems are very important, both in the real phenomenal world and in interdisciplinary research. In mathematics, a dynamical system is a system in which a function describes the time dependence of a point in a state space. A dynamical system is a triple $(\mathbf{X},\Phi,G)$ where $\mathbf{X}$ denotes the state space, $\Phi$ represents the flow of the system as a continuous map from $G \times \mathbf{X}$ into $\mathbf{X}$, and $G \subseteq \mathbf{R}$ stands for the time. When $G$ is the set of integers $\mathbf{Z}$ or $G = \mathbf{Z} + \cup\{0\}$ then the dynamical system is called discrete. When $G = \mathbf{R}$ then the system is called continuous. Let $\mathbf{X}$ be a metric space and $\Phi(n,x) = f^{\,\mathrm{on}}(x)$, where $f^{\,\mathrm{on}} = f \circ f^{n-1}, n \geq 1, f^{\,0} =$ identity, with $f \circ g$ denoting the composition $f \circ g\ (x) = f(g(x))$. Such a dynamical system is denoted by the pair $(\mathbf{X}, f)$ (Balibrea 2006). The main question is given a function $f$ and an initial value $x_0$, what happens to the sequence of iterates $\{x_0, f(x_0), f(f(x_0)), f(f(f(x_0))), \ldots\}$? Often, the evolution function $f$ is deterministic. Very simple functions, such as $f(x) = 4x(1 - x)$ for $x_0$ in [0,1], lead to bizarre and unpredictable results when iterated.

There are some similarities between notions of a dynamical system and Iterated Function System (IFSS), given in the preceding section. A dynamical system $(\mathbf{X},\Phi,G)$, with $\Phi(n,x) = f^{\mathrm{on}}(x)$ and $G = \{0,1,2,\ldots,N-1\}$ is an IFS on $\mathbf{X}$, denoted by $(\mathbf{X},f)$. So, dynamical systems are sources of deterministic fractals, because of that they are some kind of IFS. Consider the logistic map, where $\mathbf{X} = [0, 1] \subset \mathbf{R}$ and

$$\begin{gathered} f : [0, 1] \to [0, 1] \\ x \to f(x) = rx\,(1 - x)\,, \quad (0 < r \leq 4) \end{gathered}$$

$$\begin{gathered} f^n : [0, 1] \to [0, 1] \\ x \to f^n(x) = \underbrace{f\,(f\,(\ldots f(x)))}_{n\ times}. \end{gathered}$$

Let $\mathbf{H}(\mathbf{X})$ be the set of non-empty compact subsets of the closed interval $\mathbf{X} = [0,1]$. For $A = [0.8,0.9]$ in $\mathbf{H}(\mathbf{X})$ and $r = 1, 2.5, 3, 3.5, 3.75, 4$ the images $f$

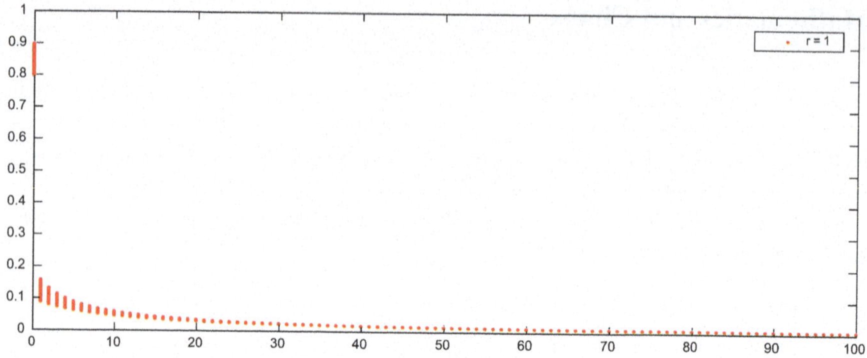


**Fig. 7** The images of $f^{\circ n}([0.8,0.9])$ for $n = 0,1,2,3,\ldots,100$, when $r = 1$

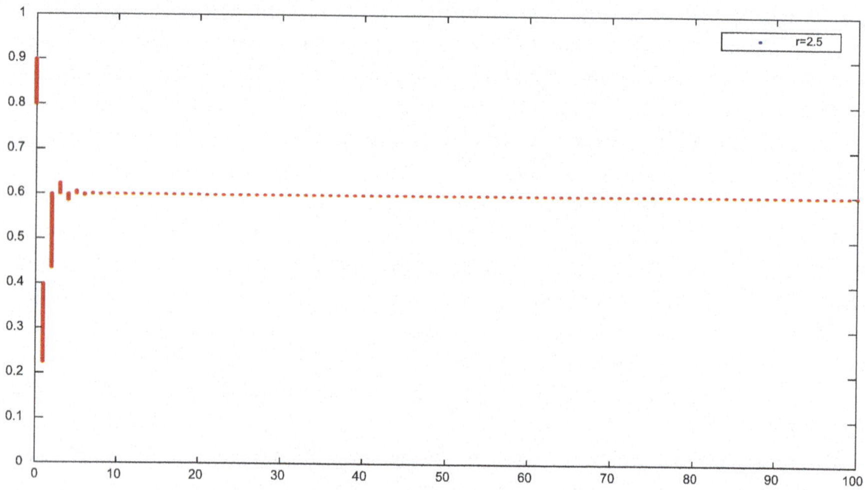


**Fig. 8** The images of $f^{\circ n}([0.8,0.9])$ for $n = 0,1,2,3,\ldots,100$, when $r = 2.5$

${}^{\circ n}(A)$ in $\mathbf{H}(\mathbf{X})$, $n = 0,1,2,\ldots,100$ are as in figures below. Different types of attractors occur.

As is seen above, the flow of the set $A = [0.8,0.9]$, that is $\Phi(n,A) = f^n(A)$, $n = 1,2,3,\ldots$ "ends" in a set, which is a single point, attracting the set $A$ to itself, for $r \leq 3$ (Figs. 7, 8, and 9). Sometimes the attractor is a finite set (Fig. 10), and sometimes infinite, seemingly, as for $r = 3.75$ and $r = 4$, for example (Figs. 11 and 12).

Now, let us see the flow of $\mathbf{X} = [0,1]$, looking at images $\Phi(n,\mathbf{X}) = f^n(\mathbf{X})$, $n = 1,2,3,\ldots,100$ in figures below. As can be seen, $f(\mathbf{X})$ is a subset (proper subset) of $\mathbf{X}$, for $0 < r < 4$ and $f(\mathbf{X}) = \mathbf{X}$, for $r = 4$ (Figs. 13, 14, 15, 16, 17, and 18).

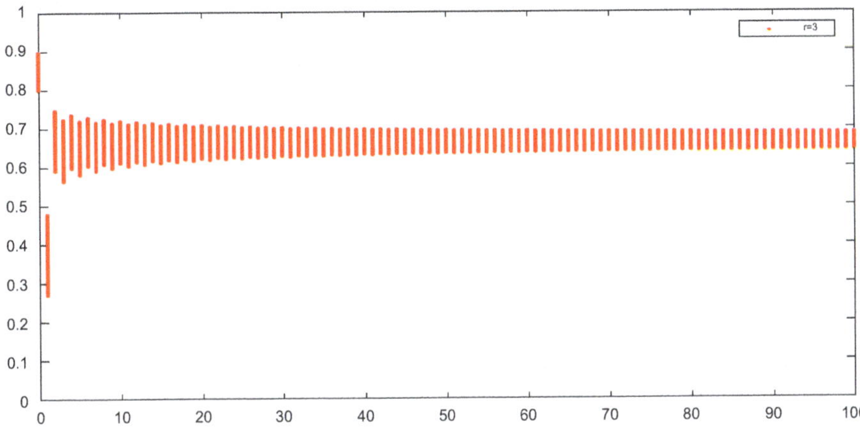


**Fig. 9** The images of $f^{\circ n}([0.8,0.9])$ for $n = 0,1,2,3,\ldots,100$, when $r = 3$

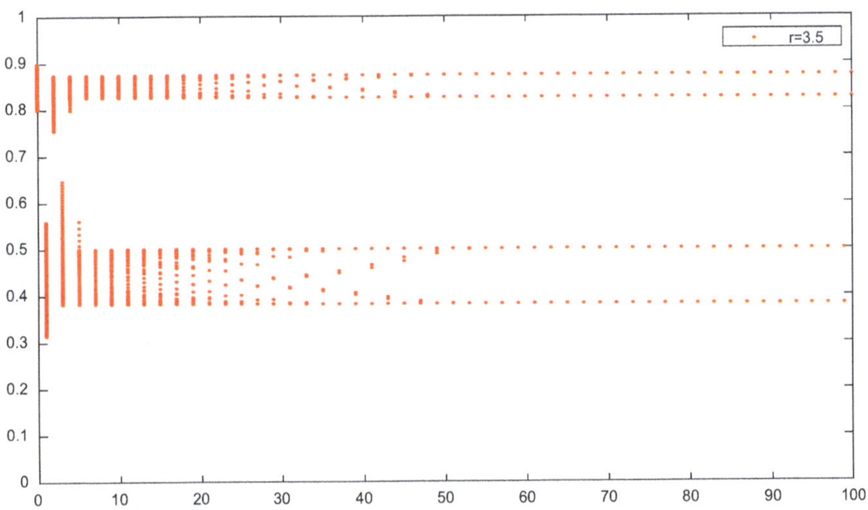


**Fig. 10** The images of $f^{\circ n}([0.8,0.9])$ for $n = 0,1,2,3,\ldots,100$, when $r = 3.5$

For some values of $r$, the flows obey "complicated" behavior, which can be more easily observed on the orbits or trajectories $(f^n(x))_{n=0,1,2,\ldots}$. When $f(x) = x$, then $x$ is called a fixed point of the system. When $f^m(x) = x$ for some $m$, then the minimal number $m$ is called the period of $x$. If $x$ is periodic of period $m$, then it is also periodic of periods $k.m$ where $k$ is any positive integer.

Let us have a look on the behavior of the trajectories of the logistic map, that is, have a look on the set of points $(n,x_n)$, $n = 0,1,2,\ldots$ connected with line segments in the coordinate $x - y$ plane, where $x_n = f^n(x_0)$. The sequence of $N = 100$ iterations of

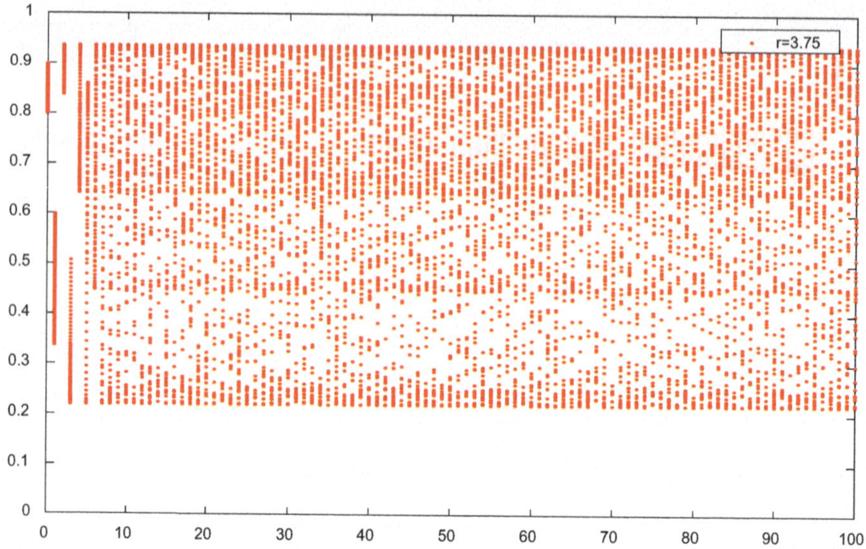


**Fig. 11** The images of $f^{\circ n}([0.8,0.9])$ for $n = 0,1,2,3,\ldots,100$, when $r = 3.75$

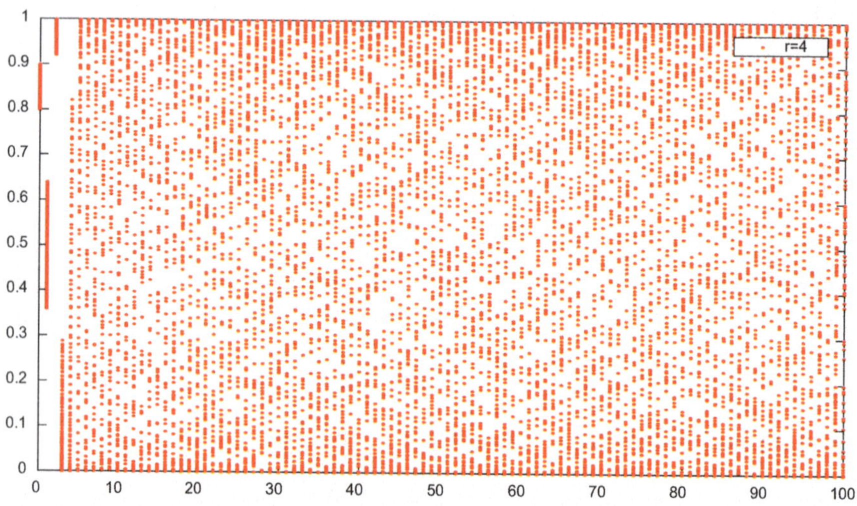


**Fig. 12** The images of $f^{\circ n}([0.8,0.9])$ for $n = 0,1,2,3,\ldots,100$, when $r = 4$

the logistic equation, which is also used as a model for population growth (Murray 1993), starting from $x_0 = 0.9$, is as in Figs. 19, 20, and 21, for the values of $r = 2.5$, 3.5, 4. The situations of steady-state, periodic oscillations and complicated ones are easily observed.

$$x_{n+1} = rx_n - rx_n^2 = rx_n\,(1 - x_n)$$

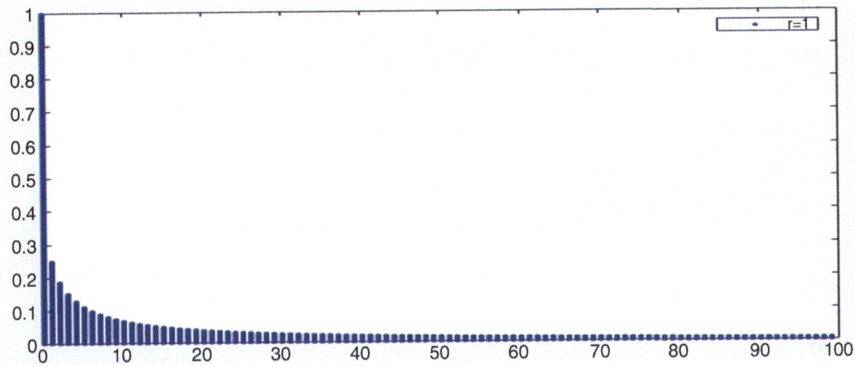


**Fig. 13** The images of $f^{\circ n}([0,1])$ for $n = 0,1,2,3,\ldots,100$, when $r = 1$

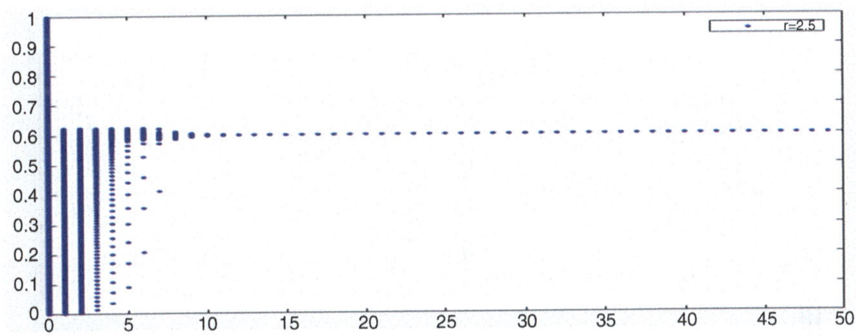


**Fig. 14** The images of $f^{\circ n}([0,1])$ for $n = 0,1,2,3,\ldots,100$, when $r = 2.5$

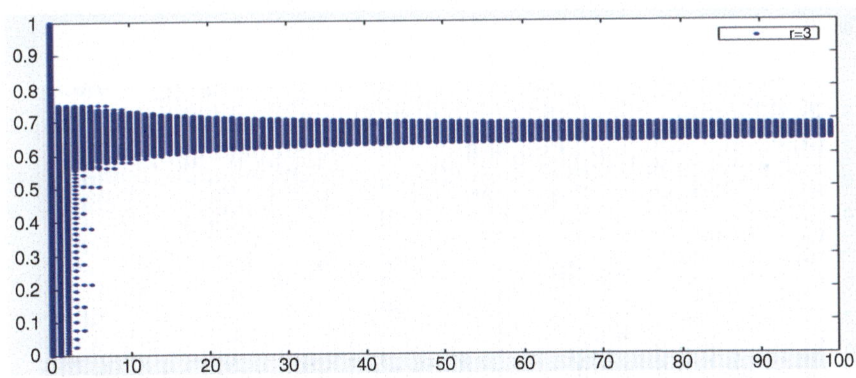


**Fig. 15** The images of $f^{\circ n}([0,1])$ for $n = 0,1,2,3,\ldots,100$, when $r = 3$

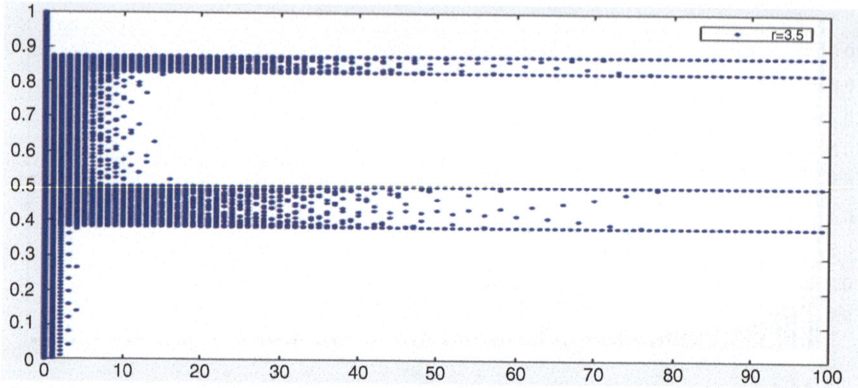


**Fig. 16** The images of $f^{on}([0,1])$ for $n = 0,1,2,3,\ldots,100$, when $r = 3.5$

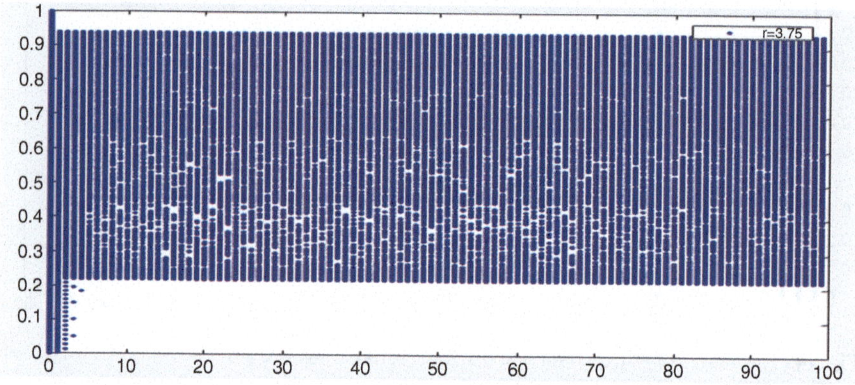


**Fig. 17** The images of $f^{on}([0,1])$ for $n = 0,1,2,3,\ldots,100$, when $r = 3.75$

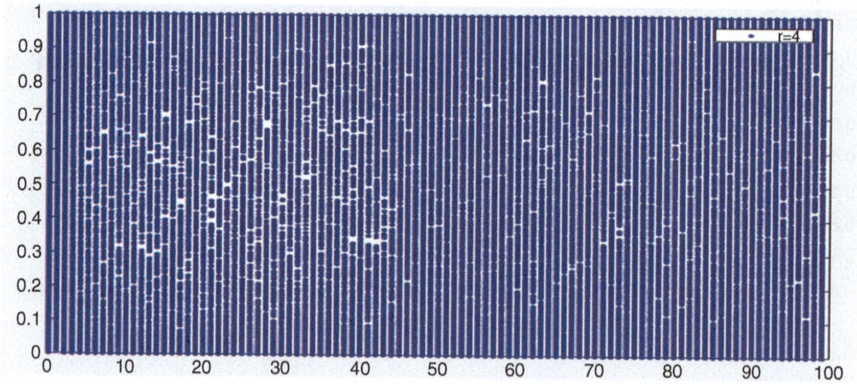


**Fig. 18** The images of $f^{on}([0,1])$ for $n = 0,1,2,3,\ldots,100$, when $r = 4$

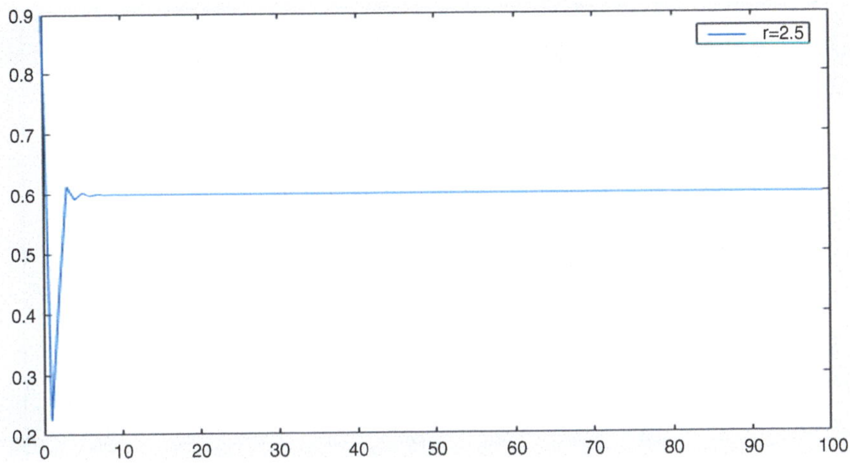


**Fig. 19** The trajectory of logistic growth for $r = 2.5$ and $x_0 = 0.9$

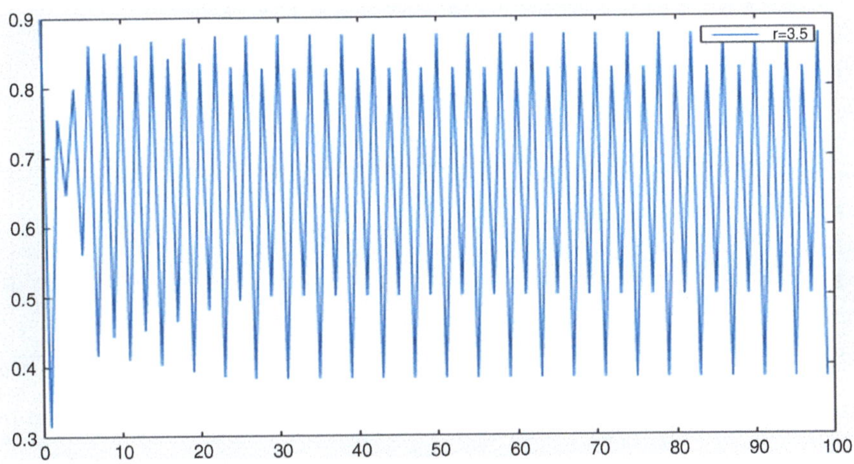


**Fig. 20** The trajectory of logistic growth for $r = 3.5$ and $x_0 = 0.9$

## Chaos

Robert L. Devaney (2003) recall that the three ingredients of chaotic dynamics are (1) sensitive dependence on initial conditions, (2) topological transitivity (mixing property), and (3) dense periodic points. The sensitivity to initial conditions is the prominent property that illustrates the chaotic nature of a dynamical system. Sensitivity to initial conditions means that each point in a chaotic system is arbitrarily closely approximated by other points with significantly different future paths, or trajectories. Starting with a limited amount of information, as is usually the

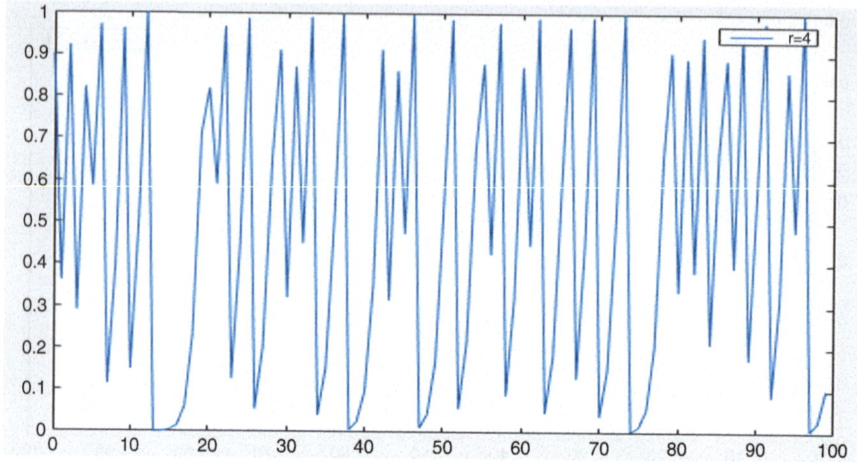


**Fig. 21** The trajectory of logistic growth for $r = 4$ and $x_0 = 0.9$

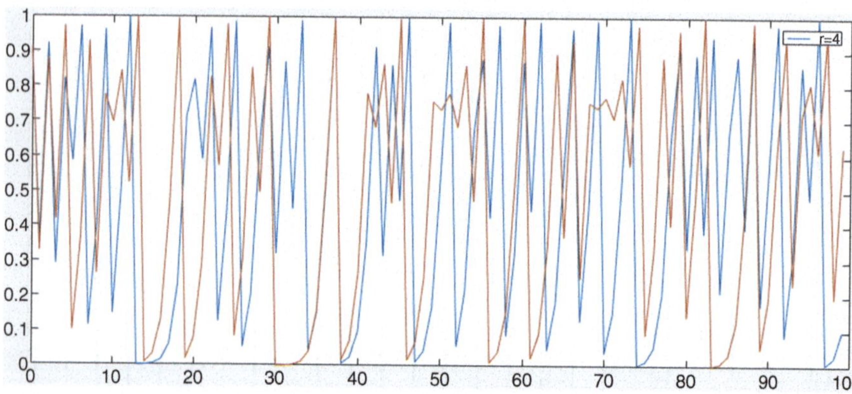


**Fig. 22** The trajectories of logistic growth for $r = 4$ and initial values $x_0 = 0.9$ (blue), $x_0 = 0.91$ (red)

case in practice, the system is no longer predictable. A small change, or perturbation, of the current trajectory may lead to significantly different future behaviors (Fig. 22).

Li and Yorke (1975) showed that if an interval or line map into itself has periodic points of period three, then the map has periodic points of all periods, considered as chaos. They proved also that, if there exists a periodic point of period three, then there is an uncountable invariant set $S \subset \mathbf{X} = [0,1]$, such that for all distinct $x, y \in S$,

(a) $\limsup |f^n(x) - f^n(y)| > 0$,
(b) $\liminf |f^n(x) - f^n(y)| = 0$.

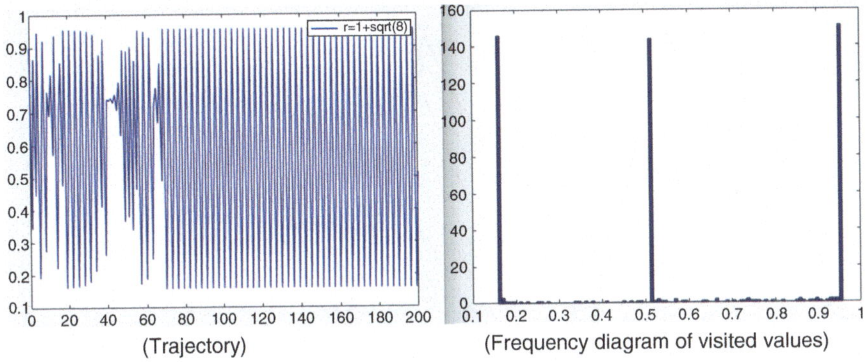


**Fig. 23** Logistic growth with parameter $r = 1 + \text{sqrt}(8)$. Population oscillates on three values

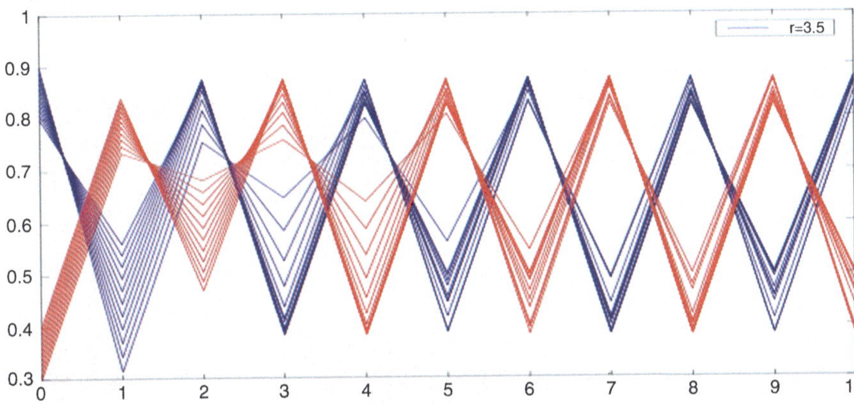


**Fig. 24** Ten stages in flow course of the sets $A = \{x_0{:}x_0 = 0.8{:}0.01{:}0.9\}$ (blue) and $B = \{x_0{:}x_0 = 0.3{:}0.01{:}0.4\}$ (red). There is no mixing, for $r = 3.5$

Conditions (a) and (b) mean that the orbit of different points $x$ and y separate and are close at infinite iterations, which is sensitivity to initial conditions. For $r = 1 + \sqrt{8}$, the logistic growth shows oscillation among three values (Fig. 23).

The mixing property ingredient of the chaos in the logistic growth model can be easily visualized, as in Figs. 24 and 25.

The property of "dense periodic points" in logistic growth, also, can easily be visualized geometrically. In the logistic growth dynamical system $f(\mathbf{X}) = \mathbf{X}$, for $r = 4$, where $\mathbf{X} = [0,1]$. The case is easily visualized by the well-known cobweb diagram in Fig. 26. Now, in order to visualize the "dense periodic points" ingredient of chaos, let us visualize the points (Fig. 27) visited at the evolution process up to last step (vertically located points in Fig. 28) with respect to step number (horizontal axe). The set of points visited up to "infinite" step in a single trajectory is a dense set in $\mathbf{X} = [0,1]$; that is, its closure is $\mathbf{X}$. The set of accumulating points in the growth sequence, that is, the attractor of the dynamical system, is a dense set in $\mathbf{X}$.

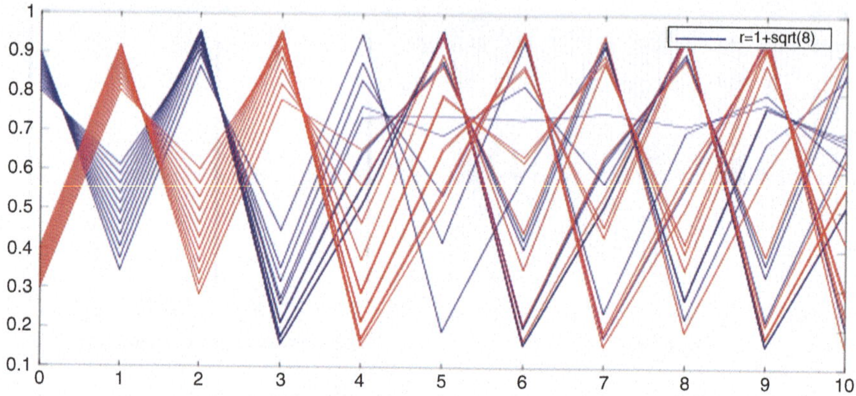


**Fig. 25** Ten stages in flow course of the sets $A = \{x_0:x_0 = 0.8:0.01:0.9\}$(blue) and $B = \{x_0:x_0 = 0.3:0.01:0.4\}$ (red). Mixing, for $r = 1 + \text{sqrt}(8)$

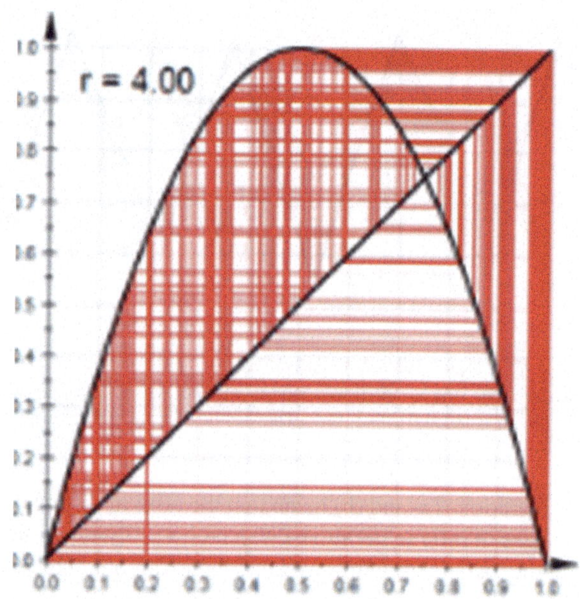


**Fig. 26** The cobweb diagram (Wikipedia)

## *Bifurcation Diagram of the Logistic Map*

The bifurcation diagram in Fig. 29 illustrates the attractor sets of logistic growth model as a function of the values of parameter $r$. Attractor sets are visualized as points on the vertical lines passing through the points ($r$, $0$) for $2.4 \leq r \leq 4$. At the bifurcation points, the system changes behavior, doubles the points of attraction, or may go to erratic behavior. For most values of $r$ beyond 3.56995, the growth exhibits chaotic behavior. There are isolated ranges of $r$, called islands of stability, where

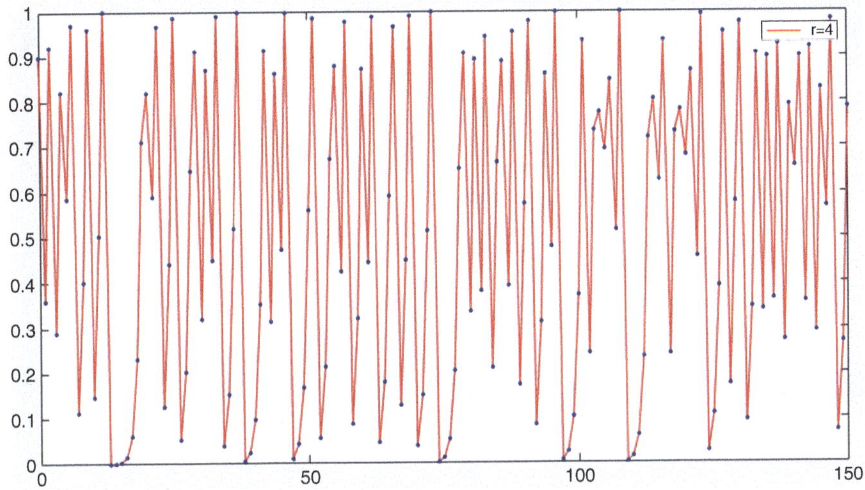


**Fig. 27** Logistic growth (150 steps), $r = 4$. The points visited at the evolution process up to last step are visualized

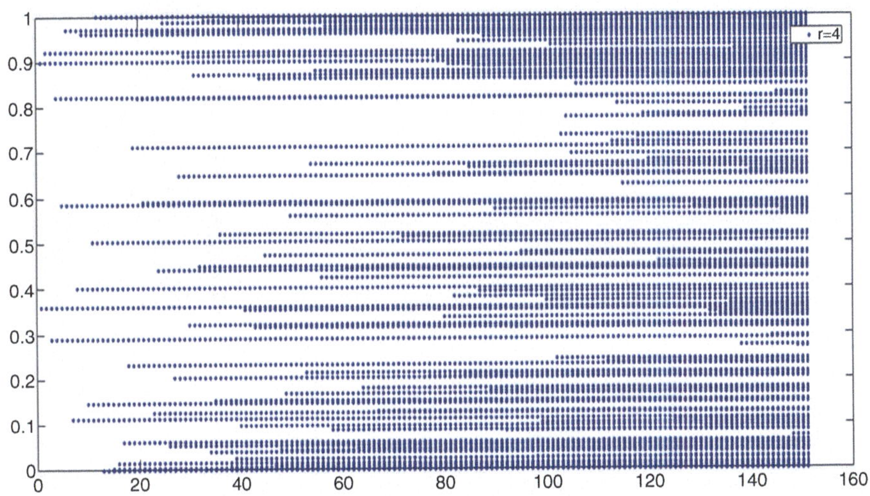


**Fig. 28** One hundred and fifty stages, showing the visited values (points up to last step) in the flow course of logistic growth, $r = 4$. Notice the emergency of new points

the system has non-chaotic behavior. The bifurcation diagram is self-similar. If you zoom in $r = 1 + \text{sqrt}(8)$ and focus on one arm of the diagram tree, the situation nearby looks like a shrunk and slightly distorted version of the whole diagram. The bifurcation diagram for the logistic map is full of interesting fractal subsets. The Feigenbaum attractor (between arrows in Fig. 30) is the set of points generated by successive iterations, for $r = 3.57$, where the period doubling is infinite. For

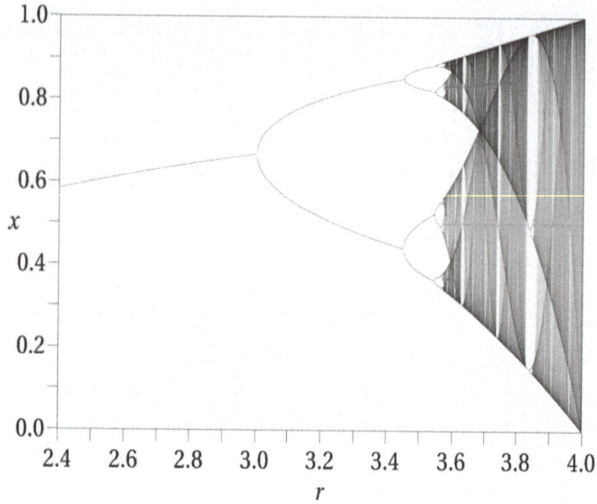


**Fig. 29** Bifurcation diagram for the logistic map for $2.4 \le r \le 4$ (Wikipedia)

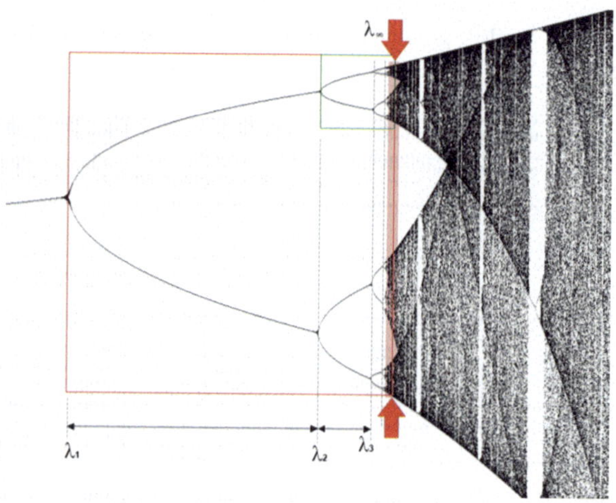

**Fig. 30** Feigenbaum Fractal (Wikipedia)

the Feigenbaum's attractor, the Hausdorff dimension is equal to 0.538 (Wikipedia). The logistic growth model is a nice example for complexity in dynamical systems, especially with emergence of the bifurcation diagram.

## *Lyapunov Exponent*

Aleksandr Lyapunov created the modern theory of the stability of dynamical systems. Lyapunov exponent of a dynamical system is a quantity that characterizes the rate of separation of infinitesimally close trajectories. In a discrete-time dynamical system $x_{n+1} = f(x_n)$, for an orbit starting from $x_0$, the Lyapunov

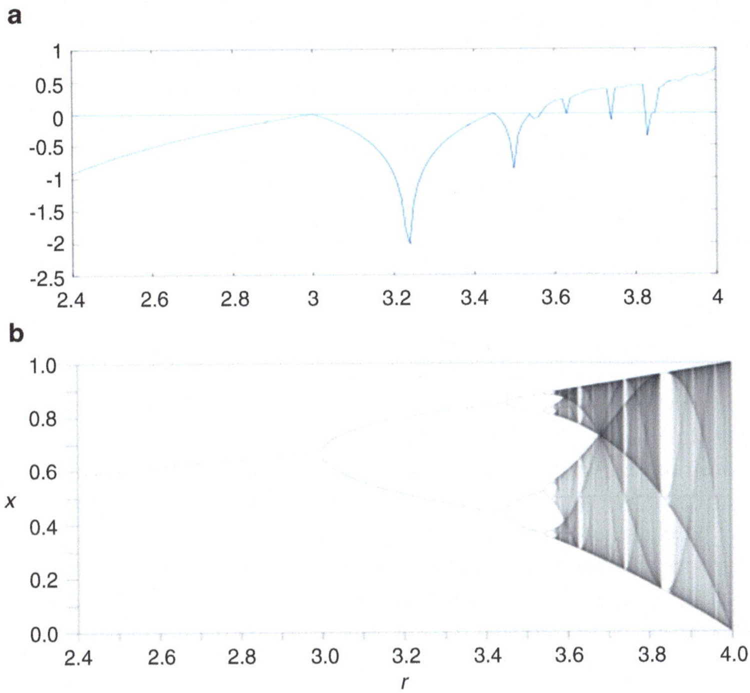


**Fig. 31** (**a**) The graph of Lyapunov exponent values versus *r*. (**b**) Bifurcation diagram

exponent is defined as

$$\lambda(x_0) = \lim_{n\to\infty} \frac{\sum_{i=0}^{n-1} \ln\left(\left|f'(x_i)\right|\right)}{n}.$$

The logistic growth evolution function $f(x_n) = rx_n(1 - x_n)$, $(0 < r \leq 4)$ can be considered as a function of the parameter $r$. So, the Lyapunov exponent is, also, a function of $r$.

$$\lambda(x_0, r) = \lim_{n\to\infty} \frac{\sum_{i=0}^{n-1} \ln\left(\left|f'(x_i, r)\right|\right)}{n}.$$

The graph of the Lyapunov exponent function $\lambda(0.9,r)$ with respect to $r$ is as in Fig. 31a, which is the same for other initial $x_0$ values. The chaotic behaviors of the logistic growth occur for positive Lyapunov exponent values, as is seen in Fig. 31b. Chaotic landscapes can be detected in the Lyapunov exponent graph.

## *Phase Space*

Given a dynamical system, it is customary to plot the state variables in a phase space, collapsing the time information in the process. The attractors can easily be observed on phase diagrams. Consider the discrete coupled logistic model for symbiotic interaction of two species, (Lopez-Ruiz et al. 2002; Lopez-Ruiz and Fournier-Prunaret 2004). The coupling depends on the population size of both species and on a positive constant $\lambda$, called the mutual benefit. Different dynamical evolutions occur, for different values of $\lambda$, as in Fig. 32 ($\lambda = 1$) and Fig. 33 ($\lambda = 1.031$).

$$\begin{aligned} x_{n+1} &= \lambda\,(3y_n + 1)\,x_n\,(1 - x_n)\,, \quad n = 0, 1, 2, \ldots \\ y_{n+1} &= \lambda\,(3x_n + 1)\,y_n\,(1 - y_n) \end{aligned}$$

Lopez-Ruiz and Fournier-Prunaret (2004) investigated thoroughly the attractive closed invariant curves and the fractal structure in the neighborhood of the point $(x,y) = (0.69,0.69)$ for $\lambda = 1.031$. Dynamical systems often possess attractors and may be classified on the basis of the effects of the dynamics on a region of the phase space (Alligood et al. 2000). Some systems have fractal attractors, also known as strange attractors. Chaos and fractals are not synonymous, although the two concepts are often conflated (Lowen and Teich 2005).

## Mathematics and Randomness

A special session on "Managing arrow, bow, and complexity: Traditional Turkish Archery" included in the program of the sixth International Symposium on Chaos, Complexity, and Leadership 2018 puts the issue of randomness into scene. The flight of an arrow toward the target is a highly complicated ballistic motion (transverse oscillations, vibrations), chaotic (sensitivity to initial conditions), and random, depending on many factors, some of which are external natural effects and others physiological and muscular effects of archers. Besides the historical research on the role of the arrow and the bow as a weapon, there is a vast literature on archery research as a sport branch. Dart games and archery are as important as dice experiments during the understanding of randomness in probability courses, especially in case of continuous random variables. As is well known, a stochastic phenomenon can be modeled with a probability space. The randomness that causes an event to occur is modeled with a suitable probability measure. The connection between randomness inherent in a real-world phenomenon and an abstract probability measure is based on Kolmogorov's Axioms (Kolmogorov 1956). Our knowledge about the randomness of real-world phenomenon is only that, what an accepted appropriate model (probability distribution) is saying to us. (Atakan et al. 2017).

The introduction of probability theory and statistics to natural and social sciences has been through demography, thermodynamics, and quantum mechanics. In thermodynamics, the Shannon measure of entropy of a system with finite or

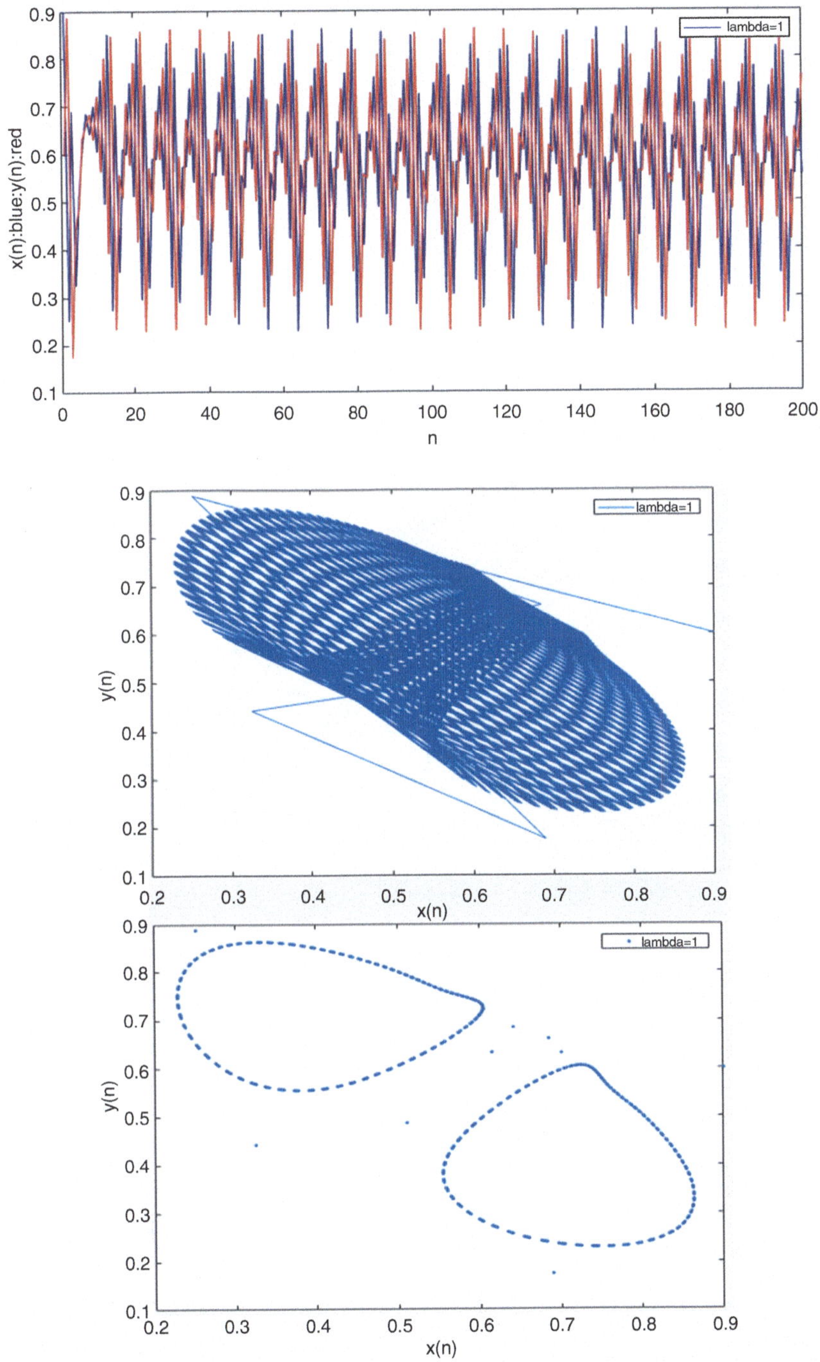


**Fig. 32** Logistic growth trajectories (top) and phase diagrams (bottom) for $\lambda = 1$. Time course of logistic growth for both species: $x_n$ (blue), $y_n$ (red), $x_0 = 0.9$, $y_0 = 0.6$. The phase diagram (successive states connected). The phase diagram (the set of visited states)

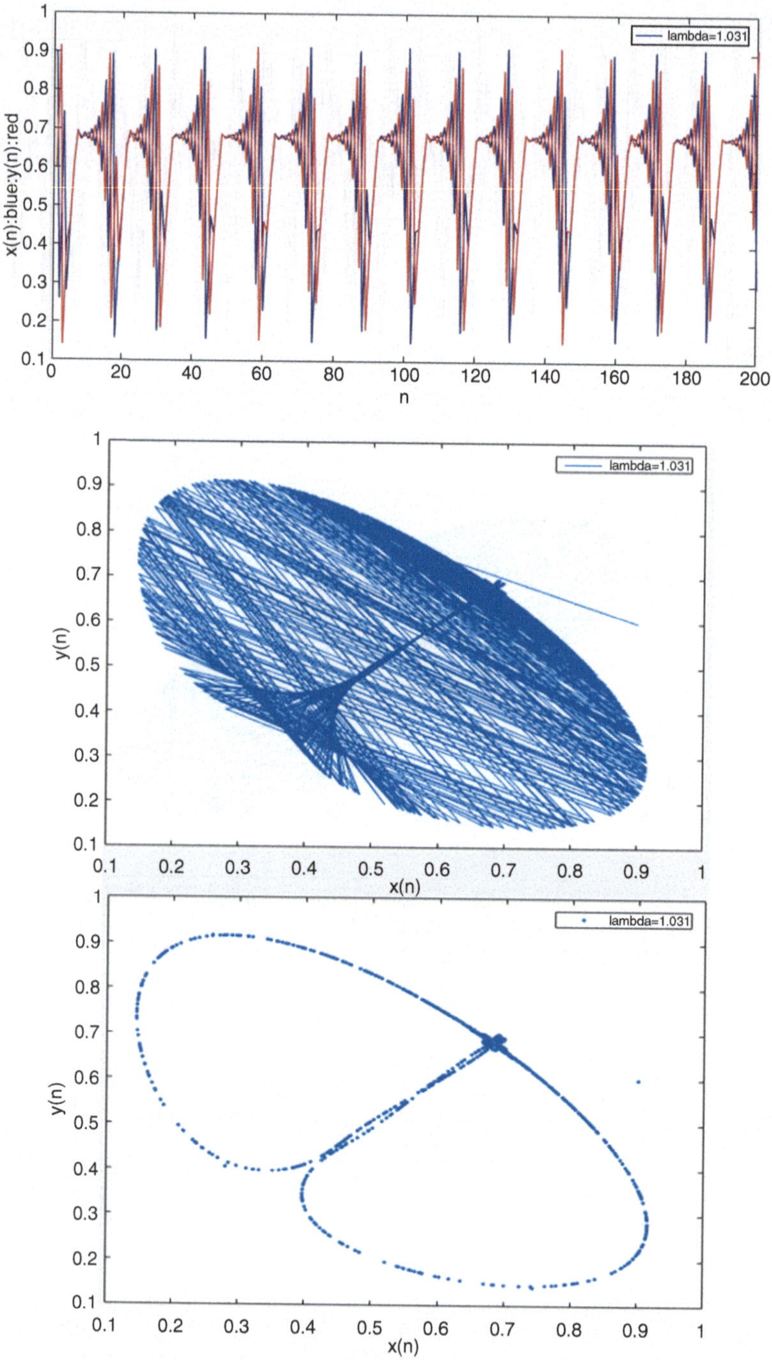


**Fig. 33** Logistic growth trajectories (top) and phase diagrams (bottom) for $\lambda = 1.031$. Time course of logistic growth for both species: $x_n$ (blue), $y_n$ (red), $x_0 = 0.9$, $y_0 = 0.6$. The phase diagram (successive states connected). The phase diagram (the set of visited states)

infinite possible states $x_n$ ($n = 1,2,\ldots,N$ or $n = 1,2,\ldots$), having probabilities $p(x_n)$, is defined as $\sum p(x_n)\log_2(p(x_n))$. The Shannon measure of entropy is a measure of disorder. As a result, disorder has its size, which is measured by Shannon entropy measure. The maximum entropy occurs under uniform probability distribution of states. Perhaps, because of that, the randomness often is thought as uniformly distributed probability measure. Remind that the probability measure acts on events and don't measure the degree of randomness. Probability distributions can be degraded by Shannon measure of entropy, with regard to randomness, indeed with regard to disorder and complexity.

Brownian motion is a nice real-world example for randomness, although the trajectory of the motion exhibits a chaotic behavior. In mathematics, Brownian motion is described by the Wiener process. A Markov process whose finite-dimensional distributions are normal distributions is called Wiener process. The Wiener process $(W_t)_{t \geq 0}$ is characterized by the following four facts:

1. $W_0 = 0$ almost surely ($P(W_0 = 0) = 1$).
2. Trajectories are continuous (almost surely).
3. $(W_t)_{t \geq 0}$ has independent increments.
4. $W_t - W_s \sim N(0,t - s)$, for $0 \leq s \leq t$.

The Wiener process can be constructed as the scaling limit of a random walk. Let $\xi_1$, $\xi_2$, ... are independently and identically distributed random variables with mean 0 and variance 1. For each $n$,

$$W_n(t) = \frac{1}{\sqrt{n}} \sum_{1 \leq k \leq \lfloor nt \rfloor} \xi_k$$

is a step function of $t$. The increments $W_n(t) - W_n(s)$ $(t > s)$ of $W_n$ are independent. For large $n$, the distribution of $W_n(t) - W_n(s)$ is close to $N(0, t - s)$ by the central limit theorem, and as $n \to \infty$, $W_n$ approaches a Wiener process by the Donsker's theorem (Wikipedia). Figure 34 shows a simulated trajectory of a such constructed Wiener process on [0,10].

The trajectory of a Wiener processes is scale invariant, meaning that $c^{-1}W_{c^2t}$ is a Wiener process for any nonzero constant $c$. The most prominent property of sample paths is that they are continuous everywhere but differentiable nowhere. A Wiener process trajectory is a fractal of Hausdorff dimension 1.5. The set of zeros of a Wiener process, that is, the set of cut points of the trajectory with the time axis, is a nowhere dense set of Lebesgue measure 0, with a fractal structure of Hausdorff dimension 0.5 (Wikipedia, Random and Natural Fractals).

The Wiener process is a stochastic process. Its trajectory is a fractal, called random fractal. Some deterministic dynamical systems have fractal attractors, called deterministic fractals. Although deterministic dynamical systems can exhibit a chaotic behavior, which might seem to be random, chaos and randomness are different concepts. C.Radhakrishna Rao is saying: "Chance deals with order in disorder while chaos deals with disorder in order" (Rao 1989, 1997). The

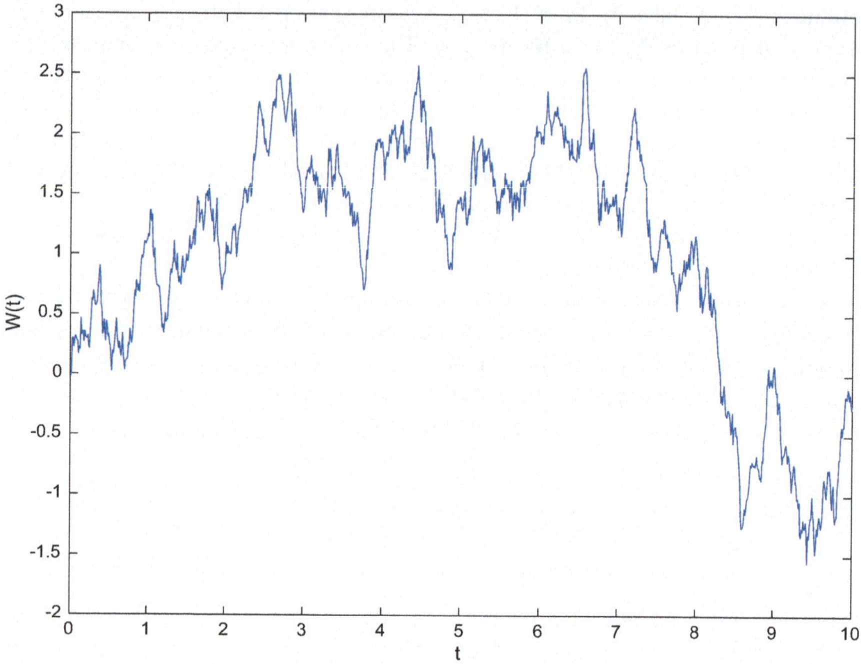


**Fig. 34** A simulated Wiener process trajectory at discrete 0, 0.01, 0.02, . . . , 9.99 times in [0,10]

time evolution of a deterministic dynamical system may appear as if random. Distinguishing deterministic dynamical systems from stochastic ones, based on observed data, is a difficult statistical problem. This problem became an important and highly studied research topic. The studies on this topic continue intensively. A great deal has been reached in this area. Criteria used to differentiate sensual randomness caused by deterministic chaotic behavior from natural randomness are based on some notions of nonlinear time series analysis.

## Conclusion

Understanding the concept of complexity and developing measures for complexity is a very complicated and difficult task. The applications of the complexity theory in sociology, as well as in all scientific areas are rising every day, gaining a new momentum in understanding complicated events. The contribution of complexity theory to the theory of international relations is thoroughly demonstrated by Tomé and Açıkalın (2017). Without forgetting the role of probability theory, we can say that chaos and complexity theory provide a new insight in understanding real-world phenomena. Easily observable events such as laminar and turbulent flow in fluids and the developments in fluid dynamics theory recently have been included in the analysis of time series as concepts and measures of complexity and chaos. Time series analysis is a widely used modeling device in many areas such econometrics, finance, meteorology, geophysics, signal processing, etc. The chaos and complexity inherent in deterministically evolving systems should not be confused with randomness in stochastic processes or time series. Everything in the real world is very complex. Nothing is simple. Mathematics and statistics are our primary tools in the way to understand the truth. Sometimes mathematical models include difficult parts, such as in the field equations of Einstein's general relativity theory. The notions of complexity and difficulty should not be confused. Complexity, chaos, and randomness are inherent in the nature, increasing toward the deep and fine foundations of the reality. How can we capture dynamics in the real world and place them in deterministic or stochastic mathematical models? Perhaps, it is useful to use C.R.Rao's idea: "If we were to speak of any rational principle in nature, then that principle can only be chance: for, it is chance, acting in collaboration with selection, that constitutes nature's reason" (Rao 1989, 1997).

## References


Alligood, K. T., Sauer, T. D., & Yorke, J. A. (2000). *Chaos, an introduction to dynamical systems* (3rd ed.). New York: Springer.

Atakan, C., Dağalp, R., Potas, N., & Öztürk, F. (2017). Randomness and chaos. In Ş. Ş. Erçetin & N. Potas (Eds.), *Chaos, complexity and leadership* (pp. 621–646). Cham: Springer.

Balibrea, F. (2006). Chaos, periodicity and complexity on dynamical systems. In A. Sengupta (Ed.), *Chaos, nonlinearity, complexity, the dynamical paradigm of nature*. Berlin: Springer.

Barnsley, M. F. (1988). *Fractals everywhere*. Boston: Academic Press.

Bensoudane, H., Gentil, C., & Neveu, M. (2008). The local fractional derivative of fractal curves. *IEEE International Conference on Signal Image Technology and Internet Based Systems*, pp. 522–529.

Cabello, A., Severini, S., & Winter, A. (2014). Graph-theoretic approach to quantum correlations. *Physical Review Letters, 112*, 040401.

Cernenoks, J., Iraids, J., Opmanis, M., Opmanis, R., & Podnieks K. (2014). Integer complexity: Experimental and analytical results II. *arXiv:1409.0446v1* [math.NT] 1 Sep 2014.

Cordwell, K., Epstein, A., Hemmady, A., Miller, S. J., Palsson, E., Sharma, A., Steinerberger, S., & Vu, Y. N. T. (2018). On algorithms to calculate integer complexity. *arXiv:1706.08424v3* [math.NT] 5 Aug 2018.

Devaney, R. L. (2003). *An introduction to chaotic dynamical systems*. New York: Westview Press.

Falconer, K. J. (1982). Hausdorff dimension and the exceptional set of projections. *Mathematika, 29*, 109–115.

Falconer, K. J. (2003). *Fractal geometry mathematical foundations and applications*. Chichester: Wiley.

Iraids, J., Balodis, K., Cernenoks, J., Opmanis, M., Opmanis, R., & Podnieks, K. (2012). Integer complexity: Experimental and analytical results. *Scientific Papers University of Latvia, Computer Science and Information Technologies, 787*, 153–179.

Kolmogorov, A. N. (1956). *Foundations on the theory of probability*. New York: Chelsea Publishing Company.

Li, T. Y., & Yorke, J. (1975). Period three implies chaos. *American Mathematical Monthly, 82*, 985–992.

Lopez-Ruiz, R., & Fournier-Prunaret, D. (2004). Complex behavior in a discrete coupled logistic model for the symbiotic interaction of two species. *Mathematical Biosciences and Engineering, 1*(2), 307–324. https://doi.org/10.3934/mbe.2004.1.307.

Lopez-Ruiz, R., Mancini, H., & Calbet, X. (2002). A statistical measure of complexity. *Physics Letters A*https://doi.org/10.1016/0375-9601(95)00867-5.

Lowen, S., & Teich, M. C. (2005). *Fractal-based point processes*. New York: Willey-Interscience.

Mandelbrot, B. B. (1982). *The fractal geometry of nature*. San Francisco: WH Freeman & Co.

Murray, J. D. (1993). *Mathematical biology*. New York: Springer.

Rao, C. R. (1989). *Statistics and truth Puting chance to work*. Dordrecht: International Co-operative Publishing House.

Rao, C. R. (1997). *Statistics and truth Puting chance to work*. Singapore: World Scientific Publishing Co.

Rovelli, C. (2017). *Fizik Üzerine Yedi Kısa Ders*. Can Sanat Yayınları. (Translation to Turkish from: Rovelli, C. (2014) Seven Brief Lessons on Physics.)

Rovelli, C. (2018). *Gerçeklik Göründüğü Gibi Değildir*. Can Sanat Yayınları. (Translation to Turkish from: Rovelli, C. (2014) Reality Is Not What It Seems: The Journey to Quantum Gravity.)

Tomé, L., & Açıkalın, Ş. N. (2017). Complexity theory as a new Lens in IR: System and change. In Ş. Ş. Erçetin & N. Potas (Eds.), *Chaos, complexity and leadership* (pp. 621–646). Cham: Springer.

Wikipedia, free encyclopedia, https://en.wikipedia.org/.

# Relationships Between Stock Markets: Causality Between G8 Countries and Turkey

**Kamil Demirberk Ünlü, Nihan Potas, and Mehmet Yılmaz**

**Abstract** This study investigated relationships between stock markets in the Group of Eight (G8) countries (Canada, France, Germany, Italy, Japan, Russia, the UK, and the USA) and the İstanbul Stock Exchange (ISE) by estimating eight different vector autoregressions (VARs). We applied the Johansen and Juselius cointegration test to identify the long-run relations between the indices. The modified Granger causality test proposed by Toda and Yamamoto was conducted to identify the causality, then forecast variance decomposition and impulse response analysis were employed to explore the impacts of unexpected shocks in the G8 countries' stock markets on the ISE. The results showed that there was no cointegration between the ISE and the G8 countries' markets, but they still affected the ISE to different degrees, and the DAX–ISE 100, CAC 40–ISE 100, and FTSE MIB–ISE 100 causal relationships were bidirectional.

**Keywords** İstanbul stock exchange · G8 countries · Cointegration · Causality · Generalized variance decomposition · Impulse response

## Introduction

Investors in developing countries eagerly follow the price movements of developed countries' stock markets; these movements are interpreted as the future movements of the emerging markets. Also, the psychological effects of the highly reputational

K. D. Ünlü (✉)
Atılım University, Ankara, Turkey
e-mail: demirberg@gmail.com

N. Potas
Ankara Hacı Bayram Veli University, Ankara, Turkey
e-mail: nihan_potas@hotmail.com

M. Yılmaz
Ankara University, Ankara, Turkey
e-mail: mehmetyilmaz@ankara.edu.tr



Ş. Ş. ERÇETİN, Ş. N. AÇIKALIN (eds.), *Chaos, Complexity and Leadership 2018*, Springer Proceedings in Complexity, https://doi.org/10.1007/978-3-030-27672-0_5

markets such as the US and UK markets cannot be ignored. Emerging markets such as that in Turkey have a high percentage of foreign investors. The decisions made by these investors can be affected by other stock markets' movements. For that reason, we examined cointegration and causality between stock markets in the Group of Eight (G8) countries (Canada, France, Germany, Italy, Japan, Russia, the UK, and the USA) and Turkey to find out (1) which markets had the largest and the longest influence on the İstanbul Stock Exchange (ISE), and (2) whether the ISE also influenced the G8 countries' stock markets. Few studies in the literature have studied the relationship between Turkey and other developed markets. Berument and Ince (2005) investigated the effects of the S&P500's returns on the ISE 100 (the Turkish stock market index) and found that the S&P500 affected the ISE 100 positively for up to 4 days. The aim of our work was to fill this gap in the literature.

Eun and Shim (1989), Kasa (1992), Smith et al. (1993), Richards (1995), and Francis and Leachman (1998) investigated the linkages between developed countries' stock markets. Also, studies have investigated the effects of the US markets on Asian countries, and some have investigated the effects on Middle East and North African (MENA) and Balkan countries. Cointegration between 13 emerging markets and the USA was studied by DeFusco et al. (1996), while Ghosh et al. (1999) found that some Asia–Pacific countries were influenced by the US and Japanese markets, and Huang et al. (2000) studied causality and cointegration among the USA, Japan, and the South China Growth Triangle. Yang et al. (2003) investigated cointegration between emerging Asian countries and the USA, while Yu and Hassan (2008) examined the integration between MENA and developed countries (the USA, UK, and France). The relation of Pacific Basin countries' markets to Japan and the USA was studied by Phylaktis and Ravazzolo (2004). Onay (2007) researched two developing countries (Brazil and Turkey) and found that the Brazilian market influenced the Turkish stock exchange. Finally, Syriopoulos (2011) explored causality between Balkan equity markets and two developed equity markets (those in the USA and Germany). In the next section, we explain our methodology and results, while the last section concludes this chapter.

## Methodology and Results

We examined the daily stock market indices of the G8 countries and Turkey. The data set consisted of the period between January 8, 2003, and September 24, 2012. We used the CAC 40, FTSE 100, DAX, FTSE MIB, Nikkei 225, RTSI, S&P/TSX Composite, and S&P500, which are indices in France, the UK, Germany, Italy, Japan, Russia, Canada, and the USA, respectively. The data on these markets was taken from Bloomberg, while the data on the ISE 100 were taken from the official website of the ISE. There were some missing data in each series, and we assigned values to the missing ones by the moving average technique. We used the natural logarithm of the series, and all series are expressed in local currency except for the RTSI series, which is expressed in US dollars. The first step in the analysis was to examine the stationarity of the data. We implemented the augmented Dickey–Fuller (ADF) unit root test (Dickey–Fuller 1979), detrended Dickey–Fuller generalized least squares (DF-GLS) test, Phillips and Perron (PP) test (Phillips and Perron 1998), Kwiatkowski–Phillips–Schmidt–Shin (KPSS) test (Kwiatkowski et al. 1992), Elliott–Rothenberg–Stock point optimal (ERS-PO) test (Elliott et al. 1996), and Ng and Perron (NP) test (Ng and Perron 2001). Tables 1 and 2 represent the results for the levels and the first differences, respectively.

The results showed that all series were stationary in the first difference; thus, the series were integrated of order one I(1) and appropriate for the Johansen and Juselius cointegration test.

At the next step, cointegration tests were used to investigate the long-run equilibrium between nonstationary time series. We employed the test of cointegration suggested by Johansen and Juselius (1990) and Johansen (1991, 1995) based on the following vector autoregression (VAR) model:

$$\Delta Y_t = c + \beta_1 \Delta Y_{t-1} + B_2 \Delta Y_{t-2} + \cdots + \beta_m \Delta Y_{t-m} + \varepsilon_t,$$

where $Y$ is a $(2 \times 1)$ vector containing the ISE 100 and the investigated stock indices of the G8 countries, $c$ is a $(2 \times 1)$ vector of constant terms, $\beta$ represents $(2 \times 2)$ coefficient matrices, and $\varepsilon$ is a $(2 \times 1)$ vector of the error term. We estimated eight different VARs, and in each of them we kept the ISE 100 while the G8 countries' series were replaced with each other. The optimal lag length of each VAR was determined by the Akaike information criterion (AIC) and the final prediction error (FPE). We could not find any evidence of cointegration between the series according to $\lambda_{trace}$ and $\lambda_{max}$ tests. The results are reported in Table 3.

At the third step we utilized the modified Granger causality test proposed by Toda and Yamamoto (1995) (TY). We utilized TY because Granger (1988) requires cointegration between the series but TY does not (Zapata and Rambaldi 1997).

The results are presented in Table 4. Panels A, B, F, G, and H indicate unidirectional Granger causality running from the S&P/TSX, FTSE 100, Nikkei 225, RTSI, and S&P500 to the ISE, while the results presented in panels C, D, and E indicate bidirectional causality between the CAC 40 and ISE 100, between the DAX and ISE 100, and between the FTSE MIB and ISE 100.

At the last step we applied the generalized variance decomposition (VDC) and impulse response (IR) methods of Koop et al. (1996) and Pesaran and Shin (1998). Here, we used the first differences of the data series. The VDC showed the forecast error of a variable caused by its own innovation and another variable in the same VAR system. Granger causality is valid only in the sample period so we employed the VDC as an out-of-sample causality test. The results confirmed the Granger causality test results.

The results of each case are summarized below:

- S&P/TSX: The amount explained by the S&P/TSX went to 12% from 0% and stabilized in 13 days.

**Table 1** Unit root test results (levels)

| | | ADF | | DF-GLS | | | | ERS-PO | | NP-Zα | |
|---|---|---|---|---|---|---|---|---|---|---|---|
| | | Statistics | Lag | Statistics | Lag | PP | KPSS | Statistics | Lag | Statistics | Lag |
| CAC 40 | Intercept | −1.6331 | 5 | −0.9559 | 5 | −1.6866 | 1.2016[a] | 12.9040 | 5 | −1.8905 | 5 |
| DAX | | −2.1699 | 11 | 0.3433 | 11 | −1.9525 | 3.3934[a] | 54.8449 | 11 | 0.3822 | 11 |
| FTSE 100 | | −2.1902 | 6 | −0.3567 | 6 | −2.2252 | 1.6393[a] | 22.6084 | 6 | −0.5891 | 6 |
| ISE 100 | | −2.2426 | 14 | 1.0815 | 14 | −2.2546 | 4.6106[a] | 134.5202 | 14 | 0.3346 | 19 |
| FTSE MIB | | −0.8148 | 17 | −08820 | 17 | −0.7260 | 3.3430[a] | 10.0743 | 17 | −2.1750 | 17 |
| Nikkei | | −1.5482 | 4 | −0.9547 | 4 | −1.5537 | −1.7294[a] | 13.3817 | 4 | −1.8410 | 4 |
| RTSI | | −2.2633 | 26 | −0.0476 | 26 | −2.2118 | 3.0409[a] | 44.3542 | 26 | −0.0557 | 26 |
| S&P500 | | −2.1659 | 18 | −0.3016 | 18 | −2.0155 | 0.7262[a] | 20.2635 | 18 | −0.5356 | 18 |
| S&P/TSX | | −2.2233 | 8 | 0.1402 | 8 | −2.2365 | 3.0216[a] | 58.4013 | 8 | 0.1358 | 8 |
| CAC 40 | Intercept and trend | −1.8342 | 9 | −1.1191 | 5 | −1.8507 | 1.0194[a] | 35.2406 | 9 | −2.7699 | 5 |
| DAX | | −2.2869 | 11 | −1.4506 | 11 | −2.0737 | 0.8492[a] | 20.7233 | 11 | −4.3671 | 11 |
| FTSE 100 | | −2.1358 | 6 | −1.6009 | 6 | −2.2240 | 0.7013[a] | 17.2218 | 6 | −5.2816 | 6 |
| ISE 100 | | −2.5254 | 14 | −1.2563 | 14 | −2.3348 | 0.6996[a] | 29.6785 | 14 | −2.9464 | 14 |
| FTSE MIB | | −2.1618 | 17 | −1.1001 | 17 | −1.9032 | 1.1000[a] | 31.9908 | 17 | −2.7420 | 17 |
| Nikkei | | −2.0542 | 4 | −1.0077 | 4 | −2.0516 | 1.0206[a] | 36.5088 | 4 | −2.3305 | 4 |
| RTSI | | −2.0385 | 26 | −1.2822 | 26 | −1.7905 | 0.7651[a] | 22.1680 | 26 | −4.0275 | 26 |
| S&P500 | | −2.1058 | 18 | −1.4884 | 18 | −2.0083 | 0.5503[a] | 20.2937 | 18 | −4.4530 | 18 |
| S&P/TSX | | −1.8866 | 8 | −1.1108 | 8 | −1.9428 | 0.8492[a] | 29.2556 | 8 | −2.9054 | 8 |

The lag lengths are determined by the Akaike information criterion (AIC)

*ADF* augmented Dickey–Fuller unit root test, *DF-GLS* detrended Dickey–Fuller generalized least squares test, *ERS-PO* Elliott–Rothenberg–Stock point optimal test, *KPSS* Kwiatkowski–Phillips–Schmidt–Shin test, *NP* Ng and Perron test, *PP* Phillips and Perron test

[a]Significance level 1%

**Table 2** Unit root test results (first differences)

| | | ADF | | DF-GLS | | | | ERS-PO | | NP-Zα | |
|---|---|---|---|---|---|---|---|---|---|---|---|
| | | Statistics | Lag | Statistics | Lag | PP | KPSS | Statistics | Lag | Statistics | Lag |
| CAC 40 | Intercept | −18.6930[a] | 8 | −1.2985 | 25 | −53.6481[a] | 0.2135 | 0.0017[a] | 8 | −1.6512 | 25 |
| DAX | | −152928[a] | 10 | −1.8308[c] | 25 | 51.6622[a] | 0.1289 | 0.0415[a] | 10 | −2.8729 | 25 |
| FTSE 100 | | −22.5462[a] | 5 | −6.6980[a] | 25 | −54.5756[a] | 0.0977 | 0.0074[a] | 5 | −14.0705[a] | 25 |
| ISE 100 | | −12.9271[a] | 13 | −0.9069 | 25 | −50.4484[a] | 0.2319 | 0.3206[a] | 13 | −0.6762 | 25 |
| FTSE MIB | | −11.0414[a] | 16 | −1.9928[b] | 25 | −51.0290[a] | 0.2502 | 0.2661[a] | 16 | −3.7242 | 25 |
| Nikkei | | −26.0074[a] | 3 | −11.3989[a] | 12 | −53.4219[a] | 0.2588 | 0.0185[a] | 3 | −118.065[a] | 12 |
| RTSI | | −8.7281[a] | 25 | −3.8119[a] | 26 | −47.6601[a] | 0.2541 | 0.2855[a] | 25 | −7.6738[c] | 26 |
| S&P500 | | −12.1093[a] | 17 | −1.0724[c] | 26 | −59.2845[a] | 0.1091 | 0.0442[a] | 17 | −0.9954 | 26 |
| S&P/TSX | | −18.5516[a] | 7 | −2.6710[a] | 25 | −53.7581[a] | 0.2099 | 0.0175[a] | 7 | −4.1385 | 25 |
| CAC 40 | Intercept and trend | −18.7336[a] | 8 | −2.9592[b] | 25 | −53.6816[a] | 0.0743 | 0.0027[a] | 8 | −3.3680 | 25 |
| DAX | | −15.3165[a] | 10 | −3.7520[a] | 25 | −51.6679[a] | 0.0655 | 0.1010[a] | 10 | −4.9810 | 25 |
| FTSE 100 | | −22.5533[a] | 5 | −21.9540[a] | 5 | −54.5757[a] | 0.0591 | 0.0266[a] | 5 | −2293.03[a] | 5 |
| ISE 100 | | −12.9793[a] | 13 | −2.2048 | 25 | −50.4623[a] | 0.0822 | 0.3910[a] | 13 | −3.3860 | 25 |
| FTSE MIB | | −11.1297[a] | 16 | −4.0336[a] | 25 | −51.0563[a] | 0.0675 | 0.6394[a] | 16 | −6.7081 | 25 |
| Nikkei | | −31.5332[a] | 2 | −25.2060[a] | 3 | −53.4699[a] | 0.0779 | 0.0565[a] | 2 | −1168.32[a] | 3 |
| RTSI | | −8.7976[a] | 25 | −6.4143[a] | 26 | −47.6972[a] | 0.0736 | 0.8549[a] | 25 | −18.1203[b] | 26 |
| S&P500 | | −12.1162[a] | 17 | −2.6492[c] | 26 | −59.2742[a] | 0.0989 | 0.0710[a] | 17 | −3.1131 | 26 |
| S&P/TSX | | −118.593[a] | 7 | −5.2608[a] | 25 | −53.8145[a] | 0.0594 | 0.0467[a] | 7 | −8.4253 | 25 |

The lag lengths are determined by the Akaike information criterion (AIC)
*ADF* augmented Dickey–Fuller unit root test, *DF-GLS* detrended Dickey–Fuller generalized least squares test, *ERS-PO* Elliott–Rothenberg–Stock point optimal test, *KPSS* Kwiatkowski–Phillips–Schmidt–Shin test, *NP* Ng and Perron test, *PP* Phillips and Perron test

[a]Significance level 1%
[b]Significance level 5%
[c]Significance level 10%

**Table 3** Multivariate cointegration test results

| Panel A (ISE 100 and S&P/TSX) lag 7 | | | | | |
|---|---|---|---|---|---|
| $H_0$ | $\lambda_{trace}$ | 1% | $\lambda_{max}$ | 1% | Eigenvalues |
| $r = 0$ | 6.7243 | 20.04 | 6.2096 | 18.63 | 0.002443 |
| $r \leq 1$ | 0.5147 | 6.65 | 0.5147 | 6.65 | 0.000203 |
| Panel B (ISE 100 and FTSE 100) lag 18 | | | | | |
| $H_0$ | $\lambda_{trace}$ | 1% | $\lambda_{max}$ | 1% | Eigenvalues |
| $r = 0$ | 10.3288 | 20.04 | 7.8594 | 18.63 | 0.003118 |
| $r \leq 1$ | 2.4694 | 6.65 | 7.8594 | 6.65 | 0.000981 |
| Panel C (ISE 100 and CAC 40) lag 7 | | | | | |
| $H_0$ | $\lambda_{trace}$ | 1% | $\lambda_{max}$ | 1% | Eigenvalues |
| $r = 0$ | 6.7243 | 20.04 | 6.2096 | 18.63 | 0.002443 |
| $r \leq 1$ | 0.5147 | 6.65 | 0.5147 | 6.65 | 0.000203 |
| Panel D (ISE 100 and DAX) lag 17 | | | | | |
| $H_0$ | $\lambda_{trace}$ | 1% | $\lambda_{max}$ | 1% | Eigenvalues |
| $r = 0$ | 14.8928 | 20.04 | 10.9857 | 18.63 | 0.004352 |
| $r \leq 1$ | 3.9071 | 6.65 | 3.9071 | 6.65 | 0.001550 |
| Panel E (ISE 100 and FTSE MIB) lag 7 | | | | | |
| $H_0$ | $\lambda_{trace}$ | 1% | $\lambda_{max}$ | 1% | Eigenvalues |
| $r = 0$ | 6.5943 | 20.04 | 6.5938 | 18.63 | 0.002594 |
| $r \leq 1$ | 0.0006 | 6.65 | 0.0006 | 6.65 | 0.000002 |
| Panel F (ISE 100 and Nikkei) lag 6 | | | | | |
| $H_0$ | $\lambda_{trace}$ | 1% | $\lambda_{max}$ | 1% | Eigenvalues |
| $r = 0$ | 6.5660 | 20.04 | 6.2004 | 18.63 | 0.002437 |
| $r \leq 1$ | 0.3656 | 6.65 | 0.3656 | 6.65 | 0.000144 |
| Panel G (ISE 100 and RTSI) lag 18 | | | | | |
| $H_0$ | $\lambda_{trace}$ | 1% | $\lambda_{max}$ | 1% | Eigenvalues |
| $r = 0$ | 14.5815 | 20.04 | 8.2726 | 18.63 | 0.003255 |
| $r \leq 1$ | 6.3089 | 6.65 | 6.3089 | 6.65 | 0.002484 |
| Panel H (ISE 100 and S&P500) lag 7 | | | | | |
| $H_0$ | $\lambda_{trace}$ | 1% | $\lambda_{max}$ | 1% | Eigenvalues |
| $r = 0$ | 8.9887 | 20.04 | 5.9673 | 18.63 | 0.002347 |
| $r \leq 1$ | 3.0214 | 6.65 | 3.0214 | 6.65 | 0.001189 |

Each panel represents the pairwise cointegration analysis between the relevant market indices. Critical values were obtained from Osterwald-Lenum (1992).

- FTSE 100: The proportion explained by the FTSE 100 increased from 0% to 22% and stabilized in 31 days.
- DAX: The effect of the DAX started from 0%, went to 14%, and balanced in 37 days.
- CAC 40: The amount explained by CAC 40 increased from 0% to 22% and stabilized in 13 days.
- S&P500: The effect of the S&P500 went to 14% from 0% and offset in 10 days.
- FTSE MIB: The innovation in the FTSE MIB influenced the ISE 100 from 0% to 20% and balanced in 9 days.

**Table 4** Granger causality test results

| Panel A | | |
|---|---|---|
| Dependent variable | ISE 100 | S&P/TSX |
| ISE 100 | – | 60.95[a] |
| S&P/TSX | 1.56 | – |
| Panel B | | |
| Dependent variable | ISE 100 | FTSE 100 |
| ISE 100 | – | 50.50[a] |
| FTSE 100 | 1.29 | – |
| Panel C | | |
| Dependent variable | ISE 100 | CAC 40 |
| ISE 100 | – | 120.92[a] |
| CAC 40 | 1.96[c] | – |
| Panel D | | |
| Dependent variable | ISE 100 | DAX |
| ISE 100 | – | 24.13[a] |
| DAX | 1.53[c] | – |
| Panel E | | |
| Dependent variable | ISE 100 | FTSE MIB |
| ISE 100 | – | 102.69[a] |
| FTSE MIB | 2.25[b] | – |
| Panel F | | |
| Dependent variable | ISE 100 | Nikkei |
| ISE 100 | – | 40.78[a] |
| Nikkei | 1.33 | – |
| Panel G | | |
| Dependent variable | ISE 100 | RTSI |
| ISE 100 | – | 46.03[a] |
| RTSI | 1.34 | – |
| Panel H | | |
| Dependent variable | ISE 100 | S&P500 |
| ISE 100 | – | 75.70[a] |
| S&P500 | 1.62 | – |

Significance implies that the column variable Granger caused the row variable
[a]Significance level 1%
[b]Significance level 5%
[c]Significance level 10%

- Nikkei: The proportion explained by the Nikkei started from 5% and balanced at 10% in 8 days.
- RTSI: The effect of the RTSI went to 20% from 0% and stabilized in 29 days.

The IR showed how a variable was influenced in the event of a shock in another variable in the same VAR system. Figure 1 shows the generalized IR effects of one-standard-deviation innovations in the G8 countries' stock prices on the ISE 100, with confidence intervals.

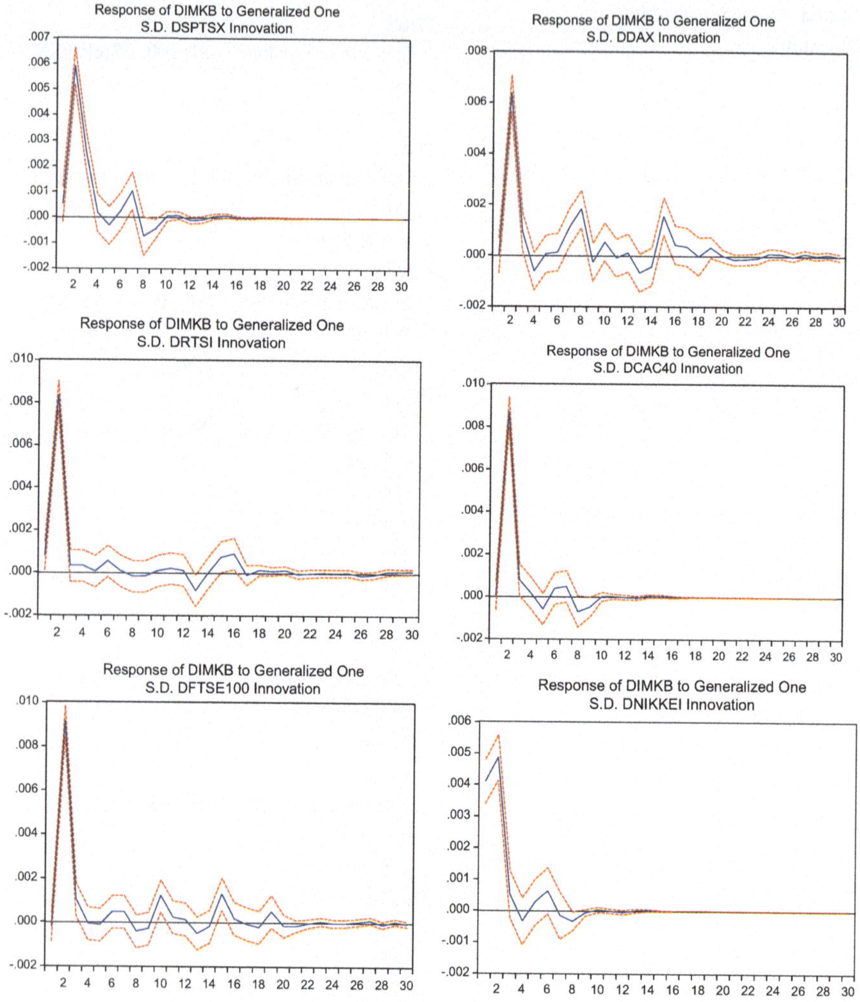


**Fig. 1** Generalized impulse responses of the ISE 100 to one-standard-deviation innovations

Shocks in the G8 countries had positive and significant impacts on the ISE 100. The impacts of each country's shocks are summarized below:

- S&P/TSX: The initial impact was positive; it increased until the third period, was statistically significant for 7 days, and died out in 10 days.
- FTSE 100: The initial impact was positive; it increased until the second period, was statistically significant for 15 days, and died out in 20 days. The FTSE 100 had the largest impact on the ISE 100.

- DAX: The initial impact was positive; it increased until the second period, was statistically significant for 15 days, and died out in 20 days. The DAX and FTSE 100 had the longest impacts on the ISE 100.
- CAC 40: The initial impact was positive; it increased until the second period, was statistically significant for 3 days, and died out in 10 days.
- S&P500: The initial impact was positive; it increased until the second period, was statistically significant for 7 days, and died out in 10 days.
- FTSE MIB: The initial impact was positive; it increased until the second period, was statistically significant for 3 days, and died out in 10 days.
- Nikkei: The initial impact was positive; it started to decrease in the second period, was statistically significant for 3 days, and died out in 8 days. The Nikkei had the smallest impact on the ISE 100.
- RTSI: The initial impact was positive; it increased until the second period, was statistically significant for 3 days, and died out in 16 days.

## Concluding Remarks

In this chapter we have examined the causality relation between the G8 countries' stock markets and the ISE 100. The data series for each country was nonstationary in terms of the levels but stationary in terms of the first differences. We could not find any cointegration between the variables, but a modified Granger causality test showed that the market in each G8 country influenced the ISE 100. The VDC and IR results showed that the largest impact on the ISE 100 was caused by the FTSE 100, while the smallest impact was caused by the Nikkei 225. Also, we found that the CAC 40–ISE 100, DAX–ISE 100, and FTSE MIB–ISE 100 causality relations were bidirectional. The results of the VDC and IR methods supported the results of the modified causality test.

## References

Berument, H., & Ince, O. (2005). Effects of S&P500's return on emerging markets: Turkish experience. *Applied Financial Economics Letters, 1*, 59–64.

DeFusco, A. R., Geppert, J. M., & Tsetsekos, G. P. (1996). Long-run diversification potential in emerging stock markets. *The Financial Review, 31*, 343–363.

Dickey, D. A., & Fuller, W. A. (1979). Distribution of the estimators for autoregressive time series with a unit root. *Journal of the American Statistical Society, 74*, 427–431.

Elliott, G., Rothenberg, T. J., & Stock, J. H. (1996). Efficient tests for an autoregressive unit root. *Econometrica, 64*, 813–836.

Eun, S. C., & Shim, S. (1989). International transmission of stock market movements. *Journal of Financial and Quantitative Analysis, 24*, 241–256.

Francis, B. B., & Leachman, L. L. (1998). Superexogeneity and the dynamic linkage among international equity markets. *Journal of International Money and Finance, 17*, 475–492.

Ghosh, A., Saidi, R., & Johnson, K. H. (1999). Who moves the Asia–Pacific stock markets—us or Japan? Empirical evidence based on the theory of cointegration. *The Financial Review, 34*, 159–170.

Granger, C. W. J. (1988). Causality, cointegration and control. *Journal of Economic Dynamics and Control, 12*, 551–559.

Huang, B. N., Yang, C. W., & Hu, W. S. (2000). Causality and cointegration of stock markets among United States, Japan and the South China Growth Triangle. *International Review of Financial Analysis., 9*, 281–297.

Johansen, S. (1991). Estimation and hypothesis testing of cointegration vectors in Gaussian vector autoregressive models. *Econometrica, 59*, 1551–1580.

Johansen, S. (1995). *Likelihood-based inference in cointegrated vector autoregressive models*. Oxford: Oxford University Press.

Johansen, S., & Juselius, K. (1990). Maximum likelihood estimation and inference on cointegration: With applications to the demand for money. *Oxford Bulletin of Economics and Statistics, 52*, 169–210.

Kasa, K. (1992). Common stochastic trend in international stock markets. *Journal of Monetary Economics, 29*, 95–124.

Koop, G., Pesaran, M. H., & Potter, S. M. (1996). Impulse response analysis in nonlinear multivariate models. *Journal of Econometrica, 74*, 119–147.

Kwiatkowski, D. P., Philips, C. B., Schmidt, P., & Shin, Y. (1992). Testing the null hypothesis of stationary against the alternative of a unit root. *Journal of Econometrics, 1992*, 159–178.

Ng, S., & Perron, P. (2001). Lag length selection and the construction of unit root tests with good size and power. *Econometrica, 69*, 1519–1554.

Onay, C. (2007). *Cointegration analysis of Bovespa and İstanbul stock exchanges*. Paper presented at the Oxford Business & Economics Conference, Oxford.

Osterwald-Lenum, M. (1992). A note with quantiles of the asymptotic distribution of the maximum likelihood cointegration rank test statistics 1. *Oxford Bulletin of Economics and statistics, 54*(3), 461–472.

Pesaran, M. H., & Shin, Y. (1998). Generalized impulse response analysis in linear multivariate models. *Economics Letters, 58*, 17–29.

Phillips, P. C. B., & Perron, P. (1998). Testing for a unit root in time series regressions. *Biometrica, 75*(2), 335–346.

Phylaktis, P., & Ravazzolo, F. (2004). Stock market linkages in emerging markets: Implications for international portfolio diversification. *International Financial Markets, Institutions & Money, 15*, 91–106.

Richards, J. A. (1995). Comovements in national stock market returns: Evidence of predictability, but not cointegration. *Journal of Monetary Economics, 36*, 631–654.

Smith, L. K., Brocato, J., & Rogers, J. E. (1993). Regularities in the data between major equity markets: Evidence from Granger causality tests. *Applied Financial Economics, 3*, 55–60.

Syriopoulos, T. (2011). Financial integration and portfolio investments to emerging Balkan equity markets. *Journal of Multinational Financial Management, 21*, 40–54.

Toda, H. Y., & Yamamoto, T. (1995). Statistical inference in vector autoregression with possibly integrated process. *Journal of Econometrics, 66*, 225–250.

Yang, J., Koları, J. W., & Min, I. (2003). Stock market integration and financial crises: The case of Asia. *Applied Financial Economics, 13*, 477–486.

Yu, J. S., & Hassan, M. K. (2008). Global and regional integration of the Middle East and North African (MENA) stock markets. *The Quarterly Review of Economics and Finance, 48*, 482–504.

Zapata, H. O., & Rambaldi, A. N. (1997). Monte Carlo evidence on cointegration and causation. *Oxford Bulletin of Economics and Statistics, 59*, 285–298.

# The Color Revolutions in the Former USSR Countries, Viewed in the Light of Chaos Theory

**Özlem Demirkıran**

**Abstract** Since the beginning of the 2000s, color revolutions have emerged in the former Soviet geography. Essentially, a pattern was applied in the development of these movements: external support provided to the opposition, youth organizations, and the media; nonviolent actions undertaken; etc. Right after elections, the people started protesting against alleged irregularities in those elections. The revolutions succeeded in Georgia (the 2003 Rose Revolution), Ukraine (the 2004 Orange Revolution), and Kyrgyzstan (the 2005 Tulip Revolution), but although the same pattern was implemented in some other countries, the revolutions did not succeed there. This chapter attempts to explain why a revolution did not happen in the case of Belarus, in the light of chaos theory.

**Keywords** Chaos theory · Former USSR countries · Color revolutions · Belarus

## Introduction

Classical physics theory states that certain reactions will occur in response to certain effects and that everything that happens can be measured precisely. The newer quantum physics theory was introduced by scientists such as Max Planck, Albert Einstein, Niels Bohr, and Werner Heisenberg, and this new approach is based on the uncertainty principle rather than on the aforementioned understanding of classical physics.

With the development of quantum theory in physics, the perception of nature and natural phenomena has changed, and the understanding that everything is in relation to other things and that these relations are constantly changing has started to develop. At this point, it is concluded that nature is complex as it cannot be explained by classical scientific understanding, and it is understood that science is

Ö. Demirkıran (✉)
International Relations Department, Süleyman Demirel University, Isparta, Turkey
e-mail: ozlemdemirkiran@sdu.edu.tr



Ş. Ş. ERÇETİN, Ş. N. AÇIKALIN (eds.), *Chaos, Complexity and Leadership 2018*, Springer Proceedings in Complexity, https://doi.org/10.1007/978-3-030-27672-0_6

not linear. Nonlinear science does not address events in a purely cause-and-effect relationship as predicted by linear science. It is neither balanced nor periodic, and it is very difficult to measure objectively. It attaches importance to small effects because they can cause big problems (Çıraklı et al. 2017: 331)

Chaos and complexity theories were born from a nonlinear science approach. In chaos theory, the fundamental understanding is that all things are in relation to each other and these relations are constantly changing; small things can create large effects by creating a domino effect. Because human beings do not have adequate analysis methods to account for all variables related to these systems, predictions about these systems can give incorrect results (Çıraklı et al. 2017: 331).

In the 2000s, color revolutions occurred in former USSR countries. They started nearly from the same point and had almost the same infrastructure, but the revolutions were successful in some countries, whereas others failed. This study is an attempt to explain the results of the color revolutions in the framework of chaos theory.

## Chaos Theory

The term "chaos" was first used by the scientist Henri Poincaré in 1900. He tried to prove whether the solar system is stable or not. As a result of this study, it was concluded that the solutions to the system of equations that determine the movement of the solar system are sensitive to the initial conditions of the solution, but that the initial conditions cannot be determined correctly. This result shows that it is not possible to determine whether the solar system is stable or not. Poincaré used the term "chaos" for this unpredictable and indeterminable situation (Ertürk 2012: 853).

However, it was Edward Lorenz who mainly drew attention to chaos theory. In an experiment with a weather forecasting machine, he discovered that a 0.1% difference caused large fluctuations in the system. This variability is called "sensitive dependence on initial conditions." This sensitive dependency occurs when small differences in input cause large changes in output; this is known as the "butterfly effect" (Bayramoğlu 2016: 51).

Important features of chaotic systems can be listed as follows:

The first feature is sensitive dependence on the initial conditions. As mentioned above, some minor changes cause unforeseen and unpredictable consequences; this is known as the butterfly effect. Lorenz described this as follows: "Does the flap of a butterfly's wings in Brazil set off a tornado in Texas?"

The second feature is nonlinearity. Chaotic systems are not linear, because their movements cannot be calculated and predicted in advance. Chaotic systems do not always respond to the same input in the same way, whereas the output is always proportional to the input in linear systems. Because of the complexity of the conditions that determine the fate of the system and the effects of the other aforementioned features, chaotic systems exhibit unpredictable behavior (Canan 2016: 186).

The third feature is incalculability/unpredictability. Although chaotic systems appear to move in an order, it is not possible to predict how they will behave after a while. Here, the concept of randomness comes into play. Randomness indicates that a small difference in the initial conditions will create unpredictable differences. As a result, unpredictable behavior can occur as many components interact in different ways (Lale 2018: 35–36).

The fourth feature is self-similarity. In 1975, Benoit Mandelbrot discovered what he called a fractal. Mandelbrot's fractals seem to be a ready visual tool for understanding the dimensional consequences of chaos and complexity theory in one of the more important aspects of this theory, self-similarity.

Bütz (1997: 17) wrote, "Self-similarity is just as the term implies. Mandelbrot found that objects and organisms grow in a self-similar scheme. The fractal self-similarity that Mandelbrot described deals with the self-similar geometric structure of trees, lungs and vascular systems in the human body, with their branching and bifurcating shapes. These distinctly tree-like fractal shapes run throughout nature and represent the notion from fractal geometry that many adaptations are self-similarly based on previous structures."

The fact that the initial conditions of the system can never be determined and formulated with complete clarity reveals the uncertainty of the science of chaos. However, it is not possible to estimate the starting point and all influential conditions. Therefore, when not all factors are taken into account, the accuracy of the results will be reduced and the system will be complicated.

## The Color Revolution Process

Series of events leading to color revolutions have been staged in many countries. In this process, where social mobility was placed at the center of revolutions, especially university students and large masses were used (Topak 2014: 235–236). As can be seen in Table 1, large numbers of people were involved in the social movements in some countries, and it was planned to make management changes through mandatory elections, but the desired administrative changes could not be implemented, except in a few countries (Topak 2014).

According to Ó Beacháin and Polese, the revolutions that started in Eastern Europe and spread to the former Soviet geography and even to the Middle East today have followed a similar pattern: "in the framework of an electoral contest a civic campaign to guarantee free and fair elections is set up. Normally, local nongovernmental organizations (NGOs) take the lead (often coordinated with political forces) and benefit from 'know how' acquired through international trainings and manuals. Through a joint effort, civil society and political actors follow a two-pronged strategy: they seek to discredit the regime (negative campaign) while pushing people to go to the polls (positive campaign). The assumption is that, where the regime is sufficiently unpopular, a high turnout will allow a resourceful opposition to win the elections. The second part of this strategy relies on the assumption that the

**Table 1** Color revolutions accomplished and attempted

| No. | State | Success | Time frame (dates)[a] | Election date[a] | Crowd size | Victims |
|---|---|---|---|---|---|---|
| 1. | Azerbaijan | No | 16.09–20.10.2003 | 15.10.2003 | 25,000 | Few |
| 2. | Georgia | Yes | 15.11–23.11.2003 | 02.11.2003 | 100,000 | None |
| 3. | Ukraine | Yes | 22.11–04.12.2004 | 21.11.2004 | 500,000 | None |
| 4. | Kyrgyzstan | Yes | 18.03–24.03.2005 | 13.03.2005 | 25,000 (in Bishkek) | Few |
| 5. | Azerbaijan | No | 08.08–26.11.2005 | 06.11.2005 | 20,000 | Few |
| 6. | Belarus | No | 19.03–23.03.2006 | 19.03.2006 | 35,000 | 1 |
| 7. | Georgia | No | 28.09–08.11.2007 | 28.03.2004 | 75,000 | Few |
| 8. | Armenia | No | 20.02–02.03.2008 | 19.02.2008 | 100,000 | Few |
| 9. | Georgia | No | 09.04–24.07. 2009 | 21.05.2008 | 50,000 | None |
| 10. | Moldova | Yes | 06.04–07.04.2009 | 05.04.2009 | 15,000 | 3 |
| 11. | Kyrgyzstan | Yes | 06.04–15.04.2010 | 23.07.2009 | 5000 (in Bishkek) | 90–100 |
| 12. | Belarus | No | 19.12–20.12.2010 | 19.12.2010 | 40,000 | Few |

Source: Baev (2011)
[a]The dates are expressed in day.month.year format

authorities might not play fair with the election results. Once the regime refuses to acknowledge the election results (by falsifying them or simply refusing to step down), people are called on to the streets and a general strike is called until the status quo changes (this may mean that the authorities step down or that they crush the protesters)" (Ó Beacháin and Polese 2010).

In the wake of the revolution in Nicaragua, the USA launched a new initiative, which was named the "Democracy Project." With this new project, the USA, which aims to export democracy, became involved in activities aimed at attracting target countries' intellectuals with radio, books, TV, funding, scholarships, and awards. According to the project, these activities were aimed at "building take control of the brain." As a result of these policies, the USA tried to create a new type of intellectuals and new leaders. On the other hand, the USA began to play with the internal dynamics of the target countries while looking for ways to bring to power the young leaders who the USA had been investing in for a long time. The first revolutions began to take place in Eastern Europe in the Cold War period before the USSR split (Oğan 2005). But the subject of this analysis—the color revolutions in the former Soviet geography—emerged in the 2000s.

The same organizational scheme, the same sources of finance, and even similar slogans were used in these revolutions. In Yugoslavia, where the first NGO revolution was held, it was seen that the motor power of the revolution was a youth organization called Otpor (which means "Resistance"). Similar organizations called Kmara (which means "Enough") in Georgia, Pora (which means "It Is Time") in Ukraine, Kel-Kel (which means "Renaissance" and "shining of the good") in Kyrgyzstan, and Zubr (which means "Bison" (the heraldic symbol of the national groups used to represent Belarus)) in Belarus were among the most important forces next to the revolution leaders (Oğan 2005).

Such youth organizations were financed by organizations such as the Open Society Foundation organizations founded by George Soros, the National Endowment for Democracy (NED), the National Democratic Institution (NDI), the International Republican Institute (IRI), the Albert Einstein Institute, Freedom House, and the Organization for Security and Co-operation in Europe (OSCE). In addition, the World Bank, representatives of humanitarian organizations, human rights groups, media development groups, and educational institutions such as Fiona Hill, the Brookings Institution, and various American universities—as well as religion-based groups operating in Central Asia (such as Mormons and Protestant groups)—contributed to civil revolutions (Akşar 2009: 28–29).

Youth organizations organized demonstrations with opposition parties and supported them in the organization and mobilization of society. In the days following the elections, the strategies implemented by these organizations were effective in gathering hundreds of thousands of people, whether or not they were related to parties or these youth organizations. The basic strategy they followed was to encourage young people to participate in protests. Their methods were to organize street demonstrations and street theater, to give concerts, to fill squares with colorful streamers, and to mobilize the "whisper network," which had a significant impact on the masses (Akşar 2009: 27).

These youth organizations were established independently, but later international NGOs and other countries' youth organizations provided financial support and supported the training of local youth and NGOs on how to conduct nonviolent mass demonstrations in general. The Open Society Foundation in Ukraine organized and financed students' travel abroad and the meetings between different groups (e.g., Otpor and Kmara members). NGOs trained tens of thousands of volunteers for election observations, sent organization members as election observers to these countries, and supervised and monitored elections. For example, the trainers of Otpor members were NGOs. Moreover, Kmara received training support from Otpor volunteers, Pora members were trained by Kmara and NGOs (Akşar 2009: 27–28), and Zubr received support from Otpor, Pora, and Kmara (Donnari 2006).

It must be mentioned that street demonstrations and the idea of nonviolent action of these youth organizations operating in civil revolutions are embodied in the thoughts of Prof. Gene Sharp (Topak 2014: 240–241), whose publication *From Dictatorship to Democracy* has become almost a "handbook" of civil revolutions. It describes in detail how to create civil revolutions and guides civil revolutionaries. It has been translated into the languages of the countries where civil revolutions have been created or could be created. Almost all relevant youth organizations' websites host the original text or translations of it. Sharp argues that "civil disobedience and international pressure" constitute the Achilles heel of dictatorships. Prof. Sharp, who was nominated for the Nobel Peace Prize in 2009 and has been accused by some groups of being a Central Intelligence Agency (CIA) front man (Arrow 2011), recommends 189 different civil disobedience action methods (Oğan 2006). Some of these are methods of nonviolent protest and persuasion such as letters of opposition or support, declarations by organizations and institutions, display of flags and symbolic colors, wearing of symbols, singing, and demonstrative funerals; methods of social noncooperation such as social boycotts and stay-at-home campaigns; methods of economic noncooperation such as economic boycotts and strikes; methods of political noncooperation such as withdrawal from government educational institutions; judicial noncooperation of government personnel; and methods of nonviolent intervention such as hunger strikes and stand-ins (Sharp 2010: 79–86).

Media play a very important role in civil revolutions. The press and media organs attract the attention of the West by broadcasting live broadcasts from the squares to keep the interest alive. Also, they issue broadcasts that support the civil revolution, call for support, and contribute to increasing the number of participants at demonstrations. The TV channels Rustavi-2 in Georgia and Canal 5 in Ukraine, the newspaper *Res Publica*, and MSN in Kyrgyzstan played this role. Generally, the West gave financial support to these opposition media organs (Oğan 2005).

The names and the elements that symbolized these revolutions were carefully selected. Generally a name and a color that were extremely influential in public opinion, easy to memorize, and could be translated into all world languages was preferred, such as the Georgian Rose (or Velvet) Revolution, the Ukrainian Orange Revolution, the Kyrgyzstan Tulip (or Lemon) Revolution, or the Belarus Jeans (or Blue) Revolution. Similarly, many of the youth organizations formed in these countries used similar emblems. The sign of the clenched fist was first used by the Otpor organization in Yugoslavia, and later it was used by the Kmara organization in Georgia and by some other organizations in other countries (Oğan 2005).

All civil revolutions emerged on the eve of the elections. International NGOs educated local volunteers about how to conduct the elections correctly and in a proper manner, sent professional election observers to these countries, and made it possible to spread the opinion that irregularities would occur in the elections. During the election and vote counting, international and local NGOs collaborated, conducted parallel vote counts, and announced the results immediately after the elections, and they believed that the international legitimacy of this parallel vote count was greater (Akşar (2009: 32–33). For example, in the November 2003 elections in Georgia, according to the results of the parallel vote counts done by the Fair Elections organization, Saakashvili received 26.60% of the votes and incumbent president Shevardnadze received 18.92%. However, according to the Central Election Commission (CEC) results, Shevardnadze received 21.32% of the votes and Saakashvili received 18.08% (Companjen 2010: 14). In Ukraine, the preliminary presidential ballot was held on October 31, 2004. Yanukovych and Yushchenko received very similar support—approximately 41% each. The second ballot on November 21 triggered mass protests when the CEC declared Yanukovych to be the winner on November 22. The official results were published 2 days later, giving Yanukovych 49.46% and Yushchenko 46.61% (Gerlach 2014: 11). On the basis of exit polls funded by various Western government embassies and four Western NGOs (the NED, Charles Stewart Mott Foundation, Eurasia Foundation, and George Soros Renaissance Foundation) it was alleged that the election results were largely falsified in favor of Yanukovych. Following the February and March parliamentary elections in 2005 in Kyrgyzstan, allegations of irregularities in the election and manipulation of a large number of votes by opposition groups led to mass protests (Kakışım 2017: 243). Similarly, when it was announced that Lukashenka had received more than 80% of the votes in the first ballot, major protests began in Belarus. People who thought that the ruling party cheated in the elections went out on the street and protested against the government. While civil disobedience was gradually expanding, international pressure on the current powers was increased. Then, the people who revolted seized strategic points such as the parliament, the presidency, and the radio and TV buildings, and seized power. After the revolution, the previous opposition won the elections and came to power (Oğan 2005).

As seen above, attempts have been made to realize color revolutions in many countries in the former Soviet geography. However, in Belarus, for example, the same practices have not resulted in a change of government as they did in Georgia, Ukraine, and Kyrgyzstan.

After the incumbent president Lukashenka was announced to have won the first ballot by more than 80%, with alleged irregularities in the election, large-scale protests started in Belarus on March 19, 2006. The opposition, civil groups, and Zubr mobilized activists and managed to establish a tent city. This so-called Jeans Revolution ended when the police used force and uplifted the tents and demonstrators. Many of them were jailed (Gerlach 2014: 22).

Although (as in Georgia, Ukraine, and Kyrgyzstan) youth organizations, the opposition, and media were supported from outside and the protests started with allegations of irregularities in the elections, in Belarus the color revolution was not successful. Some of the factors that were ultimately effective can be identified, such as the Lukashenko administration's intense pressure on individuals and civil society organizations, and its willingness to use force against demonstrators. In addition to suppressing the politically active critics of the regime, the media were highly censored and legal barriers made it difficult for independent media to work in the country. Besides, there was a divided opposition in Belarus. The West tried to enter Belarus with NGOs, but such organizations had been under extreme pressure in Belarus since 1997 (Marcus 2010: 124–128). Polls have since shown that Lukashenka remains popular (Gerlach 2014: 21). Other factors were the lack of a negative perspective of Russia or Putin, the lack of nationalist feelings, and the use of the economy as a weapon by Lukashenka (Marples 2006: 358–360).

## After the Color Revolutions...

In the 2000s, pro-Western governments came to power in Georgia, Ukraine, and Kyrgyzstan with color revolutions, with support from the West for youth organizations, the opposition, and media. However, this situation did not bring about developments in favor of the West in these countries after the revolutions.

It can be said that the West's interests have been least damaged in Kyrgyzstan, among these three countries. Bakiyev was brought to power after the opposition demanded Akayev's resignation, claiming that he cheated in the elections in 2005. Bakiyev was dismissed in a bloody coup 5 years later, accused of corruption and authoritarianism (Milliyet 2010). In October 2011, Almazbek Atanbayev became president. Kyrgyzstan tried to pursue a balanced foreign policy with Russia, China, and the USA in his term, and continued to maintain both Russian and American bases (Halidov 2013).

The pro-Western strategy and policies of Saakashvili, who came to power with the revolution in Georgia, attracted increasing attention to the region from Moscow and, after 1992, calm gave way to increasing tension, conflict, and instability. The summer of 2004 saw the start of heated conflicts between the South Ossetians and the Georgians. Because it shared the same culture and history, South Ossetia wanted to be part of the Russian Federation (RF) and complained that Georgia was unwilling to resolve this issue. In response, Georgia deployed 2000 troops to South Ossetia (Atıcı Köktaş 2015: 99).

The other disputed territory, Abkhazia, demanded unification with Soviet Russia in the USSR period in 1978. The Abkhazian Supreme Soviet declared Abkhazia to be an independent republic in the USSR in a session without participation from Georgian representatives in 1990. However, the decision was immediately annulled by the Georgian Supreme Soviet. The war began on August 14, 1992, when the Georgian army entered Abkhazia, and the leadership of the Georgian leader Zviad Gamsakhurdia ended in the same year. Abkhazians succeeded in removing the Georgian army from their lands in 1993. After Kosovo declared its independence on February 17, 2008, the South Ossetian Parliament called on the RF, the Commonwealth of Independent States (CIS), the United Nations (UN), and the European Union (EU), to recognize its independence on March 3, 2008. On August 7–8, 2008, Saakashvili chose to use military force against Abkhazia and South Ossetia. Then Moscow intervened and a war between the two countries started. After the war, Russia recognized the independence of South Ossetia and Abkhazia (Atıcı Köktaş 2015: 99–100).

In Ukraine, Yushchenko declared that Yanukovych had cheated in the elections and canceled the elections. Yushchenko won the new elections. After the Orange Revolution, the Yushchenko administration pursued a pro-Western policy. However, because of pressures imposed by Russia and lack of adequate support from the West, Yushchenko lost the presidential election in 2010. Viktor Yanukovych, who managed to appeal to both the Russian and Western sides in his campaign, won the election. During the rule of Yanukovych, Ukraine gave up its North Atlantic Treaty Organization (NATO) membership and extended the duration of use of Russia's Sevastopol naval base until 2042. At the same time, Russian ships were given the freedom to cross the Kerch Strait. The pro-Russian foreign policy of Yanukovych's government caused reactions from the populations in the western and central regions of the country. Long protests began upon the decisions of Yanukovych to abandon the Eastern Partnership Agreement with the EU in November 2013 and to embark on political, economic, and military rapprochement with Russia. In the western cities of the country, buildings and belonging of the state and the ruling party were captured by the demonstrators. As a result, Yanukovych left the country, but—more importantly—Russia took advantage of this confusion and annexed Crimea. Under the influence of the developments in Crimea, the ethnic Russian minority in the eastern part of Ukraine and the pro-Russian separatist forces initiated popular uprisings against the Kiev administration in Luhansk, Donetsk, Slavyansk, and Kharkiv (Sandıklı and İsmayılov 2015: 6–21; Cabbarlı 2016: 105–113). The uncertainty in Ukraine continues.

## Conclusion

After the end of the Cold War, semiauthoritarian governments were established in the former Soviet countries. In the 2000s, nonviolent protests emerged against these governments, claiming that the elections held in many of these countries were unfair. Essentially, a pattern was applied. The opposition, youth organizations, and media organizations were supported financially and differently by international NGOs; the administrations changed in Georgia, Ukraine, and Kyrgyzstan, and the pro-Western opposition came to power. However, in some countries, such as Belarus, protests were suppressed and the revolution failed.

The developments in the countries where the revolutions were successful and pro-Western governments came to power were particularly detrimental to the regional and global interests of the West.

Chaos theory emphasizes that there are many different variables that can cause an event and that small effects can cause large changes in the results of events. Therefore, because not all factors could be taken into account, different factors in Belarus prevented the success of the revolution despite the same pattern being applied that was successful in other countries. Although the starting point was the same, the result was different because of the effects of different factors.

With regard to the influence of the West in the region, it can be said that the implementation of color revolutions through Western support for internal factions has had a negative effect on the USA with the butterfly effect. Instead, it has contributed to the increase in Russia's international influence, especially with its annexation of Crimea.

## References

Akşar, K. (2009). *Soğuk Savaş Sonrasi Dönemde Büyük Güçlerin Kafkasyaüda Nüfuz Kazanma Mücadeleleri-Kirgizistan Devrimi*, (Unpublished master dissertation), Trakya University, Edirne.

Arrow, R. (2011, February 22). Devrimin kitabı yazılır mı? *BBC News*. https://www.bbc.com/turkce/haberler/2011/02/110222_gene_sharp. Accessed 2 Mar 2019.

Atıcı Köktaş, N. (2015). Saakaşvili Dönemi Gürcistan: İç ve Dış Politika Üzerine Bir Değerlendirme. *Ardahan Üniversitesi İktisadi ve İdari Bilimler Fakültesi Dergisi, 2*, 95–110.
Baev, P. K. (2011). "A Matrix for Post-Soviet 'Color Revolutions': Exorcising the Devil from the Details", *International Area Studies Review, 14*(2), 3–22.
Bayramoğlu, G. (2016). Karmaşıklık Paradigması Işığında Örgüt Teorilerinin Yeniden Değerlendirilmesi. *Selçuk Üniversitesi Sosyal Bilimler Ensstitüsü Dergisi*, (35), 49–63.
Bütz, M. R. (1997). *Chaos and complexity implications for psychological theory and practice*. Washington, DC: Taylor and Francis.
Cabbarlı, H. (2016). İki Güç Arasındaki Ukrayna. *Karadeniz Araştırmaları*, (50), 95–123.
Canan, S. (2016). *Kimsenin Bilemeyeceği Şeyler*. İstanbul: Tuti Kitap.
Çıraklı, Ü., Dalkılıç, S., & Hacıhasanoğlu, T. (2017). Kaos Teorisi, Karmaşıklık Teorisi, Karmaşık Uyarlamalı Sistemler: Sağlık Hizmetleri Açısından Bir Derleme. *International Journal of Academic Value Studies (Javstudies), 3*(16), 330–343.
Companjen, F. J. (2010). Georgia. In D. Ó Beacháin & A. Polese (Eds.), *The colour revolutions in the former Soviet republics: Successes and failures* (pp. 13–29). Abingdon: Routledge.
Donnari, E. (2006, March 13). *Zubr opposition movement: Strong like bison*. https://cafebabel.com/en/article/zubr-opposition-movement-strong-like-bison-5ae004c3f723b35a145dba41/. Accessed 2 Mar 2019.
Ertürk, A. (2012). Kaos Kuramı: Yönetim ve Eğitimdeki Yansımaları. *Kastamonu Eğitim Dergisi, 20*(3), 849–868.
Gerlach, J. (2014). *Color revolutions in Eurasia. SpringerBriefs in Political Science*. Cham: Springer.
Halidov, İ. (2013). Kırgızistan Devrimlerinde Son Durum. *Akademik Bakış Dergisi, 38*, 1–19.
Kakışım, C. (2017). Küreselleşen Dünyada Devlet Egemenliği ve Ülke Toprağı Kavramları: "Renkli Devrimler" Örneği. *Sobider (Sosyal Bilimler Dergisi The Journal of Social Sciences), 4*(10), 236–246.
Lale, A. (2018). *Kaos Teorisi ve Uluslararası İlişkiler: Arap Baharı Örneği* (Unpublished master's thesis). Gazi University Institute of Social Sciences, Ankara.
Marcus, U. (2010). Belarus. In D. Ó Beacháin & A. Polese (Eds.), *The colour revolutions in the former Soviet republics: Successes and failures* (pp. 118–135). Abingdon: Routledge.
Marples, D. R. (2006). Color revolutions: The Belarus case. *Communist and Post-Communist Studies, 39*, 351–364.
Milliyet. (2010, April 8). Kırgızistan'da Halk Darbesi. *Milliyet*. http://www.milliyet.com.tr/kirgizistan-da-halk-darbesi/dunya/haberdetay/08.04.2010/1222128/default.htm. Accessed 7 Mar 2019.
Ó Beacháin, D., & Polese, A. (2010). Introduction. In *The colour revolutions in the former Soviet republics: Successes and failures* (pp. 1–12). Abingdon: Routledge.
Oğan, S. (2005). *Kırgızistan'da "Yağma" Devrimi*. http://turksam.org/kirgizistan-da-yagma-devrimi. Accessed 20 Mar 2019.
Oğan, S. (2006). *Turuncu Devrimler*. http://turksam.org/turuncu-devrimler-kitabi-birinci-bolum. Accessed 23 Feb 2019.
Sandıklı, A., & İsmayılov, E. (2015). *2014–2015 Yılında Harp Akademileri Komutanlığında İcra Edilen Paneller*. İstanbul: Genelkurmay Başkanlığı.
Sharp, G. (2010). *From dictatorship to democracy*. Boston: Albert Einstein Institution.
Topak, S. T. (2014). Sivil Toplum Kuruluşları ve Sosyal Medya Bağlamında "Renkli Devrimler" ve "Arap Baharı" Süreçlerinin Karşılaştırmalı Analizi. *Eskişehir Osmangazi Üniversitesi İİBF Dergisi, 9*(3), 233–254.

# Brexit in the Light of Chaos Theory and Some Assumptions About the Future of the European Union

**Erjada Progonati**

**Abstract** Chaos theory is one of the physics revolutions of the twentieth century. Although the use of chaos theory in the natural sciences is a very useful tool, it has become a controversial issue for the social sciences. To make chaos theory more understandable within the framework of the international relations discipline, this work attempts to explain the future of the European Union (EU) as a case study. In this context, although the EU project has embodied the common vision of the Western European countries for political unification, this plan for integration has been delayed by the possibility of an even more pessimistic outcome such as disintegration of the EU. This uncertainty about the EU could lead to deep chaos needing to improve the prospects and a reconstruction of the European idea.

**Keywords** Chaos theory · EU · Greece · Brexit

## Introduction

The concept of chaos includes irregularity and uncertainty. In an environment of irregularity and uncertainty, parties encourage chaos in order to dominate the existing system or to obtain a more privileged position. In political and social systems where the balance of power is disturbed, chaos continues until a new geopolitical equilibrium and a predictable atmosphere are formed. It is possible to say that the end of the Cold War and the inability to create a new geopolitical equilibrium are the main reasons for the current international crises.

In the social sciences, although there is an emphasis on the concept of chaos as "the butterfly effect," there is no clear definition of this term, but there are many phenomena that can be evaluated as chaotic. The assassination of Prince Ferdinand, leading to the start of World War I; the emergence of the biggest spate of looting

E. Progonati (✉)
Süleyman Demirel University, Isparta, Turkey
e-mail: erjadaprogonati@sdu.edu.tr



Ş. Ş. ERÇETİN, Ş. N. AÇIKALIN (eds.), *Chaos, Complexity and Leadership 2018*, Springer Proceedings in Complexity, https://doi.org/10.1007/978-3-030-27672-0_7

and anarchy in the USA after the shooting of a black teenager; and the beginning of the Arab Spring by self-immolation of a young Tunisian can be explained in this context.

To make chaos theory more understandable within the framework of the international relations discipline, we will try to explain the future of the European Union (EU) as a case study. In this study, we will discuss the political situation in Greece from 2008 to the present and events and consequences that might shed light on the future of the EU. The fact that the Greek crisis has spread to other EU countries such as the UK, which is one of the most powerful members of EU, means that many balances can change in the future of the EU. All of these events will create a new socioeconomic situation in the EU and in the rest of the world.

## Chaos Theory in International Relations

Chaos theory is one of the physics revolutions of the twentieth-century. Although the use of chaos theory in the natural sciences is a very useful tool, it has become a controversial issue for the social sciences. This theory examines irregular fluctuations, susceptibility to distortions, and long-term unpredictable systems. The two common features of chaotic behavior are its sensitive and mutant characteristics. The first step in trying to apply chaos theory to a given system is to define the system in question. When we design the system, we define the rule of how the elements of the system change (Kantemnidis 2015: 19–20).

Political scientist Dylan Kissane proposes three assumptions to make a series of predictions about the international system in the short and medium terms. The first assumption is that the nature of the international system is chaotic and that the system is sensitive to the nature of the initial conditions, complex long-term behavior, and unpredictability. Secondly, in systems with chaotic behavior, actors seek security. The third assumption is that these actors that seek security tend to interact with other actors in the international system (Kantemnidis 2015: 20).

The problems in these assumptions are that each concept is not abstract and measurable, and that the variables that may affect the system are not defined. The "chaotic international system" statement is ambiguous. Social scientists are forced to imitate physicists to produce models that will demonstrate the existence of chaos, with which they can inform decision makers for the purpose of policy making. Glenn emphasizes that learning about chaos is not difficult. The only difficulty learning about it quickly, because not everyone is familiar with it. This may bring about failure to recognize chaotic events in physical and social systems. He also argues that the practice of chaos is broad and that decision makers must be familiar with the fundamental consequences and insights of chaos theory (James 1996, p. xii).

In international relations and political science, most cases are based on their sensitivity to initial conditions; therefore, many events in these disciplines can be explained using the chaos approach. Thus, the results of political parties' activities and elections can be evaluated in terms of chaos theory because some small incidents in an election campaign can change the final results altogether.

The chaos theory used in international relations helps us to model the real world and events that have happened. "Modeling" means creating a representation in your mind, on paper, or in a laboratory of the world in which you are interested. This is because we cannot combine all of the real world and its dynamics in our minds. In this case, after analyzing certain events and dynamics in our minds, and maybe manipulating things, it will be easier to see how the rest of the world has changed. Thus, we need to select "a piece of the world" that represents the entire world and analyze its dynamics. In this analysis, we determine future events and results in the real world about whether we are right. Application of chaos theory in international relations should be done with models. Models are theoretical structures to capture the basic features of real systems. Models represent the specific joints of general theories, so the creation of good realistic models is not a trivial task. The actual systems are always open to some extent; that is, their boundaries are not well defined and are so intrinsically complex that they can challenge a definite characterization. So maybe we cannot use models as solutions to everything, and noncritical usage of them can lead to incorrect estimates. However, it would be very useful for decision makers to examine the isolated subsystems of the real world, which are limited to a small number of variables.

A moderate solution for the use of chaos theory in international relations comes from Saperstein, according to whom the theories of nonlinear dynamic systems deal with mathematical models. He says we cannot use modeling always and without making a critical decision, because we can get the wrong results. Real-world systems consist of undefined boundaries.

The complexity of real-world systems makes it impossible to gather all of the information necessary to create a model that is sufficient to predict the future of international relations among all actors. Thus, Kamminga thinks it is not appropriate to use mathematical theories in the social sciences, and describes human social systems as important but extremely problematic in how to identify and study them. The construction of mathematical models includes significant simplifications that can have tremendous consequences for understanding of real dynamic systems (Kamminga 1990: 52). Nevertheless, it may be useful for decision makers to examine isolated subsystems in the real world that are limited to a small number of variables (Saperstein 1999, p. 10).

From the perspective of this work, the discussion about Grexit, Brexit, and other potential "-exits" raises the matter of the EU's future, as well as the future of Greece and the UK. The "-exit" terminology was coined in February 2012 by two economists from the US financial giant Citigroup to explain the possibility of Greece's departure from the single European currency (Straits Times 2016).

The word "Grexit" has been used for a long time. The word "Brexit," was derived from it to describe the 2016 referendum in the UK. Meanwhile, other derived "-exit" terms have been coined to refer to the possible exit scenarios of other countries from the European integration project. In this context, referring to the European integration project, the EU's enlargement process is somewhat ambiguous.

## Dangerous Fragmentation Inside the European Union

### *Grexit and Its Consequences Inside and Outside Greece*

In 2007, the world faced a new economic crisis. This crisis, triggered by turbulence in the US financial system, led to a significant redefinition of economic power worldwide and completely changed the roles, relationships, and agenda of world power centers.

The stagnation in Greece is not directly linked to the main causes of the crisis in the USA and Europe, but it is directly related to the effects of the globalized crisis on the imbalances that had already accumulated in Greece before 2007–2009 and caused an unfavorable financial situation in that country (Hrisovalantis 2015, pp. 40–41).

After 2 years of saving measures in accordance with the conditions of the lenders, the Greek government's failure to drive the country toward economic growth led it to a new election in January 2015. The Syriza Party, headed by Alexis Tsipras, won the election and promised to put an end to memoranda and austerity. One feature of this effort was the holding of a referendum on whether to adopt the new austerity measures implemented by the European Commission, the European Central Bank, and the International Monetary Fund at the Eurogroup on June 25, 2015. In the referendum, 62.5% of votes were cast against the proposed austerity measures. A week after the referendum, however, Greece's Syriza-led government took even tougher measures. Amid fears of Grexit, EU leaders had to sit at the negotiating table to prevent it (Chatzinikolaou 2017) and, as a result, Greece has remained part of the EU and the eurozone. If the result of the negotiations had been different, Greece could have withdrawn from the eurozone and a Grexit process would have begun.

During the crisis, Greece was described as the "chronic patient" of the eurozone. In this period, Greek pensions and wages were reduced by up to 40%, with aggressive taxation and the collapse of the middle classes. Despite the great efforts of the Greek government and after all of the sacrifices of the people, the future of the country depends on continuous growth of the economy and attainment of a budget surplus for years. Greece's debt repayments will continue until 2060 (Prentoulis 2018).

At the beginning of 2017, the risk of Grexit reappeared. In particular, the financial cost of delays in the evaluation procedures—that is, a final agreement on the 2018 correction program—was greater than the potential benefit for Greece. Today, Greece seems to be entering into a more normal process with the end of its official debt agreement with the troika (the International Monetary Fund, European Central Bank, and European Commission) (Alcidi et al. 2012). The crisis costs Greece 25% of its gross domestic product (GDP), and the unemployment rate is 20% even after emigration (Prentoulis 2018). Although debt arrangements have been made, Greece is not entirely independent from its creditors, because a series of audits and reforms must continue for a healthy economy. As long as Greece does not implement the reforms demanded by the institutions, the possibility of it leaving the eurozone will persist.

A possible outflow of any country from the eurozone and the EU will mark many developments in its economy in all sectors, affecting both businesses and individuals, and increasing the unemployment rate, household bankruptcies, stagnation, etc. Grexit is a big challenge for other countries in the eurozone. Even if most private creditors have already made deposits from the countries facing a debt crisis, the citizens of these countries (e.g., citizens of Italy and Spain) can transfer their money to more secure bank accounts abroad. Such an event would lead to a crisis in the banks, especially those in Southern Europe, because the countries that borrow money from those countries would be affected and this would result in an international banking crisis.

Another reality is that although Greece represents only 0.3% of the global GDP, it will represent the failure of European integration if it withdraws from the EU and will also lead to large global balance changes. In particular, this will undermine the confidence of the EU project, because the efforts to revive Greece will have failed and this situation will result in the withdrawal of a large number of countries from the eurozone, seriously harming the integration of the EU and the progress of the project.

Finally, a possible Grexit might affect the political structure of the region and the role of other major economies at the international level. If left alone, Greece will likely receive financial support from strong countries outside the European sphere of influence, such as Russia, China, or Middle Eastern countries. These factors will pose a financial risk not only for Greece but also for all of the other actors in the Balkans. Such a situation would make it clear that the second-largest reserve of money in the world, and therefore the investment levels in Europe, would be reduced and the eurozone would bring back the recession. This would also undermine the EU's political role because its lack of capacity to guarantee its own currency would diminish its power in any political negotiations.

## *Brexit and Its Echoes in the Eurozone*

In 2009, the economies of the eurozone countries were in recession. More specifically, since 2008 the GDP of France experienced a recession of 2.6%; Spain, 3.5%; the Netherlands, 4%; Denmark, 4.9%; Germany, 5%; Sweden, 5.2%; Ireland, 7%; and Finland, 7.8%. However, in the last four years, following the implementation of concrete measures, the aforementioned countries show a marked increase in the stagnation (Hrisovalantis 2015, p. 4).

The emergence of the debt crisis in Greece since 2009 and then in other eurozone countries, especially in the southern region of Europe, raised concerns about the sustainability of the EU and the preservation of its basic structure. In the last 2 years, many things changed in the EU: the refugee crisis intensified, nationalism was significantly strengthened, and, most importantly, the UK decided to leave the EU.

The UK has long been recognized as one of the most "Euro-skeptic" members of the EU. The former British prime minister David Cameron, faced with increasing pressure from tough Euro-skeptic lines inside and outside the British Conservative Party, called for a review of Britain's requirements for the accession to the EU. Thus, on February 23, 2016, after an agreement was reached between Cameron and other EU governments, a date was fixed for a public referendum on whether UK would or would not stay in the EU.

On June 23, 2016, that referendum was held in the UK. The result of the referendum, in which the participation rate was 72%, was 51.9% in favor of leaving the EU and 48.1% against leaving. As a result, David Cameron resigned and Theresa May assumed the position of prime minister on July 13, 2016.

On March 29, 2017, the UK government accepted the referendum results and triggered Article 50 of the Lisbon Treaty (a so-called EU exit clause; this article describes the procedures that a member state of the EU must perform to leave the union).

David Cameron and the then finance minister, George Osborne, foresaw the unfavorable situation and the consequences of a Brexit process for the UK. According to estimates, the growth rate in the UK would be 3.6% lower in the 2 years following the referendum, unemployment would reach half a million, wages would fall by 2.8%, and the value of the pound would drop significantly. Brexit is likely to have great political and economic repercussions for the UK and the union itself, because the UK is the EU's second-largest economy and a major diplomatic and military force.

It should be noted also that after the possible exit of the UK from the EU, much of the uncertainty about the future of the European Union stems from concerns about British trade. According to various studies and analyses, Brexit may have a direct and long-lasting effect on the economies of the UK and the EU. On the basis of these arguments, the financial markets reacted negatively to the result of the Brexit referendum. However, a meeting of the seven finance ministers and the central banker expressed confidence that the UK was well prepared for the results of Brexit and would maintain its economy's flexibility. The International Monetary Fund concluded that those eurozone members with strong trade and financial ties with the UK (Ireland, Malta, Cyprus, Luxembourg, the Netherlands, and Belgium) would be more affected by Brexit.

It seems that the Brexit process has been underestimated by British politicians. The short-term effects of Brexit may not be too bad, but its long-term impact will make room for negative scenarios. This is because British politicians still fail to understand how the EU will react to the crisis.

In 2017, with the re-election of Theresa May, the UK faced a surprise because, instead of strengthening the UK's future Brexit negotiations, the area of bargaining had become much narrower. The vote of the people of Greece for its own government was a vote of confidence, and so Greece continued to negotiate with the EU, whereas the vote of the people of the UK gave a result to leave the EU.

Naturally, a question arises: Why did the UK choose not to make the same choice as Greece (despite its lack of severe pressures in 2015), but to leave the

EU instead? In fact, there is the fear that the UK will look for other partners in the post-Brexit period and weaken (Bennet 2017). From 2010 onward, it appears that the EU has faced a number of problems, including slow growth and high unemployment, economic and political pressures, and the rise of populist parties in many European countries. Such problems hamper efforts to solve many internal and external problems that are very important for the future of the European Union and Europe in general, such as the long-lasting Greek debt crisis and the possibility of Greece leaving the eurozone (Grexit), the possible exit of the UK from the EU (Brexit), the ongoing flows of migrants and refugees, the continuous increase in terrorism and the Islamophobia phenomenon, or the attitude of extreme-right-wing European parties toward the EU. These factors seem to have had a major effect on the UK's decision to leave the union. In the face of these important and challenging issues, the future, existence, and consequent survival of the EU are constantly being questioned. These issues cause concern for supporters of the European project.

All of the above issues have echoed in other member states of the EU. The populist nationalist parties of both the right and the left base, with their policy lines on Europe, think it is better to leave the EU. The Nationalist Front Party in France, the nationalist parties in Austria and Germany, and leftist populists such as Podemos in Spain have repeated their rhetoric about leaving the EU. This rhetoric takes place in important member countries of the EU and even in the founder members of the union. Thus, the French extreme-right leader Marine Le Pen called for "Frexit" shortly after the results of the UK's membership referendum were announced. "Victory of freedom! As I've asked for years, now we must have the same referendum in France and other EU countries," she declared on Twitter. The extreme-right party headed by Marine Le Pen is the third largest party in France behind the ruling Socialist Party and the center-right Republicans. In addition, after the referendum in the UK, the leader of the far-right extremist Freedom Party in Netherlands, Geert Wilders, said, "Now it's our turn," and in his party's election campaign he promised a referendum called "Nexit." Moreover, in Austria the extreme-right party leader Norbert Hofer proposed the idea of "Oexit" (named after the Austrian name Österreich), and in Sweden, where the level of support for leaving the EU was 31% in opinion polls, the right-wing Swedish Democrats introduced the idea of "Swexit." A petition called "Fixit" (although the English version of this name does not have the proper connotations), requesting Finland's exit from the EU, collected thousands of signatures. The populist Danish People's Party (DPP) said that they want to renegotiate the EU agreements, otherwise they are eager for "Dexit," which means the exit of Denmark from the EU. "Gerxit" was also discussed in the French and English media; however, this idea wasn't too adaptable on official platforms in Germany. Nevertheless, Frauke Petry, the chairman of the right-wing populist Alternative Germany Party (AfD), emphasized that "Brexit" was a warning to the EU. "If the EU does not abandon the semisocialist experiment on continuous integration, Europeans will follow the British way and take back their sovereignty," Petry said. Italy's leading right-wing politician Matteo Salvini also supports "Italexit," and this is very strange because Italy is one of the founder countries of the EU. This whole process does not seem certain, but it is clear that the EU is not the same as it was before 2008.

## The Russian Factor in the Chaos Within the European Union

Since the end of the Cold War, the USA has not had the same contact with the military and political alliances that helped it to prevail in the conflict. Although Russia remains weak in comparison with the USA, it continues to pursue its goal of dividing political alliances to regain its strategic position in Europe. While the institutions and the vision of the EU are being taken seriously after the shock of Brexit, and reform scenarios are being discussed, a candidate who will overthrow this order will win the upcoming critical elections in Moscow, which does not want a strong EU. Meanwhile, the close relations of the EU's strengthening right-wing parties with Moscow and their support have led Russia's upmanship toward the EU (Valansi 2017).

Why is the Moscow factor so apparent in EU issues? The link between the USA and the European democracies over the last 100 years has been based on common values such as freedom, human rights, and the rule of law. Putin's Russia rejects these values (thus, Putin puts his political opponents under pressure or imprisons them, sends assassins to kill Russia's dissidents, and uses similar tactics against them). Putin is now trying to export his leadership brand. The Russian leader has formed alliances with many right-wing European political parties and leaders who offer consistent commitment to Russian interests. The European right-wing parties benefit from the economic and security crises in Europe to ensure public support, and they are thus trying to undermine the Western liberal order (Gude 2017).

Another argument that is connected with the Russian presence in EU issues is the neo-Russian imperialist idea known as Eurasianism. The main architect of Eurasianism is Alexander Dugin. In many articles and books, he has discussed the "revival of fascism" in Russia. In the late 1980s and early 1990s, Dugin began to establish a network with far-right groups in Europe (Laruelle 2015, p. 82). In 1989 he went to France, and in 1992 he contacted Jean-Marie Le Pen's National Front. It is known that Le Pen has long been a fan of Putin. During the 1990s and 2000s, Dugin hosted the conferences of the National Front authorities. During the tenure of Italian Prime Minister Silvio Berlusconi, Dugin also had strong ties with right-wing groups in Italy.

When Ukraine elected a pro-Western leadership in 2014, Moscow supported an uprising in its Russian-speaking eastern provinces (Jenkins 2018). Russia's relations with Europe are not the same, till the time that Putin began trying to bring Crimea back to Russia. Representatives of far-right parties served as election observers in Crimea, acknowledged the annexation of Crimea, and spoke against Western governments that were against Russia's interests (Polyakova 2016). One of Putin's goals is to use these extremist right-wing parties to undermine the political consensus in the West and the institutions that support the liberal international order, which Russia sees as a threat. Russia seems to have succeeded at this point because it has gained sympathy in many countries of Europe. The pro-Russian campaigns by the former leader of the UK Independence Party (UKIP) Nigel Farage have also been important in ensuring the UK's withdrawal from the EU. On the basis of these arguments, it can be said that Russia is in opposition to the EU's political and economic development; also, it fuels the chaos that has become present in Europe together with Grexit.

## Consequences

It can be deduced that if social scientists have a background in mathematics and physics, or if they collaborate with a team of such scientists, chaos theory can be successfully applied to international relations studies. Use of systems such as cooperation, similar concepts of attractors, nonlinearity, equilibrium points, stability, control of chaos, synchronization, etc., helps researchers to make models that predict chaotic behavior. Predictions are a qualitative approach to the unpredictability and imbalance of a system without violating mathematics or physics.

In this aspect, this work emphasizes some assumptions about the EU's future. Democracy seems not to have taken place in the policies of the EU, as Monnet thought. Brussels continues to manipulate people through distrust. Especially in the last 20 years, the EU has been unable to earn the trust of the European people, and its geography has become depressed. It seems that it will take many years for Europe to cope with unemployment and especially to cope with unemployment among its young population. Comprehensive institutional restructuring of the eurozone may provide healthy growth rates. However, the need for such extensive reforms has never been accepted in Brussels. It seems that the EU is no longer a ray of hope. The idea behind the eurozone project is that a single currency requires common economic policies; thus, it ultimately pushes for the formation of a political union. This leads to questions about the existence of the EU.

The European project is at risk of collapse, and the main source of this risk is the failure of economy and policy. The Greek uprising in 2015 (the "Athens Spring") represented an enlightened effort to challenge the dynamism of separation in Europe. The pressure of the Athens Spring contributed to the EU's rate of shift toward the Brexit vote and the rises of the National Front in France and other populist parties in other EU member countries.

Supporters of the European project are worried that for the first time in the EU's 60-year history, some aspects of European integration may stop, or even the whole EU structure may collapse. However, some think that the recent crises in the EU could lead to beneficial reforms and promotion of political and economic integration, ultimately making it a more efficient and consistent community.

## References


Alcidi, C., Giovannini, A., & Gros, D. (2012). *Grexit: Who would pay for it?* Centre for European Policy Studies. https://www.ceps.eu/ceps-publications/grexit-who-would-pay-it/.

Bennet, J. (2017). From Grexit to Brexit: A view from Athens. *British Academy Review*, no. 31. The British Academy. https://www.thebritishacademy.ac.uk/grexit-brexit-view-athens.

Chatzinikolaou, P. (2017). Could Grexit follow Brexit? *Euro Crisis in the Press*. https://blogs.lse.ac.uk/eurocrisispress/2017/04/12/could-grexit-follow-brexit/.

Gude, K. (2017). How Putin undermines democracy in the west, chapter and verse. *Newsweek*. https://www.newsweek.com/how-putin-undermines-democracy-west-chapter-and-verse-568607.

Hrisovalantis, T. G. (2015). *Το Μέλλον της Ευρωπαϊκής Ένωσης: Προς Ομοσπονδία ή Προς Διάλυση* [*The future of the European Union: Towards a federation or dissolution*] (Postgraduate diploma thesis). University of Thessaly. http://ir.lib.uth.gr/bitstream/handle/11615/47303/16659.pdf.

James, G. E. (1996). *Chaos theory: The essentials for military applications*. Newport: Naval War College Press.

Jenkins, S. (2018). Forget Brexit, war in Ukraine is the biggest threat to Europe. *The Guardian*. https://www.theguardian.com/commentisfree/2018/nov/26/forget-brexit-ukraine-europe-russia.

Kamminga, H. (1990). What is this thing called chaos? *New Left Review, 181*, 49–59.

Kantemnidis, D. (2015). Chaos theory and international relations. *Journal of Mediterranean and Balkan Intelligence, 6*(2), 19–28. http://www.academia.edu/20398467/Chaos_Theory_and_International_Relations.

Laruelle, M. (2015). *Eurasianism and the European far right: Reshaping the Europe–Russia relationship*. Lexington: Lanham.

Polyakova, A. (2016). *Putinism and the European far right*. Institute for Modern Russia. https://imrussia.org/en/analysis/world/2500-putinism-and-the-european-far-right.

Prentoulis, M. (2018). Greece may still be Europe's sick patient, but the EU is at death's door. *The Guardian*. https://www.theguardian.com/commentisfree/2018/aug/21/greece-europe-eu-austerity.

Saperstein, A. M. (1999). *Dynamical modeling of the onset of war*. Singapore: World Scientific.

Straits Times. (2016). From Grexit to Brexit: Eurosceptics claim their exit. *Straits Times*. https://www.straitstimes.com/world/europe/from-grexit-to-brexit-eurosceptics-claim-their-exit.

Valansi, K. (2017). Rusya ve İsrail'in Avrupa Sağıyla Dansı [Dance of Russia and Israel with the European right]. *Şalom Gazetesi*. http://www.salom.com.tr/arsiv/haber-102571-rusya_ve_Israilin_avrupa_sagiyla_dansi.html.

# Intra-Specific Competition in Prey Can Control Chaos in a Prey-Predator Model

**Md Saifuddin and Santanu Biswas**

**Abstract** In this article, we propose a general prey-predator model with the presence of Allee effect in prey. We have considered different competition coefficients within the prey population, which leads to the emergent carrying capacity. The stability analysis of the system is discussed. Further, the dynamical behavior of the system is analyzed, taking delay and emergent carrying capacity as bifurcation parameters. Time delay can turn a stable equilibrium into an unstable one. It was shown that our system experiences the Hopf bifurcation, as the delay parameter crosses some critical values. Further increases in delay produce chaos, which can be controlled by the emergent carrying capacity.

**Keywords** The Allee effect · Delay · Stability analysis · Eco-epidemiological system

## Introduction

In recent times, significant research has been done on the development of the concept for the Allee effect, which corresponds to the positive correlation between population size/density and per-capita growth rate at low population density (Allee 1931). Ecologists paid significant attention to this topic, as it relates to species extinction. Furthermore, disease has been considered to be one of the main cause for species disappearance, and if it is connected with the Allee effect, the interaction between them has substantial biological importance in nature (Hilker et al. 2007). On the other hand, many ecological and manmade activities in biology and medicine can be better interpreted with the help of time delays. In classic books such as (MacDonald 1989; Gopalsamy 1992; Kuang 1993) topics on the relevance of time

M. Saifuddin (✉)
Department of Mathematics, Bidhan Chandra College, Hooghly, India

S. Biswas
Department of Mathematics, Adamas University, Barasat, India



Ş. Ş. ERÇETİN, Ş. N. AÇIKALIN (eds.), *Chaos, Complexity and Leadership 2018*, Springer Proceedings in Complexity, https://doi.org/10.1007/978-3-030-27672-0_8

**Table 1** Variables and parameter description of different parameters for the model (1)

| Variables/ & Parameters | Biological meaning |
|---|---|
| $S$ | Density of prey |
| $P$ | Density of the predator |
| $\theta$ | Individual searching efficiency |
| $\beta$ | Attack rate of the predator |
| $\alpha = c\beta$ | Net gain to P by consuming S |
| $b$ | Intra-specific competition |
| $a$ | Intrinsic growth rate of S |
| $d$ | Natural death rate of P |
| $\tau$ | Gestation period of P |

delays in practical models are discussed in detail. To the best of our knowledge, there are very few works on time-delayed population dynamics in the presence of the Allee effect (Yan et al. 2005; Celik et al. 2008; Pal et al. 2012; Zhang and Zang 2014; Biswas et al. 2015). In this chapter, a strong Allee effect with emergent carrying capacity is discussed.

## The Model

$$\begin{aligned} \frac{dS}{dt} &= S(a - bS(t-\tau))(S-\theta) - \beta SP \\ \frac{dP}{dt} &= P\,[\alpha S - d]. \end{aligned} \tag{1}$$

All the variables and parameters are positive. The variables and parameters used in Model (1) are presented in the Table 1.

The initial conditions for the system (1) take the form

$$S(\phi) = \psi_1(\phi), \quad P(\phi) = \psi_2(\phi), \qquad -\tau \le \phi \le 0,$$

where $\psi = (\psi_1, \psi_2)^T \in C_+$ such that $\psi_i(\phi) \ge 0$ $(i = 1, 2)$, $\forall\, \phi \in [-\tau, 0]$, and $C_+$ denotes the Banach space $C_+\left([-\tau, 0], \mathbb{R}^2_+\right)$ of continuous functions mapping the interval $[-\tau, 0]$ into $\mathbb{R}^2_+$ and denotes the norm of an element $\psi$ in $C_+$ by

$$\|\psi\| = \sup_{-\tau \le \phi \le 0} \Big\{ \mid \psi_1(\phi) \mid, \mid \psi_2(\phi) \mid \Big\}.$$

For biological feasibility, we further assume that $\psi_i(0) > 0$, for $i = 1, 2$.

## Mathematical Analysis of the System (1) with No Time Delay

We first study the system (1) with no time lag. The system (1) without time delay can be written as

$$\begin{aligned}\frac{dS}{dt} &= S(a - bS)(S - \theta) - \beta SP \\ \frac{dP}{dt} &= P\left[\alpha S - d\right].\end{aligned} \tag{2}$$

The system (2) has the following boundary equilibria: $E_0 = (0, 0)$, $E_\theta = (\theta, 0)$ and $E_1 = (1, 0)$. Here, $E_0$ is always stable, $E_\theta$ is stable if $\frac{a}{b} < \theta$ and $E_1$ is stable if $\frac{a}{b} > \theta$.

The system has a unique interior attractor $E^* = (S^*, P^*)$, where $S^* = \frac{d}{\alpha}$ and $P^* = \frac{(a-b\frac{d}{\alpha})(\frac{d}{\alpha}-\theta)}{\beta}$. The interior equilibria exist if $\frac{a}{b} > \frac{d}{\alpha} > \theta$.
$E^*$ is stable if $S^* > \frac{a+b\theta}{2b}$.

## Mathematical Analysis of the Time Delay Model

In this section we study our delay model (1). Here, we study local stability analysis of equilibria, the permanence and existence of switching the stability of the delay differential equation (1). We analyze the delay system with respect to the interior equilibria $E^*$ only.

### *Local Stability Analysis*

Let $\widetilde{E} = (\widetilde{S}, \widetilde{P})$ be any equilibrium point of the system (1). The linearized system of the system (1) at $\widetilde{E} = (\widetilde{S}, \widetilde{P})$ is

$$\begin{aligned}\dot{x}(t) &= ((a - \widetilde{S}b)(2\widetilde{S} - \theta) - \beta\widetilde{P})x(t) - \beta\widetilde{S}y(t) - \widetilde{S}b(\widetilde{S} - \theta)x(t - \tau), \\ \dot{y}(t) &= x(t)\alpha\widetilde{P}\end{aligned} \tag{3}$$

Then the characteristic equation of the delayed system (1) around any equilibrium point $\widetilde{E} = (\widetilde{S}, \widetilde{P})$ is given by

$$det\begin{bmatrix}((a - \widetilde{S}b)(2\widetilde{S} - \theta) - \beta\widetilde{P}) - e^{-\lambda\tau}\widetilde{S}b(\widetilde{S} - \theta) - \lambda & -\beta\widetilde{S} \\ \alpha\widetilde{P} & -\lambda\end{bmatrix} = 0. \tag{4}$$

The characteristic equation at the interior equilibrium $E^* = (S^*, P^*)$ of the dynamical system with positive delay reduces to the following transcendental equation:

$$\lambda^2 - C_1\lambda + C_2 = [C_3\lambda]e^{-\lambda\tau}. \tag{5}$$

Where

$$\begin{aligned} C_1 &= (a - \widetilde{S}b)(2\widetilde{S} - \theta) - \beta\widetilde{P}, \\ C_2 &= \alpha\beta\widetilde{S}\widetilde{P}, \\ C_3 &= \widetilde{S}b(\widetilde{S} - \theta). \end{aligned} \tag{6}$$

For the delay-induced system (1), it is known that the equilibrium point $E^*$ will be asymptotically stable if all the roots of the corresponding characteristic equation (5) have negative real parts. But, the classical *Routh–Hurwitz criterion* cannot be used to investigate the stability of the system. Since the Eq. (5) is a transcendental equation, it has infinitely many eigenvalues. To determine the nature of the stability, we require the sign of the real parts of the roots of the characteristic equation (5).

Let $\lambda(\tau) = \zeta(\tau) + i\rho(\tau)$ be the eigenvalue of the characteristic equation (5), substituting this value in Eq. (5) we obtain real and imaginary parts respectively as

$$\zeta^2 - \rho^2 - C_1\zeta + C_2 = [(C_3\zeta)\cos\rho\tau + \rho C_3 \sin\rho\tau]e^{-\zeta\tau}, \tag{7}$$

and

$$2\zeta\rho - C_1\rho = [C_3\rho\cos\rho\tau - (\zeta C_3)\sin\rho\tau]e^{-\zeta\tau}. \tag{8}$$

A necessary condition for a stability change of $E^*$ is that the characteristic equation (5) should have purely imaginary solutions. We set $\zeta = 0$ in (7) and (8). Then we get,

$$-\rho^2 + C_2 = [\rho C_3 \sin\rho\tau], \tag{9}$$

$$-C_1\rho = [\rho C_3\cos\rho\tau]. \tag{10}$$

Eliminating $\tau$ by squaring and adding the Eqs. (9) and (10), we obtain the algebraic equation for determining $\rho$ as

$$\rho^4 + (C_1^2 - 2C_2 - C_3^2)\rho^2 + (C_2^2) = 0. \tag{11}$$

Substituting $\rho^2 = \theta_1$ in Eq. (11), we obtain a quadratic equation given by

$$k(\theta_1) = \theta_1^2 + \sigma_1\theta_1 + \sigma_2 = 0, \tag{12}$$

where

$$\sigma_1 = C_1^2 - 2C_2 - C_3^2, \quad \sigma_2 = C_2^2.$$

Now $\sigma_2 > 0$ and $\sigma_1 < 0$ implies that (12) has at least one positive root. The following theorem gives a criterion for the switching in the stability behavior of $E^*$ in terms of the delay parameter $\tau$.

**Theorem 1** *Suppose that $E^*$ exists and is locally asymptotically stable for* (1) *with $\tau = 0$. Also let $\theta_0 = \rho_0^2$ be a positive root of* (12).

1. *Then there exists $\tau = \tau^*$ such that the interior equilibrium point $E^*$ of the delay system* (1) *is asymptotically stable when $0 \leq \tau < \tau^*$ and unstable for $\tau > \tau^*$.*
2. *Furthermore, the system will undergo a Hopf bifurcation at $E^*$ when $\tau = \tau^*$, provided that $Z(\rho)X(\rho) - Y(\rho)W(\rho) > 0$.*

***Proof*** Since $\rho_0$ is a solution of the Eq. (11), the characteristic equation (5) has the pair of purely imaginary roots $\pm i\rho_0$. From Eqs. (9) and (10), we have that $\tau_p^*$ is a function of $\rho_0$ for $p = 0, 1, 2, \ldots$; which is given by

$$\tau_p^* = \tfrac{1}{\rho_0} \arccos\left[-\tfrac{C_1}{C_3}\right] + \tfrac{2\pi p}{\rho_0}. \tag{13}$$

Now the system will be locally asymptotically stable around the interior equilibrium point $E^*$ for $\tau = 0$, if the condition $S^* > \frac{a+b\theta}{2b}$ holds. In that case by *Butler's lemma*, $E^*$ will remain stable for $\tau < \tau^*$, such that $\tau^* = \min_{p \geq 0} \tau_p^*$.

Also, we can verify the following transversality condition

$$\frac{d}{d\tau}[Re\{\lambda(\tau)\}]_{\tau=\tau^*} > 0.$$

Differentiating equations (7) and (8), with respect to $\tau$ and then putting $\zeta = 0$, we obtain

$$\begin{aligned} X(\rho)\tfrac{d\zeta}{d\tau} + Y(\rho)\tfrac{d\rho}{d\tau} &= Z(\rho), \\ -Y(\rho)\tfrac{d\zeta}{d\tau} + X(\rho)\tfrac{d\rho}{d\tau} &= W(\rho). \end{aligned} \tag{14}$$

Where

$$\begin{aligned} X(\rho) &= -C_1 + \tau(\rho C_3 \sin \rho\tau) - C_3 \cos \rho\tau, \\ Y(\rho) &= -2\rho + \tau(\rho C_3 \cos \rho\tau) + C_3 \sin \rho\tau, \\ Z(\rho) &= \rho^2 C_3 \cos \rho\tau, \\ W(\rho) &= -\rho^2 C_3 \sin \rho\tau. \end{aligned} \tag{15}$$

Solving the above system, we have $\frac{d}{d\tau}[Re\{\lambda(\tau)\}]_{\tau=\tau^*,\rho=\rho_0} = [\frac{Z(\rho)X(\rho)-Y(\rho)W(\rho)}{X^2(\rho)+Y^2(\rho)}]_{\tau=\tau^*,\rho=\rho_0}$, which shows that $\frac{d}{d\tau}[Re\{\lambda(\tau)\}]_{\tau=\tau^*,\rho=\rho_0} > 0$ if $Z(\rho)X(\rho)-Y(\rho)W(\rho) > 0$. Therefore, the transversality condition is satisfied and hence Hopf bifurcation occurs at $\tau = \tau^*$. This completes the proof of the theorem.

## *Uniform Persistence of the System*

In this section, we present conditions for uniform persistence of the system (1). We denote by $\mathbb{R}_+^2 = \{(S, P) \in \mathbb{R}^2 : S \geq 0, P \geq 0\}$ the non-negative quadrant and by $int(\mathbb{R}_+^2) = \left\{(S, P) \in \mathbb{R}^2 : S > 0, P > 0\right\}$.

**Definition** System (1) is said to be uniformly persistent if a compact region $D \subset$ int $(\mathbb{R}_+^2)$ exists such that every solution $\Xi(t) = (S(t), P(t))$ of the system (1) with initial conditions eventually enters and remains in the region $D$.

**Boundedness of the solution of the delayed system** (1):
One can write the first equation of the system (1) as

$$\frac{dS}{S} = [a\,(1 - bS)\,(S - \theta) - \beta P]\,dt.$$

Integrating between the limits 0 and $t$, we have

$$S(t) = S(0)\exp\left\{\int_0^t [a\,(1 - bS)\,(S - \theta) - \beta P]\,ds\right\}.$$

Similarly from the second equation of the system, we have

$$P(t) = P(0)\exp\left\{\int_0^t \left[\alpha \frac{S(s-\tau)P(s-\tau)}{P} - d\right] ds\right\}.$$

where $S(0) = S_0 > 0$ and $P(0) = P_0 > 0$. Therefore, $S(t) > 0$ and $P(t) > 0$.

**Lemma 1** *All the solutions of the system* (1) *starting in* $int(\mathbb{R}_+^2)$ *are uniformly bounded with an ultimate bound* $M$.

*Proof* The proof is similar to the proof of Biswas et al. (2016).

## Numerical Simulation

Several numerical experiments are performed on the system (1) to validate our theoretical findings. In the present investigation, the gestation delay ($\tau$), Allee effect ($\theta$) and intra-specific competition (b) are the key parameters. We investigate the

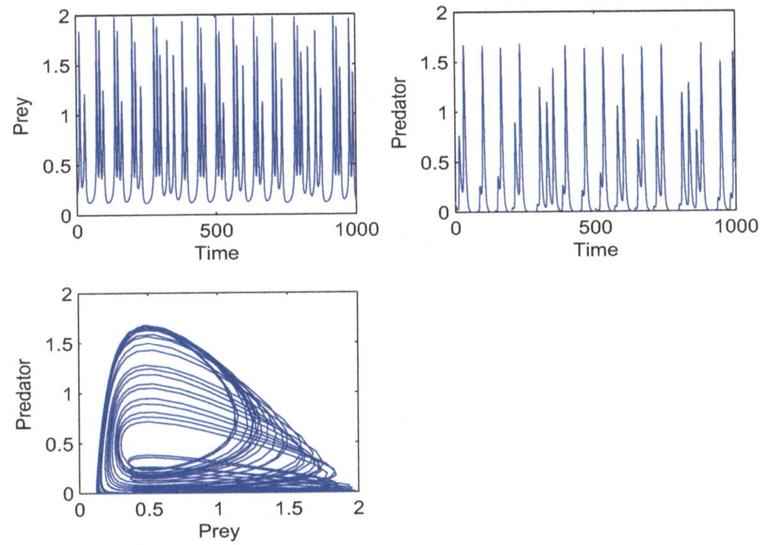


**Fig. 1** The delayed system (1) is chaotic with the time delay $\tau = 2.5$ and the other parameters $\beta = 0.2, m = 0.4, a = 0.9, \theta = 0.1, \alpha = 0.8, b = 1$, with the initial condition [S(0),I(0),P(0)] = [0.8, 0.1, 14]

system (1) for different values of the above parameters. This study demonstrates the feasibility of different complex dynamical behavior, including limit cycle, higher periodic oscillations and chaos (Fig. 1).

Consider the set of parameters as $\tau = 2.5, \beta = 0.2, m = 0.4, a = 0.9, \theta = 0.1, \alpha = 0.8, b = 1$, with the initial condition $[S(0), I(0), P(0)] = [0.8, 0.1, 14]$. In the presence of strong Allee if we increase the delay parameter, the system will be more chaotic. Now keeping the other parameter values the same, if we increase the emergent carrying capacity (b), then we observe that the system (1) shows 2-periodic solutions for $b = 1.04$ (Fig. 2). Figure 3 shows the limit-cycle oscillations for $b = 1.06$. Again, if we increase the value of $b$, then we observe that for $b = 1.16$ the system (1) converges to stable focus (Fig. 4).

To make it more clear, we draw the bifurcation diagram with respect to the emergent carrying capacity (b) for $1 < b < 1.2$. For a gradual increase in b, the system (1) switches its stability from chaotic oscillation to period doubling, period doubling to limit cycle oscillation, and limit cycle oscillation to stable focus. Figure 5 shows that for $b \in [1, 1.04)$ the interior equilibrium $E^*$ is chaotic, for $b \in [1.04, 1.06)$ it shows period doubling, for $b \in [1.06, 1.16)$ it exhibits limit cycle oscillation, and for $b \in [1.16, 1.2]$ the interior equilibrium is stable.

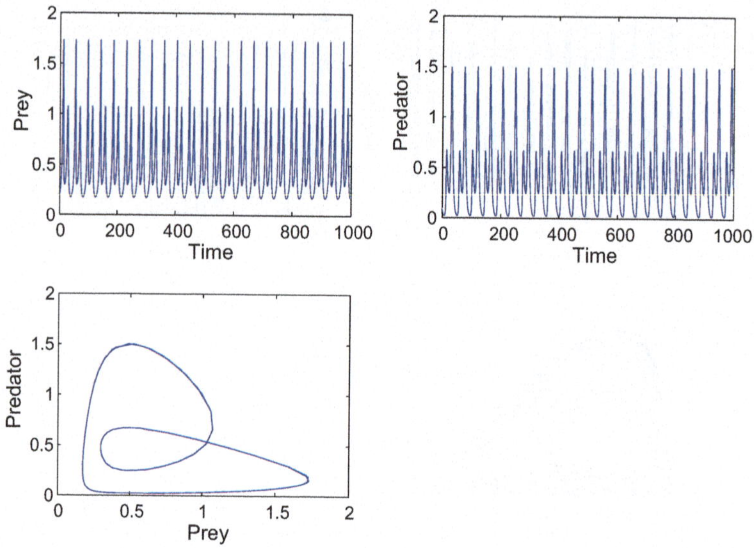


**Fig. 2** The delayed system (1) is two periodic with the time delay $\tau = 2.5$ and the other parameters $\beta = 0.2, m = 0.4, a = 0.9, \theta = 0.1, \alpha = 0.8, b = 1.04$, with the initial condition [S(0),I(0),P(0)] = [0.8, 0.1, 14]

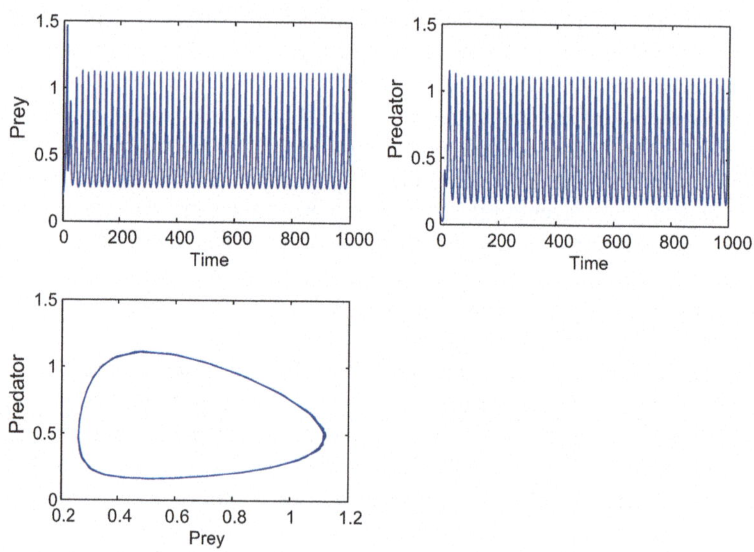


**Fig. 3** The limit cycle oscillation of the delayed system (1) with the time delay $\tau = 2.5$ and the other parameters $\beta = 0.2, m = 0.4, a = 0.9, \theta = 0.1, \alpha = 0.8, b = 1.06$, with the initial condition [S(0),I(0),P(0)] = [0.8, 0.1, 14]

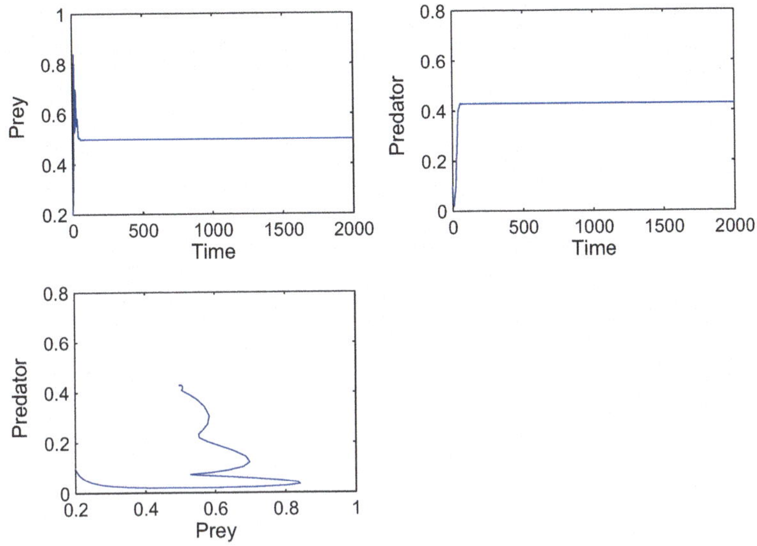


**Fig. 4** The delayed system (1) is stable with the time delay $\tau = 2.5$ and the other parameters $\beta = 0.2, m = 0.4, a = 0.9, \theta = 0.1, \alpha = 0.8, b = 1.16$, with the initial condition [S(0),I(0),P(0)] = [0.8, 0.1, 14]

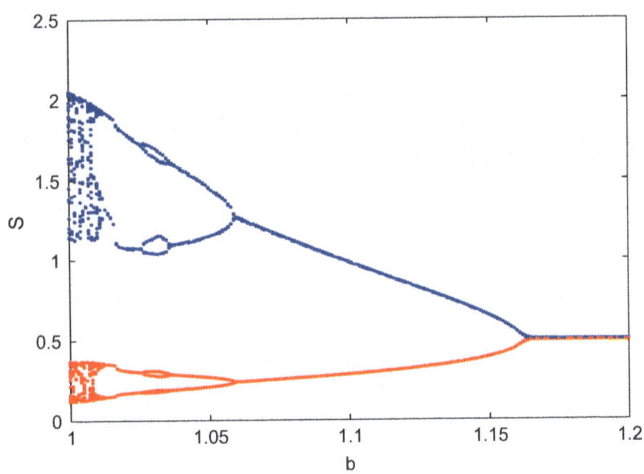


**Fig. 5** Bifurcation diagram of the system (1) with respect to b, when the other parameters are fixed as in Fig. 1

## Discussion

For our model, the time delay due to gestation plays an important role. Time delay can turn a stable equilibrium into an unstable one; that is, there is a critical value $\tau^*$, such that for $\tau^* > \tau$, the positive equilibrium $E^*$ is stable, and it loses stability as $\tau$ passes through its critical magnitude from lower to higher values. It was shown that the system (1) experiences Hopf bifurcation, as the delay parameter $\tau$ crosses some critical values $\tau^*$. A further increase in delay beyond the bifurcation point leads to complex dynamical behavior, including chaos. We have also shown numerically that chaos can be controlled by the emergent carrying capacity. In future, we would like to study a diffusive prey–predator model with emergent carrying capacity and the Allee effect to explore the possibilities of Turing instability and pattern formation.

## References

Allee, W. C. (1931). *Animal aggregations. A study in general sociology*. Chicago: University of Chicago Press.

Biswas, S., Sasmal, K. S., Samanta, S., Saifuddin, M., Khan, Q. J. A., Alquranc, M., & Chattopadhyaya, J. (2015). A delayed eco-epidemiological system with infected prey and predator subject to the weak Allee effect. *Mathematical Biosciences, 263*, 198–208.

Biswas, S., Saifuddin, M., Sasmal, K. S., Samanta, S., Pal, N., Ababneh, F., & Chattopadhyaya, J. (2016). A delayed prey-predator system with prey subject to the strong Allee effect and disease. *Nonlinear Dynamics, 84*, 1569–1594.

Celik, C., Merdan, H., Duman, O., & Akin, O. (2008). Allee effects on population dynamics with delay. *Chaos, Solitons & Fractals, 37*, 65–74.

Gopalsamy, K. (1992). *Stability and oscillation in delay differential equation of population dynamics*. Dordrecht: Kluwer Academic.

Hilker, F. M., Langlais, M., Petrovskii, S. V., & Malchow, H. (2007). A diffusive SI model with Allee effect and application to FIV. *Mathematical Biosciences, 206*, 61–80.

Kuang, Y. (1993). *Delay differential equation with applications in population dynamics*. New York: Academic.

MacDonald, N. (1989). *Biological delay systems: Linear stability theory*. Cambridge: Cambridge University Press.

Pal, P. J., Saha, T., Sen, M., & Banerjee, M. (2012). A delayed predator–prey model with strong Allee effect in prey population growth. *Nonlinear Dynamics, 68*, 23–42.

Yan, J., Zhao, A., & Yan, W. (2005) Existence and global attractivity of periodic solution for an impulsive delay differential equation with Allee effect. *Journal of Mathematical Analysis and Applications, 309*, 489–504.

Zhang, T., Zang, H. (2014). Delay-induced Turing instability in reaction-diffusion equations. *Physical Review E, 90*, 052908.

# Angela Merkel's Chancellor Democracy and Leadership in Times of Crisis

**Başar Şirin**

**Abstract** Since the time of the first chancellor of the German Federal Republic, Konrad Adenauer, the term "chancellor democracy" (*Kanzlerdemokratie*) has been used to describe the unique position of the federal chancellor as head of government in the German political system, with a similar amount of power to that of the president in a presidential democracy. In addition to basic principles such as the major decision-making role in the cabinet, indisputable control over his/her own party, personal prestige in the majority of society, and recognizable delimitation of the opposition, the chancellor's strong personal involvement in foreign policy is also accepted as an important determinant of the chancellor democracy. From that perspective, while only several German chancellors in history have embodied all of those features together, the current federal chancellor, Angela Merkel, has shown exceptional leadership since she became the chancellor in 2005. In fact, she has even been accused of exceeding her power and exhibiting self-empowerment, especially with her response to the European debt crisis and, more recently, the refugee crisis. On the other hand, both Merkel's initial "open door" policy (which let about 1 million refugees move to Germany in a short period) and her decision to reach a deal with Turkey to prevent illegal passage from Turkey to Greece through the Aegean Sea were intensively discussed in German public opinion. Correspondingly, this chapter examines Merkel's 13 years in office from the chancellor democracy perspective and aims to show that although, during her reign, Merkel successfully took many crucial decisions in times of crisis, her government, her party, and the German public disapproved of her refugee policy. As a result, Merkel's political power and chancellor democracy were severely damaged and had to come to an end in late 2018.

**Keywords** Angela Merkel · Democracy · Leadership · Crisis

---

B. Şirin (✉)
Freie Universität, Berlin, Germany



Ş. Ş. ERÇETİN, Ş. N. AÇIKALIN (eds.), *Chaos, Complexity and Leadership 2018*,
Springer Proceedings in Complexity, https://doi.org/10.1007/978-3-030-27672-0_9

## Introduction

Without a doubt, the German federal chancellor Angela Merkel has been one of the most influential political figures since the Second World War in both Germany and the European Union (EU). During her time in office of more than 13 years, she has already reached the same level as the two most influential politicians in the Federal Republic of Germany: Konrad Adenauer and Helmut Kohl. While Adenauer played an indisputable role in shaping Germany's postwar political choices, Kohl rightfully earned the name "chancellor of unity" by achieving German unity during his reign in 1990. Moreover, both chancellors actively contributed to European cooperation and integration processes.

On the other hand, in comparison with the legacies of her two predecessors, Chancellor Merkel's legacy is much more disputable. On the one hand, *Time* maga zine (http://time.com/time-person-of-the-year-2015-angela-merkel/, 2015) declared her to be "chancellor of the free world" by describing her as "Europe's most powerful leader" and, similarly, the *New York Times* referred to her as "the liberal West's last defender" with her strong commitment to Western political values, norms, and beliefs. On the other hand, her personal decision to accept refugees from Syria resulted in one of the most profound periods of political and social turbulence in the history of the EU (https://www.nytimes.com/2016/11/13/world/europe/germany-merkel-trump-election.html, 2016).

When the political performance of German chancellors during their reigns are considered, the German political science term *Kanzlerdemokratie* ("chancellor democracy") provides critical criteria to study this issue more deeply. As a journalistic idea that emerged in the 1950s in response to the dominant political leadership of Chancellor Adenauer, this concept mainly referred to the chancellor's political ability to control his/her party and to guide the policies of the government, thereby influencing the political process and public policy in the country (Helms 2011). According to Niclauß (2015), there are five elements of the chancellor democracy, which reflect the federal republic's political tradition:

- The federal chancellor is both formally and informally the central political figure in the decision-making process.
- The political competition in the country is heavily dependent on the chancellor's personal popularity and prestige in society, as well as in the media.
- The chancellor fully controls his/her own party.
- There is a clear contrast between the government and the opposition in terms of major political issues.
- The chancellor plays a leading role in the country's foreign policy decision making, along with the federal foreign minister.

Although those five elements are widely used by political science experts to compare chancellors' performances, they actually indicate an ideal situation for a powerful chancellor. Therefore, only a few former federal chancellors have met these criteria during some periods in their reigns. Accordingly, this article examines

Chancellor Merkel's political performance as the leader of Germany by focusing on these five elements of her chancellor democracy, particularly in times of crisis.

## Historical Background of the Chancellor Democracy

For leaders of postwar Germany, the "overdemocratic institutions" and "mass democracy" of Germany during the Weimar Republic was one of the significant reasons for the rise of the Nazi regime. Therefore, providing a stable political system in the country to overcome deficiencies in the liberal parliamentarism was the first aim of German politicians after the end of the Second World War. In parallel with that, the new German constitution gave the parliament (Bundestag) mainly a function of controlling the executive and representing social and economic interests of different groups in society. Promoting legislative issues, however, was not counted as a responsibility of the parliament. Moreover, the so-called constructive vote of no confidence let the parliament oust the chancellor only in the case that an immediate alternative candidate was presented as a replacement. To strengthen the executive branch, the legal power of the president was also confined with the formal nomination of a candidate for the chancellorship, and in practice, presidents have no influence on chancellor candidates and cabinet members (Mommsen 2007).

When all of these early political decisions are taken into consideration, it can be argued that the chancellor democracy in Germany was actually not the result of Adenauer's dominant personality or autocratic tendencies but a conscious decision made by federal Germany's early leaders. In fact, in the cabinets of Adenauer, ministers always had a large amount of flexibility, as long as their policies did not challenge the government's general strategy (Helms 2005). Therefore, although the emergence of the chancellor democracy concept was directly connected with Adenauer, it has since become one of the significant attributes of the political culture in the federal republic.

## Elements of the Chancellor Democracy

### *The Chancellor Principle*

Unlike the other elements of the chancellor democracy, the chancellor principle finds its legal basis in the German Basic Law, especially in Article 65. According to this article, "the federal chancellor shall determine and be responsible for the general guidelines of policy [*Richtlinienkompetenz*]." In other words, with the so-called *Richtlinienkompetenz*, the chancellor becomes the primary decision maker in the government by enforcing his/her aims in defining core issues of the government's policies. Accordingly, the chancellor also has the right to reorganize institutions of the executive branch such as ministries or other administrative offices.

When considered from this perspective, the influence of the chancellor principle emerges in German politics mostly before the formation of governments. The number of ministers appointed directly by the chancellor also reflects the weight of the chancellor in the government. In the case of grand coalitions or a slight difference between the coalition parties in terms of the popular votes they have obtained, the influence of the chancellor decreases dramatically.

The chancellor's personal initiative to implement a specific policy or support for a policy proposal from a minister—or the chancellor's objection to a policy—also directly depend on the chancellor's power in the government (Kropp 2004).

Moreover, the chancellor's right to ask parliament for a vote of confidence is one of the extraordinary powers of chancellors in Germany. According to this right, in the case of a decision of no confidence in parliament, the chancellor may ask the federal president to dissolve the Bundestag within 21 days.

## *Personal Prestige*

As the chancellor is the leader of the country, his/her personality is undoubtedly a significant determinant in politics. In other words, chancellors are generally seen (by the public) as being responsible for both the successes and the failures of the government. In that sense, the politics in the country are embodied within the personality of the chancellor and, most of the time, the policies of the government are explained and defended by the chancellor. Also, in the media, the chancellor has the most exceptional visibility. The implications of personal prestige therefore are successes in both federal and state elections, as well as in public opinion polls.

## *Party Leadership*

The party leadership element of the chancellor democracy emphasizes the chancellor's control over his/her own party. Although in the German political tradition, it is not necessary for the chancellor to be a party leader, unconditional support from the chancellor's own party plays an essential role in strong leadership, especially in the parliament. In the history of federal Germany, in addition to Helmut Schmidt and Ludwig Erhard, who did not hold party seats during their chancellorships, Gerhard Schröder, for example, was not the party leader when he became chancellor and only after a period of time did he gain the party leadership.

Because in many Western countries such as the USA and France, new political figures such as Donald Trump and Emanuel Macron have entered politics without any formal support from a political party, the "leadership concentration" issue has been intensively discussed in those countries recently. Some political observers have also argued that Chancellor Merkel's policies have been increasingly shaped independently from her party (De Mucci 2018). However, in comparison with those other examples, the political weight of the political parties in Germany is still an indisputable fact.

## *Distinction Between the Government and the Opposition*

Like other elements, the clear distinction between the opposition and the government has been influenced by Adenauer's leadership style. In contrast to the comprehensive view that in hard times an all-party government should be formed by overcoming differences among the parties, Adenauer considered that the strongest party should take the leadership in the country and the other large party should counterweight this party by taking the opposition role (Mommsen 2007).

Since, before every Bundestag election, both the Christian Democratic Union/Christian Social Union (CDU/CSU) and the Social Democratic Party (SPD) propose a candidate for the chancellorship and, at the same time, indicate their preferred coalition with different smaller parties, this dualism mostly dominates the federal election. Accordingly, parties nominate their candidates before the election and the whole campaign term is focused on these two candidates, while vast numbers of viewers follow the TV debates between candidates.

Even in the grand coalitions, as was seen during the first and third Merkel cabinets, those two parties try to differentiate themselves with their policy proposals and find their chancellor candidates outside the government.

## *Personal Engagement in Foreign Policy*

The federal chancellor's personal engagement in foreign policy issues, along with that of the federal minister for foreign affairs, has been another implication of the chancellor democracy. In Germany, traditionally, the leader of a small coalition party or an influential political figure has mostly taken the foreign affairs position, such as Hans-Dietrich Genschner, Joschka Fischer, or Frank-Walter Steinmeier. On the other hand, since Adenauer (who, along with the chancellorship, took the federal foreign ministry position between 1951 and 1955), different chancellors have come to the forefront with some crucial foreign policy issues such as Willy Brandt's *Ostpolitik* ("eastern policy"), Kohl's great contribution to the 1992 Maastricht Treaty and the EU's enlargement process, and Schröder's decisions to join wars in Kosovo and Afghanistan as Germany's first external military operations since the end of the Second World War.

Moreover, representing Germany in the international arena is seen generally by Germans from a positive point of view. Accordingly, taking active roles in international meetings—such as United Nations (UN) meetings, Group of Twenty (G20) meetings, or Group of Eight (G8) meetings—or hosting those events in Germany are also considered by Germans to be an indication of strong leadership.

## The End of the Chancellor Democracy

As there is a clear need for some elements to establish a stable chancellor democracy, according to Korte (2000) there are some indications of the end of the chancellor democracy. For him, weakening of the chancellor democracy emerges firstly as power erosion. Without a doubt, one of the essential power sources of a chancellor is his/her own party. Therefore, as a result, a lack of support from the chancellor's party may severely diminish the chancellor's power. Accordingly, an inner party or public discussion about the party leadership and emergence of new candidates for the party leadership is an essential indicator of loss of power by the chancellor.

Moreover, if the chancellor loses his/her capability to act effectively, in parallel with increasing power of other political actors such as opposition parties or other challenging political figures, this may practically deadlock the politics in the country. In that case, to overcome such blockades, voters may approach new political actors. In addition, lack of successful political communication is a sign of decreasing chancellor power. Since every policy proposal of the government should be well explained in order to get support from voters, problems in communication may result in inadequate popular support and persistently low figures in opinion polls and other polls such as elections in various federal states.

Last, but not least, if the chancellor loses his/her perception of reality and experiences increasing loneliness in the administration, these are decisive indicators. Because observation of the political situation in the country and addressing the expectations of the voters are among the primary duties of a government, problems in information gathering, as well as opinion forming, may lead to a false image of the political situation. A natural tendency to rely on useful reports and ignore negative news therefore increasingly damages the chancellor's political power, as he/she can no longer meet the expectations of the voters.

## Chancellor Merkel and the Chancellor Democracy

### *Merkel and the First Grand Coalition (2005–2009)*

Although Merkel was not a totally new figure in German politics and served as the minister for women and youth, minister for the environment, and general secretary of the CDU between 1991 and 2000, her personality, leadership style, and potential in politics were entirely unknown to many Germans when she became the leader of the CDU in 2000 and a candidate for the chancellorship in the 2005 election. However, especially in foreign policy issues, she had very clear positions such as strong opposition to Turkey's bid for EU membership (https://www.spiegel.de/international/turkey-and-the-eu-the-pros-and-cons-a-333126.html, 2004).

Merkel's first term in the chancellor's office started with the country's second grand coalition between Merkel's CDU, the CSU (their smaller Bavarian sister party), and the SPD party of the previous chancellor, Gerhard Schröder. From that perspective, Merkel's area of influence in her first term was very much limited, as the CDU/CSU parliamentary group could obtain only four more members in the parliament in comparison with the SPD and, accordingly, Merkel's own CDU party had only six positions in the government along with the chancellor's office, while the SPD gained eight and the CSU had two cabinet seats.

On the other hand, with regard to controversial issues such as the SPD's election promise of an overall national minimum wage, Chancellor Merkel could defend her position by strictly opposing this proposal, even if this dispute seriously threatened the continuation of the coalition government (https://www.dw.com/en/minimum-wage-compromise-gives-merkel-a-domestic-boost/a-2615268, 2007).

Similarly, after the resignation of the SPD leader and vice chancellor, Franz Müntefering (who was known as "Mr. Grand Coalition" and supported the chancellor in many critical governmental issues) from the government, Merkel could maintain the stability of the government (http://www.spiegel.de/international/germany/muentefering-resignation-merkel-loses-mr-grand-coalition-a-517132.html, 2007).

In her first term in the chancellor's office, Merkel's leadership in her party was quite indisputable. In the first party chair election after she became chancellor, Merkel received 93.06% of the votes at the 2006 CDU party convention in Dresden. Two years after that vote, Merkel strengthened her position in the party leadership with the support of 94.83% of votes at the 2008 party convention.

As one of the vital signs of political success in Germany, Merkel and her party's election performance in the federal state elections was somewhat mixed. While in this first term the CDU/CSU dominated the elections in one of the most populous federal states, Baden–Württemberg, they clearly underperformed in federal states such as Hessen, Hamburg, and Saarland either by losing an absolute majority or by losing a significant number of popular votes (https://www.dw.com/en/german-conservatives-suffer-losses-in-key-state-election/a-3091972, 2008).

In public opinion surveys during those years, the CDU/CSU consistently stayed ahead of any other party. However, in surveys regarding the most influential politicians in Germany in terms of performance and sympathy (such as Politbarometer), Merkel was in fourth place after Christian Wulff, Friedrich Merz, and Joschka Fischer when she became chancellor in 2005. In the following years, while Merkel climbed up to the first position in this survey from time to time, she was topped by political figures such as the SPD leader Kurt Beck or the foreign minister Frank-Walter Steinmeier. When her overall performance as chancellor is considered, however, generally more than 70% of Germans found Merkel's performance to be quite good in her first term as leader (Forschungsgruppe Wahlen: Politbarometer 2019).

Foreign policy and external issues undoubtedly became essential elements in her performance. In this regard, two significant developments in the second year of her term helped Merkel rapidly enhance her positive image in public opinion. Firstly, by taking the rotating presidency of the Council of the EU in January 2007, Chancellor Merkel took a very active role in meetings regarding the Treaty of Lisbon (initially known as the Reform Treaty). Especially with the Berlin Declaration

(2007), which was adopted during the 50th anniversary of the Treaty of Rome, the German government created a new consensus among the member states of the EU regarding the future perspective of the union. Secondly, at the 33rd G8 Summit, which was held in June 2007 in Heiligendamm, Chancellor Merkel not only attracted the attention of the German public by playing a leading role in the climate change issue but also improved German–American relations, which had seriously deteriorated during the reign of the previous chancellor, Gerhard Schröder. Accordingly, the chancellor's visit to Greenland to draw attention to the consequences of global warming and to call for the USA and China to sign the Kyoto Protocol strengthened her profile as the "climate chancellor" (https://www.nytimes.com/2007/12/21/world/asia/21transfer.html, 2007).

When all of these developments are considered together, it can be argued that Angela Merkel's first term as chancellor was not very challenging. In particular, cooperation and consensus with other coalition parties were the major characteristics of that term. Nevertheless, Chancellor Merkel developed essential features of the chancellor democracy, particularly in terms of her personal prestige, party leadership, and foreign policy involvement.

## *Merkel and the CDU/CSU–FDP Coalition (2009–2013)*

Despite a slight loss of votes in 2009 election, the weight of the CDU and CSU in the coalition was substantially increased in Merkel's second cabinet (which was formed with the liberal Free Democratic Party (FDP)) in comparison with the grand coalition. With 14.6% of the popular vote and 93 parliamentary seats, the FDP occupied only six ministerial posts. While the CSU had four ministries, as in the previous government, this time Merkel had the chance to control eight positions, including the chancellor's office, directly.

Unlike her first term, a significant crisis hit the Merkel government in the first months of that term of government. As a continuation of the 2008 global financial crisis, the 2009 European debt crisis became the most dominant political and economic issue not only for Germany but also for the whole EU. A deep government crisis between the coalition partners regarding recovery proposals and reform programs were some serious challenges for the continuation of the government as well. Nevertheless, Merkel's persistent efforts to achieve a solution to the crisis let her improve the chancellor democracy in this chaotic political situation.

In addition to the European debt crisis, the nuclear disaster at the Fukushima Daiichi nuclear plant in Japan and its direct influence on the energy transition debate in Germany, the reorientation of the German army and the suspension of compulsory conscription, and large-scale protests against a government-funded railway station project (publicly known as Stuttgart 21) in Stuttgart were controversial issues that Merkel had to face in the first half of her second term.

In that term, the election of the federal president in 2010 was another critical political issue, which gave Merkel a chance to test her political weight in German politics. As the joint candidate of the CDU, CSU, and FDP, Merkel's choice, Christian Wulff (the incumbent minister president of Lower Saxony, from the CDU) was successfully elected as the tenth federal president of Germany against a prevalent public figure, civil rights activist and also the next federal president, Joachim Gauck.

In that period, Merkel's undisputable power in the party leadership was reconfirmed on a number of occasions. Above all, in 2009, the coalition agreement between the CDU/CSU and the FDP was approved by the CDU delegates without any opposition vote. Similarly, in 2011, during the CDU party convention, Angela Merkel was elected for the sixth time as the party leader with the support of 90.44% of party delegates. In 2012, the CDU delegates once again elected Merkel, with 97.94% of the votes.

In her second term, Merkel substantially increased her personal prestige. Although the performance of the government was severely criticized in public opinion during the first year of that term, the popularity of the government increased dramatically after 2010 and reached almost 70% before the 2013 election (Forschungsgruppe Wahlen Politbarometer 2013a). Without a doubt, Merkel's personal popularity profited from her strong leadership as well. Although her disapproval rate was almost 45% and her popularity fell behind that of politicians such as the defense minister Karl-Theodor zu Guttenberg, the minister of labor and social affairs Ursula von der Leyen, the minister of finance Wolfgang Schäuble, and the SPD leader Frank-Walter Steinmeier in late 2010, she changed this trend in a very short time and carried her approval rate to almost 85% in 2 years (Forschungsgruppe Wahlen Politbarometer 2013a).

Merkel's election losses in two important federal state elections—the 2011 Baden–Württemberg and 2012 North Rhein–Westphalia state elections—were probably two notable setbacks for Merkel in her second term. In particular, as a result of public protests concerning the Stuttgart 21 project, the CDU lost the minister president post in Baden–Württemberg, which had been governed by the CDU since 1953. Similarly, in North Rhein–Westphalia (the most populous federal state in Germany), Merkel experienced a very harsh defeat, with her candidate Norbert Röttgen (the federal minister for the environment, nature conservation, and nuclear safety, and the deputy leader of the CDU) losing 8.3% of the popular vote and correspondingly the minister president's office.

Since the SDP was the major opposition in the Bundestag between 2009 and 2013, the distinction between the government and the opposition was much more visible than that in the previous legislative session. Nevertheless, the SPD's chancellor candidate Peer Steinbrück (a former minister president of North Rhein–Westphalia and a former federal minister of finance) ran quite a weak election campaign, and his public support for the chancellorship mostly stayed around 30%. On the other hand, Merkel's public support was mostly about 60% throughout the whole campaign period (Forschungsgruppe Wahlen Politbarometer 2013b).

As discussed above, the European debt crisis was probably the most crucial foreign issue in those years. Although Chancellor Merkel's proposals of harsh austerity programs for Greece, Ireland, Portugal, and Spain to reduce budget deficits were severely criticized in the EU by those arguing that these programs would cause adverse effects on the economic growth of these countries, her strong insistence showed that her personal political power, as well as Germany's institutional power, reached a very high level in the EU and that Merkel alone could force other member states to take decisions against their will (Kundnani 2011).

Another crucial foreign policy issue in this period was the 2011 military invention against the Gaddafi regime in Libya, led by the North Atlantic Treaty Organization (NATO). Although the intervention decision taken by the UN was actively supported by Germany's three NATO allies (France, the UK, and the USA), Chancellor Merkel and foreign minister Westerwelle opted to abstain—along with Brazil, Russia, India, and China—in the voting. After that decision, the German government also decided not to take part in any military operation against Libya. As a reaction to these decisions, Germany's decision to break the solidarity of the alliance and the danger of its isolation in international politics were severely criticized both domestically and internationally, but Merkel was able to avoid any negative political consequences for herself by leaving the foreign minister, Westerwelle, to justify the decision publicly. However, this difficult undertaking substantially contributed to Westerwelle's resignation from her post in August 2011 (https://www.dw.com/en/german-foreign-minister-under-fire-after-libya-u-turn/a-15349250, 2011).

When all of the political developments in Merkel's second term are considered together, probably the most apparent indication of how Merkel established a powerful chancellor democracy was the 2013 German federal elections. In the election campaign before the election, the CDU focused not on any specific election issue but mainly on the personal image of Merkel. A CDU election poster almost 2400 square meters in size near Berlin's central train station, which showed only Merkel's diamond-shaped hand gesture under the motto "Germany's future in good hands," became the symbol of Merkel's indisputable leadership in Germany (https://www.spiegel.de/international/germany/giant-merkel-campaign-poster-raises-eyebrows-in-berlin-a-920196.html, 2013). Unsurprisingly, Merkel increased her votes by 7.8% and won the election with 41.8% of the popular votes. At the same time, her party obtained 311 parliamentary seats and fell only five seats short of an overall majority of 316.

## *Merkel and the Second Grand Coalition (2013–2017)*

Despite the decisive personal triumph of Merkel in the 2013 German federal election, her junior coalition partner FDP's catastrophic election result, which meant that it fell out of the parliament for the first time since its establishment after the end of the Second World War, forced Merkel to form another grand coalition after 4 years. Nevertheless, an almost 15% difference in votes between the CDU/CSU and the SPD strengthened Merkel's hand before the coalition talks. In comparison with her first grand coalition, Merkel was able to obtain one more ministry post in the government, while the SPD was able to gain seven ministry positions.

Like the European debt crisis in her previous term, the 2015 European refugee crisis now became by far the most crucial political issue and influenced not only Merkel's political career but also the politics of the whole EU. Her personal decision to open Germany's borders to refugees from Syria without consulting any cabinet or party member, and her proposal to find a joint solution with Turkey (known as the Merkel Plan), were equally criticized by coalition parties (including her own party) and the opposition parties in the parliament (Alexander 2017). Still, Merkel persistently maintained her individual leadership style in the government.

Although Merkel clearly kept control over her own party in these years, party delegates also showed their displeasure during the CDU party leader's election in 2016. Whereas just 1 year before the beginning of the uncontrollable illegal migration to Germany, Merkel had reached her record-high 96.72% party delegate support at the 2014 party convention, this number fell to 89.50% in 2016. In fact, this result was the worst since she became chancellor in 2005 and may be accepted as the first warning for her chancellor democracy.

Unlike her support from the party, her personal prestige in society decreased very rapidly after the refugee crisis. Before the crisis, especially in late 2014, the CDU/CSU maintained strong public support in opinion polls, with about 42%, while Merkel also enjoyed high popularity, with an almost 80% approval rate. Accordingly, 65% of the population believed that Merkel should once again compete for the next federal election in 2017 (Forschungsgruppe Wahlen Politbarometer 2014). With the beginning of the refugee crisis in late 2015, however, the initial warning signs from public opinion emerged. Although nearly three quarters approved of her performance as the chancellor, her refugee policy was supported by only 50%.

Moreover, she suddenly dropped back to become the fourth most popular politician in Germany after Wolfgang Schäuble, Frank-Walter Steinmeier, and the CDU politician Wolfgang Bosbach (Forschungsgruppe Wahlen Politbarometer 2015). After mass sexual harassment incidents involving refugees in Cologne on New Year's Eve in 2015, the CDU/CSU's public support also rapidly fell to 33%, while a newly established right-wing populist party, Alternative for Germany (AfD), achieved its best results in opinion polls with a shocking 12% support. In the same period, Angela Merkel went one more step down and became only the fifth most popular politician in public polls (Forschungsgruppe Wahlen Politbarometer 2016). Nevertheless, with a dramatic reduction in the numbers of illegal immigrants after a deal was made with Turkey to prevent illegal passage, Merkel's and the CDU/CSU's public support were slightly enhanced in late 2016. Also, Merkel's announcement that she would once again stand for the chancellor's position in the next federal election was responded to positively by 64% of the population.

In a similar vein to what had happened in the previous legislative session, German voters showed their displeasure in some important federal state elections. After failing to gain the minister president post in Baden–Württemberg in the previous election, the CDU this time lost 12% of its votes in the 2016 state election and was unable to win the election for the first time since the foundation of Baden–Württemberg in 1952. In another vital state election in Berlin, the CDU lost another

5% of the popular votes and obtained only 17% support. More importantly than that, the right-wing populist party AfD easily entered the state parliament with more than 14% of the votes.

Although, as a result of the grand coalition, the distinction between the government and the opposition was not so visible in the parliament, Martin Schulz (the SPD's chancellor candidate and the incumbent president of the European Parliament) posed a severe threat to Merkel during the 2017 election campaign as a candidate from outside the government. Particularly in late 2016 and early 2017, Schulz overtook Merkel in public opinion surveys and even reached a 49% rating in February 2017, while only 38% of Germans preferred Merkel as the chancellor. In the meantime, the SPD increased its public support to 32%, while the CSU/CDU remained at 34% (Forschungsgruppe Wahlen Politbarometer 2017). This trend, however, was able to be reversed by Merkel before the federal election in September 2017.

The first major foreign policy test for Merkel in her third term was the 2014 Russian military intervention in Ukraine. Unlike Germany's hesitant approach during the Libya crisis, Merkel took a clear leading role in the EU once again and signaled that Germany and the EU would not easily accept Russian revisionist foreign policy. In that sense, Germany increased its resilience, deterrence, and defense options through economic and military instruments. However, Merkel also kept all of the communication channels open and tried to take a mediator role between the two parties involved in the crisis. Therefore, with this so-called "hybrid *Ostpolitik*," Merkel managed this severe security challenge successfully (Daehnhardt and Handl 2018).

As mentioned above, the European migrant crisis was undoubtedly one of the biggest political crises in the history of Germany and the EU. It substantially threatened German public order and changed the whole of Europe politically, economically, and socially. In particular, Merkel's decision to open German borders to refugees and to defend her choice with her famous *Wir schaffen das* ("We can do it") discourse, which was supported by European moral values as well as Christian responsibility, did not get enough support from society. On the contrary, anti-immigrant political discourses became much more visible in society, thanks to right-wing populist politicians. To achieve an urgent solution, Merkel negotiated and made decisions on behalf of the whole EU almost alone during the 2016 EU–Turkey Refugee Agreement talks (Alexander 2017). However, support from the German public and the other EU countries was quite limited.

When all of the significant political developments in Merkel's third term are considered together, it is quite safe to argue that the strength of Merkel's chancellor democracy, which reached its peak in 2014, was markedly decreased after the refugee crisis. Even though she found an urgent solution and cut illegal migration by almost 90% in the summer of 2016, neither her decision to accept refugees nor her agreement with Turkey received the approval of the majority of the population. Still, her support was sufficient to win another federal election in 2013. In terms of her chancellor democracy, however, her performance in this election was not powerful enough to change the decreasing trend.

## The End of Merkel's Chancellor Democracy

On 14 March 2018, Chancellor Merkel was sworn in for the fourth time as chancellor of Germany. In comparison with the 2013 elections, the CDU/CSU votes were diminished by 8.6% and reached only 32.9%. This result was not only her lowest election result since she became chancellor, but it was also the worst election result for her party since the first Bundestag election in West Germany after the end of the Second World War in 1949. Moreover, personally, it was also an obvious indication that Merkel's chancellor democracy was not as stable as before.

In the summer of 2018, a conflict between Merkel and the interior minister and CSU leader Horst Seehofer regarding Germany's policy on refugees and asylum-seeking, Seehofer's threat to resign, and the danger of a break in the seven-decade partnership between the CDU and the CSU revealed that it was "the end of German politics as we know it," right at the beginning of Merkel's fourth term (https://www.spiegel.de/international/germany/csu-merkel-conflict-means-german-politics-is-changing-a-1216204.html, 2018). Although, after very intensive negotiations, the two leaders reached a compromise at the last minute, further damage to Merkel's chancellor democracy came from her own party in September 2018. Despite Merkel's open support for the CDU/CSU parliamentary group leadership, Volker Kauder (known as Merkel's "right-hand man" and close ally) was ousted after 13 years of parliamentary group leadership. This radical action of the parliamentary group showed openly that Merkel's support in her parliamentary group was severely diminished (https://www.dw.com/en/merkel-ally-sent-packing-in-surprising-parliamentary-group-leadership-vote/a-45635982, 2018). At the same time, the CDU's public support reached a record low of 28%, and the grand coalition would lose its parliamentary majority in the event of a new Bundestag election. At a personal level, satisfaction with Merkel's leadership fell to 23% in September 2018 (Forschungsgruppe Wahlen Politbarometer 2018).

Shortly after these developments, in October 2018, Chancellor Merkel announced that she would not seek another term as chancellor in the 2021 federal election. On top of that, she decided not to run again for the CDU chairmanship during the next party convention in December 2018. Accordingly, she was replaced as the party leader by Annegret Kramp-Karrenbauer on 7 December 2018, and Merkel's very long-term chancellor democracy officially came to an end within a couple of months.

## Conclusion

During her long journey as the leader of Germany, Angela Merkel has successfully solved many complex political problems. As of 2019, Merkel is still the chancellor of Germany, and it is expected that she will continue in that role until the 2021 federal election. In this way, her tenure in the chancellor's office will have lasted as long as those of the two iconic ex-chancellors Konrad Adenauer and Helmut Kohl. However, it is clear that her chancellor democracy has already ended. Often named as the world's most powerful woman, leader of the free world, or the savior of the EU, Merkel could not get rid of the disputes over her refugee policy. Following her decreasing prestige in society, loss of control over her government, and, finally, evident opposition from her own party, Merkel found no other option but to accept her political defeat. In other words, after the failure of her involvement in a foreign policy issue, other elements of Merkel's chancellor democracy—namely, personal prestige, the chancellor principle, and party leadership—dissolved one by one.

## References

Alexander, R. (2017). *Die Getriebenen: Merkel und die Flüchtlingspolitik: Report aus dem Innern der Macht*. Munich: Siedler.

Daehnhardt, P., & Handl, V. (2018). Germany's eastern challenge and the Russia–Ukraine crisis: A new *Ostpolitik* in the making? *German Politics, 27*(4), 445–459.

De Mucci, R. (2018). Governmental leadership without a political party. *Open Journal of Political Science, 8*, 278–290.

Forschungsgruppe Wahlen Politbarometer. (2013a). *Ganz allgemein: Macht die Bundesregierung aus CDU/CSU und SPD ihre Arbeit alles in allem gesehen eher gut oder eher schlecht?* https://www.forschungsgruppe.de/Umfragen/Politbarometer/Langzeitentwicklung_-_Themen _ im_Ueberblick/Politik_-_Archiv/Arbeit_BR_2013_1.jpeg.

Forschungsgruppe Wahlen Politbarometer. (2013b). *Wen hätten Sie nach der Bundestagswahl im September lieber als Bundeskanzlerin oder Bundeskanzler, Angela Merkel oder Peer Steinbrück?* https://www.forschungsgruppe.de/Umfragen/Politbarometer/Langzeitent wicklung_-_Themen_im_Ueberblick/Politik_-_Archiv/Lieber_BK_2013_1.jpeg.

Forschungsgruppe Wahlen Politbarometer. (2014). *Politbarometer* Juli 2014. https://www.forschungsgruppe.de/Umfragen/Politbarometer/Archiv/Politbarometer _2014/Juli_2014/.

Forschungsgruppe Wahlen Politbarometer. (2015). *Politbarometer September II 2015*. https://www.forschungsgruppe.de/Umfragen/Politbarometer/Archiv/Politbarometer_2015/Septem ber_II_2015/.

Forschungsgruppe Wahlen Politbarometer. (2016). *Politbarometer April II 2016*. https://www.forschungsgruppe.de/Umfragen/Politbarometer/Archiv/Politbarometer_2016/April_II_2016/.

Forschungsgruppe Wahlen Politbarometer. (2017). *Politbarometer März 2017*. https://www.forschungsgruppe.de/Umfragen/Politbarometer/Archiv/Politbarometer_2017/Maerz_2017/#_.

Forschungsgruppe Wahlen Politbarometer. (2018). *Politbarometer September II 2018*. https://www.forschungsgruppe.de/Umfragen/Politbarometer/Archiv/Politbarometer_2018/September _II_2018/.

Forschungsgruppe Wahlen Politbarometer. (2019). *Was meinen Sie, macht Bundeskanzlerin Angela Merkel ihre Arbeit alles in allem gesehen eher gut oder eher schlecht?* https://www.forschungsgruppe.de/Umfragen/Politbarometer/Langzeitentwicklung_-_Themen _im_Ueberblick/Politik_II/10a_Arbeit_Merkel_3.jpg.

Helms, L. (2005). *Presidents, prime ministers and chancellors: Executive leadership in western democracies*. New York: Palgrave Macmillan.

Helms, L. (2011). Angela Merkel and the unfulfilled promise of chancellor democracy. *Current History, 110*(734), 97–102.

Korte, K. R. (2000). Konjunkturen des Machtwechsels in Deutschland: Regeln für das Ende der Regierungsmacht? *Zeitschrift für Parlamentsfragen, 31*(4), 833–857.

Kropp, S. (2004). Gerhard Schröder as "coordination chancellor": The impact of institutions and arenas on the chancellor's style of governance. In *Germany on the road to "normalcy": Policies and politics of the red–green federal government (1998–2002)* (pp. 67–88). New York: Palgrave Macmillan.

Kundnani, H. (2011). Germany as a geo-economic power. *The Washington Quarterly, 34*(3), 31–45.

Mommsen, H. (2007). The origins of chancellor democracy and the transformation of the German democratic paradigm. *German Politics and Society, 25*(2), 7–18.

Niclauß, K. (2015). *Kanzlerdemokratie: Regierungsführung von Konrad Adenauer bis Angela Merkel*. Wiesbaden: Springer.

# The Achilles' Heel of Strategic Management: Strategic Leadership in a Chaotic Environment

**Halil İbrahim Özmen**

**Abstract** The main view of strategic management thinking is analysis of external and internal environmental conditions, and determination of the enterprise's strategies in this context. In this manner, enterprises not only position themselves but also take positions. Enterprises also specify and implement their strategies according to their positions. The person who is responsible for the management process—which includes creation, implementation, and evaluation of the strategy of the enterprise or strategic business unit—is the strategic leader. The most important function of the strategic leader is that they assume a decisive role in the strategic management process. The leadership behavior they exhibit in the implementation of strategies is the main determinant in achieving success. However, it is thought that a strategic leader who adopts the mechanical point of view of the past will not be successful in today's complex and chaotic environmental conditions. The main reason for this is that the business world is a complex, dynamic, and nonlinear system. An analysis made according to circumstances does not result in complex and nonlinear systems. In the story of Achilles, whom Homer described in his work, his mother's efforts were unable to protect Achilles from death. In a strategic management process, if a strategic leader continues to think in a traditional way and exhibits modernist management behavior, he or she may meet the same fate as Achilles.

**Keywords** Achilles · Strategic leadership · Strategic management · Chaos

## Introduction: The Strategic Context of Achilles' Heel

Mother tells me,
the immortal goddess Thetis with her glistening feet,
that two fates bear me on to the day of death.
If I hold out here and I lay siege to Troy,

H. İ. Özmen (✉)
Süleyman Demirel University, Isparta, Turkey
e-mail: halilozmen@sdu.edu.tr



Ş. Ş. ERÇETİN, Ş. N. AÇIKALIN (eds.), *Chaos, Complexity and Leadership 2018*,
Springer Proceedings in Complexity, https://doi.org/10.1007/978-3-030-27672-0_10

> my journey home is gone, but my glory never dies.
> If I voyage back to the fatherland I love.
> my pride, my glory dies . . .
> true, but the life that's left me will be long . . .
> —Homer (1990, pp. 497–504)

Achilles, one of the great heroes of Greek mythology, has a tragic story although he is known for his heroism in the Trojan War. Zeus, "the father of gods and people," has lost his heart to the goddess Thetis, who will be the mother of Achilles. However, Zeus listens to a fortune teller who says that "the boy to be born from Thetis will be stronger than his father" and he stops seeing her. Then Thetis marries Peleus, a heroic king. Peleus is a human being and a mortal. Achilles is born from the marriage of Thetis and Peleus. Thetis, who does not want her child to be mortal like his father (this shows strategic intent), plunges him into the River Styx, which has magical water. Thus, no weapon can harm her son (this shows strategic foresight). Achilles becomes the greatest hero of the Achaeans thanks to his mother's efforts, and he does not lose any war he fights in. Agamemnon, who fought in front of the walls of Troy for 9 years but could not win a victory, invites Achilles to the battle in order to win. Achilles joins the war even though he knows he will die if he does, as his horse Xanthus has predicted it. He has achieved many successes (this shows strategic leadership) in the war and even kills Hector, the prince of Troy. Achilles is killed by Paris when he is shot in the heel with an arrow at the end of the war (Homer 1990; Hard 2004; Woodard 2007). Even though Achilles dies in the Trojan War, in this article on strategic leadership and chaos, we honor his name and confirm the prophecy made by Xanthus.

The basic (modernist) view of strategic management is that business strategies should be prepared by an analytical method within a rational thought system in order that they can be implemented in accordance with the desired results and targets set by the senior management (Ülgen and Mirze 2018). The thinking behind this approach is the scientific practice of objectively observing nature, formulating hypotheses about governing laws, and then testing these laws against quantitative data, thus moving toward more accurate understanding of the laws (Stacey and Mowles 2016). However, today's scientific perspective is changing. It differs from the mechanical point of view, which believes that everything can be determined in advance, by evolving into a nonlinear point of view in which uncertainty and disorder exist (Prigogine and Stengers 1984). In line with this perspective, the idea of strategic management is evolving from a modernist point of view based on rational thought and analysis to a postmodern perspective (Horwitch 1987). With the consideration that organizations are complex systems (Amagoh 2016; Thiétart and Forgues 1995), strategic management has been evaluated as being "at the edge of chaos" (Beinhocker 1997).

From the legend of Achilles, as told by Homer, we understand that strategic intentions based on the strategic foresight that is possessed may not be actualized, even if they are combined with strategic leadership. A modernist approach cannot be sufficient to explain that all of these efforts may be made for nothing. But, at this point, the basic concepts of chaos theory become more explainable as follows:

*sensitive dependence on initial conditions* (Parker and Stacey 1994; Werndll 2009) (the seeds of the Trojan War are sown at the wedding of Thetis and Peleus by the goddess of disagreement, Eris, when she drops an apple onto the table); *nonlinearity* (Bussolari and Goodell 2009) (the murder of a very powerful hero, such as Achilles, by someone who does not have warrior characteristics, such as Paris); *strange attractors* (Thiétart and Forgues 1995) (although Achilles was told he would die, he joined the Trojan War anyway); *feedback* (Gleick 1995) (Achilles killed all of the nuns who took refuge in the Temple of Apollo during the Trojan War, and the arrow that Paris fired to kill Achilles was directed by the god Apollo to hit his heel, which was the only vulnerable part of him); and *self-organization and renewal* (Jantsch 1980; Weick 1977; Murphy 1996) (while it was thought that after Achilles' death, the war could not be won, Epeius built the famous wooden horse that was left in front of the gates of Troy and proved to be the strategy that won the war).

## From *Strategos* to Chaos: Strategic Management

Conceptually, the word "strategy" is based on the word *strategos*, which was used in the ancient Greek civilization to refer to a military general. Kleisthenes, who destroyed the tyrants' power in Athens and seized power with the support of the people in 507 BCE, removed the existing tribal organization, which was based on lineage, and he divided Athens into ten tribes on a geographical basis. He also asked the tribes to elect ten *strategoi* to lead their armies in the time of war. These ten *strategoi* formed the War Council of Athens (Cummings 1993). Books written on military strategy—such as *The Art of War*, written by Sun Tzu (2014) in the sixth century BCE; *The Prince*, written by Niccolò Machiavelli (2008) and published in 1532; and *On War*, by Carl von Clausewitz (2018), who started writing it in 1816 but died before its publication in 1832—are still effective references on contemporary strategic management thought. In addition, strategies regarding the management of kings and countries are described in *Management of Managers*, written by Abu Naceb Sühreverdi (1974) for Salahuddin al Ayyubi in the eleventh century AD. The usage in the business world of the concept of strategy, which has been used in the military and political fields since the ancient Greek period, started after the Second World War. The first studies on business strategy were described in Chandler's *Strategy and Structure* (1962), Ansoff's *Corporate Strategy* (1965), and Andrews's *The Concept of Strategy* (1971). Chandler emphasized that there is a relationship between strategy and organizational structure, while Ansoff (who has been called "the father of strategy") emphasized the complexity of businesses and the complexity of the business environment. Ansoff explained this perspective in his book by saying that "Our concern in this book is with the behavior of complex organizations in turbulent environments." Ansoff emphasized that both businesses and their environments are complex structures. The first hypothesis in the book is: "When any ESO [environment serving organization] is confronted with the prospects of extinction, it focuses all of its energy on a search for a survival

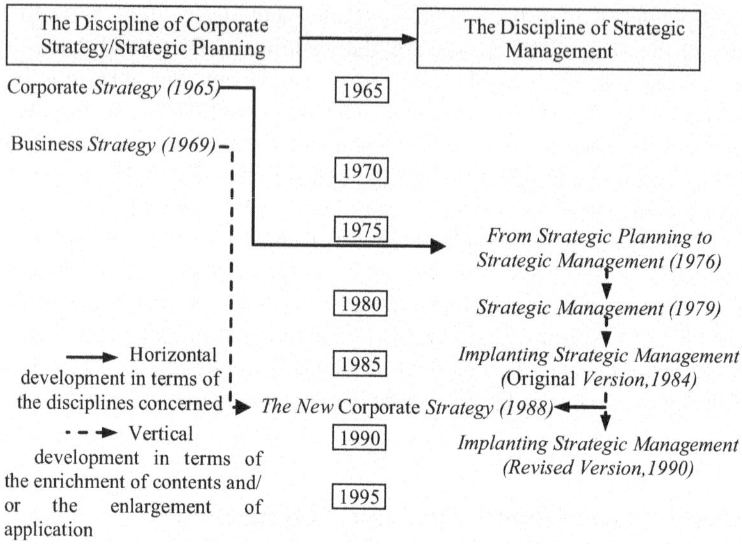


**Fig. 1** Development of Dr. Ansoff's works: disciplines and major books (Ansoff 2007)

strategy" (Ansoff 1965). In this context, the starting point of the strategy is the desire for survival of the enterprise, depending on the competition. Figure 1 shows the development of Ansoff's works.

Before the 1970s, organizations used to link strategies with the future on the basis of a planning approach. They started to link their strategies with the environment they were in with the effect of system and contingency approaches in the 1970s. In the 1980s, Michael Porter and Henry Mintzberg made significant contributions to the idea of strategic management. Porter (1979, 1985, 1996) focused on the content of strategies, while Mintzberg (1978, 1993, 1994) focused on the strategic process. It seems that the source-based view gained weight in the 1990s. In this context, it has come to the fore that enterprises can obtain a competitive advantage if they have basic capabilities that cannot be imitated by other enterprises (Prahalad and Hamel 1990) (see Table 1).

Skinner (1969) defined strategy as follows: "Strategy is a set of plans and policies by which a company aims to gain advantages over its competitors." Similarly, Henderson (1989) defined strategy as conscious study of an action plan to improve the competitive advantage of a business. He stated that this process is a repetitive process, which starts with determining the holdings and location of the enterprises. However, some modernist studies have revealed that there are not always targeted strategies applied in enterprises as "intended/intentional" strategies obtained as a result of managers' analysis; there can be "spontaneously emerged" strategies, depending on situations and conditions, that are different from the intentional ones (Mintzberg and Waters 1985). The main reason for this is that the business world is

**Table 1** Abridged history of strategic management

| Period | 1960s | 1970s | 1980s | 1990s | 2000s |
|---|---|---|---|---|---|
| Label | Definition of strategy | Conceptualization of strategic management | Industrial organization economics view of strategy | Resource-based view of strategy | New paradigm for strategic management |
| Some leading authors | Chandler (1962), Ansoff (1965), Learned et al. (1965), Andrews (1971) | Rumelt (1974), Mintzberg (1978), Ansoff (1979) | Porter (1980), Porter (1986) | Barlett and Ghoshal (1987), Wernerfelt (1984), Barney (1991), Prahalad and Hamel (1990) | Nonaka (1991), Hamel (2000), Pfeffer and Sutton (2000) |
| Dominant themes | Corporate strategy, planning, and growth | Strategic management content and process | Competitive advantage development | Resource and capability development | Learning, knowledge, and innovation |
| Rationales | Strategy as a rule for making decisions | Evaluation and implementation of critical aspects of the formulated strategy | Five-forces analysis of industry attractiveness to develop a competitive advantage through generic strategies | Valuable, rare, and costly-to-imitate resources without close substitutes can be sources of sustained competitive advantage | Dynamic strategic model by which firms obtain valuable information, create knowledge, and accumulate intangible capabilities in a process of learning |
| Strategic concepts, tools, and techniques | SWOT; experience curve; growth share matrix | Value chain | Five-forces model, strategic choice | Core competence value system; VRIO; game theory | New integrated information technology systems |

Source: Mele and Guillen Parra (2006)
*SWOT* strengths, weaknesses, opportunities, and threats; *VRIO* value, rarity, imitability, organizational aspects

a complex, dynamic, and nonlinear system. The properties of complex systems can be expressed as follows (Koçel 2014):

- The constant change of the system, which is a living organism
- The fact that this change is not always in the form of repetition
- The many internal and external factors affecting this change
- The fact that it is difficult to know all of these factors and their relationships with each other
- The nonlinearity of these relations
- The fact that the intensification of each split into unexpected directions in the system focuses itself on repeating around certain environmental characteristics
- Sensitive adherence to the initial values in the motion of the system

If industries behave like chaotic systems, then a new strategic viewpoint is needed for businesses. Levy (1994) explained the relationship between chaos theory and strategy as shown in Table 2.

**Table 2** Relevance of chaos theory to strategy

| | |
|---|---|
| Long-term planning is very difficult | In chaotic systems, small disturbances multiply over time because of nonlinear relationships and the dynamic, repetitive nature of chaotic systems. As a result, such systems are extremely sensitive to the initial conditions, which makes forecasting very difficult |
| industries do not reach a stable equilibrium | Chaotic systems do not reach a stable equilibrium; indeed, they can never pass through exactly the same state more than once. Chaos theory also suggests that changes in industry structures can be endogenous. Corporate decisions to enter or exit the market, or to develop new technologies, alter the very structure of the industry, which in turn influences future organizational behavior |
| Dramatic change can occur unexpectedly | Large fluctuations can be generated internally by deterministic chaotic systems. If economic systems are chaotic, then we do not need to search for wars or natural disasters to account for economic depressions or a crash in the stock market |
| Short-term forecasts and predictions of patterns can be made | Although the unpredictability and instability of chaotic systems have been emphasized, there is also a surprising degree of order in chaotic systems. If we imagine that strategic decisions in companies are made on a monthly or even an annual cycle, then industry simulation models might be able to make useful predictions over a time horizon of several months or possibly years |
| Guidelines are needed to cope with complexity and uncertainty | Because of the complexity of strategic interactions, one does not always know why a particular strategy is successful. While the complexity of industry systems dictates the need for broad strategies, the dynamic nature of chaotic systems mandates that strategies adapt. As industry structures evolve and competitors change their strategies, a firm clearly needs to change its own guidelines and decision rules. The problem here is that there is no simple way of deriving optimal strategies for a given system |

Source: Levy (1994)

Systems operating in a turbulent and complex environment have complex features. In such a structure, the elements in the upper and lower systems are driven by mutual interactions and the structure can continue its life with continuous radical changes that do not repeat themselves in time. The balance deteriorates and a process toward irregularity and breakdown may occur in some situations when not all necessary data can be obtained in a complex environment and small-scale variables are ignored (Ülgen and Mirze 2018). In the context of this chapter, we can say that the concept of strategy, derived from the word *strategos*, as was used more than 2500 years ago, has reached the edge of chaos.

## Strategic Leadership at the Edge of Chaos

Here lies a man who knew how to enlist in his service better men than himself.
—Andrew Carnegie's tombstone

Leadership is an essential skill for anyone who develops or implements business strategy. Therefore, it is important to understand how to lead people so they can carry a strategy to success (Kourdi 2009). Because of this importance, leadership, within the framework of strategy and management, is a field of interest that attracts a lot of attention; thus, both academicians and professionals often dwell on it. Although there are many reasons for this, the biggest one is the difference that leadership creates in terms of results (Day 2000). When most people begin to do research and read resources to learn about the concept of leadership, they come up with the question "What is leadership?" and philosophers and authors, working on and thinking about leadership, try to answer it. The answer to people trying to learn leadership is really surprising because this question does not have a single answer, nor exactly a right answer.

Starting from Socrates (Adair 2002), who was thought to be the first person to examine the nature of leadership, many people have published works on leadership. The first systematic approach to leadership was the great man theory, first described in 1840. This theory was based on the belief that leadership is congenital. The first basic approach to leadership in modern times was the approach of characteristics (1910–1950). The behavior of leaders became the focal point once it was understood from research studies that not every leader had the same features. In this context, the Iowa State University (Lewin et al. 1939), Ohio State University (Halpin and Hunt 1957), Michigan State University (Likert 1958), managerial grid (Blake and Mouton 1964), and behavioral leadership theories came into prominence.

In the 1970s, situational leadership theories were developed on the basis of the idea that leadership was influenced by circumstances. Fiedler's contingency model (Fiedler 1976), the path–goal theory (House 1971), the situational leadership model (Hersey et al. 1979), and the three-dimensional leadership model (Reddin 1977) can be considered approaches within the context of situational leadership theory. In contrast to these approaches, other theories have been developed, each of which has a separate extent such as the leader–member exchange theory (Graen and Uhl-Bien 1995), transformational leadership (Bass 1985), visionary leadership (Nanus 1995), and servant leadership (Greanleaf 1970). It is possible to develop new theories based on the nature of the leader and leadership. One of these theories is strategic leadership.

The strategic management process includes strategic thinking (see Papatya 2017), strategic formulation, strategic implementation, and strategic evaluation processes (Ülgen and Mirze 2018). The key role of strategy in business is how to ensure effective competition in the market while maintaining profitability (Porter 1985). At the focus point of strategic leadership, there is the strategy of the enterprise. The *strategic leader* is the person responsible for the management process, which includes creation, implementation, and evaluation of the strategy of the enterprise or strategic business unit (Dinçer 1998). According to Sullivan and Harper (1998), strategic leadership means management and control of purposeful and well-considered actions in terms of the organization's aims, culture, strategy, basic identity, and critical processes. A strategic leader is a person who can make strategic changes when necessary through the ability to see the future, to create a vision, to be flexible, and to empower other people (Ülgen and Mirze 2018). Thompson et al. (2015) described strategic leaders as individuals with many different leadership roles, including those of a visionary, entrepreneur, strategist, chairman of the board, strategy practitioner, process integrator, coach, crisis solver, business master, speaker, interviewer, motivator, arbitrator, negotiator, policy maker, enactor, and mentor (see Table 3).

**Table 3** Strategic leadership practices

| Twentieth-century practices | Twenty-first-century practices |
|---|---|
| Are outcome focused | Are outcome and process focused |
| Are stoic and confident | Are confident but without hubris |
| Seek to acquire knowledge | Seek to acquire and leverage knowledge |
| Guide people's creativity | Seek to release and nurture people's creativity |
| Work flows are determined by the hierarchy | Work flows are influenced by relationships |
| Articulate the importance of integrity | Demonstrate the importance of integrity by actions |
| Demand respect | Are willing to earn respect |
| Tolerate diversity | Seek diversity |
| React to environmental change | Act to anticipate environmental change |
| Serve as a great leader | Serve as a leader and as a great group member |
| View employees as a resource | View organizational citizens as a critical resource |
| Operate primarily through a domestic mind-set | Operate primarily through a global mind-set |
| Invested in employees' development | Invest significantly in citizens' continuous development |

Source: Ireland and Hitt (2005)

The focus of strategic leadership is a sustainable competitive advantage, or the enduring success of the organization. The work of strategic leadership is to drive and move an organization forward so it will thrive in the long term (Hughes and Beatty 2005). Ireland and Hitt (2005) said that competition in the global economy of the twenty-first century would be complex, challenging, and full of opportunities and dangers. In this context, effective strategic leadership practices can help businesses that have to compete under turbulent and unpredictable conditions. According to Ireland and Hitt (2005), strategic leadership practices in the twenty-first century differ from those in the previous century. After the introduction of chaos theory, it was realized that systems are unbalanced, not balanced, and that work life is removed from stability and becomes variable. It was also realized that processes are not recycled, because time is unidirectional, data cannot be measured accurately, and thus the results cannot be determined in advance. Therefore, it is difficult to maintain the validity of the strategic leadership concept, which is designed according to conditions of order, in the context of chaos. It is necessary to rethink the process of strategic leadership (see Fig. 2).

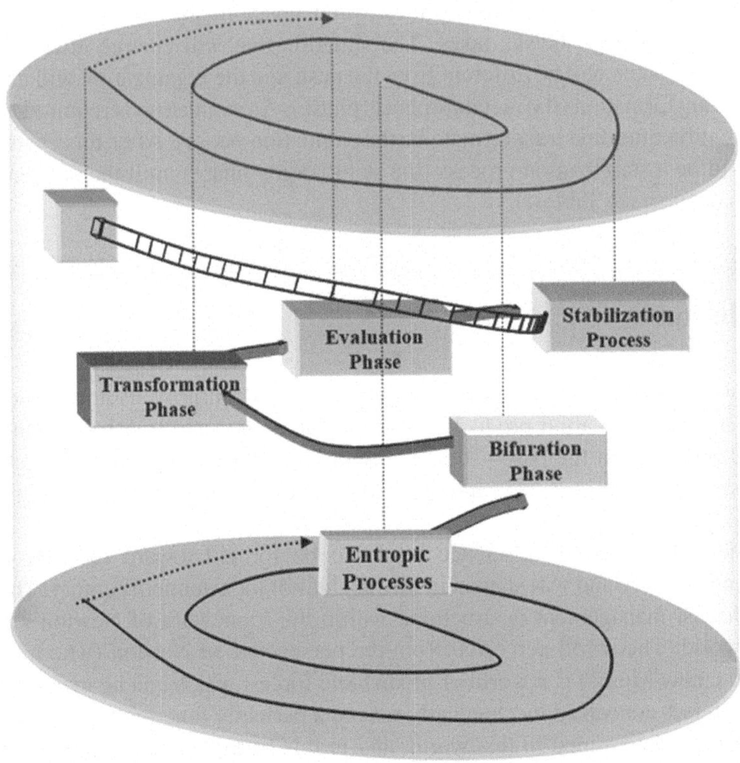


**Fig. 2** Strategic leadership processes at the edge of chaos

**Table 4** Strategic leadership processes phases

| Phase | Description |
|---|---|
| Entropic processes | This is the stage when problems and tensions occur because of the tendency of the system to deteriorate |
| Bifurcation phase | With the increase in tension caused by entropy, this is the stage at which the system comes to the point of selection. At this point, it will either turn toward the change or resist the change and tend to break down |
| Metamorphosis phase | This is the stage at which the complicated system, because of the developing bifurcation, is directed toward creating a new form that starts with nonlinear conditions in its critical structure |
| Evaluation phase | This is the stage at which an evaluation of the new situation resulting from the metamorphosis is made |
| Stabilization phase | At this stage, after the evaluation is carried out, the structure is preserved until the entropic processes phase |

Sources: Schwarz (1996) and Goldstein (1994, pp. 32–52)

Strategic leaders should first consider their situation in terms of entropic processes. The leader will reach a bifurcation phase, thanks to the decisions on strategic management he/she takes. The situation that will emerge after strategic decisions are taken will be different from the past, and the organization will undergo change/transformation (the metamorphosis phase). An evaluation is required for this new situation after this transformation (the evaluation phase). After this evaluation, there will be a stabilization process (the stabilization phase) until the next entropic process occurs (see Table 4).

## Conclusion

When we look at our environment, dizzying facts can surprise us at any moment. The speed of information production, new technologies, and transformation make today different from the past—so much so that when we say the world we live in is different from the past, we cannot say that our perception of the changing world remains the same.

Modernist management science began to be formed toward the end of the nineteenth century and was shaped in the first half of the twentieth century. The idea of modernist management is structured within the framework of Newton's vision of the world. The world perceived from the perspective of Newton (who has been called the new Moses) is a world of predictable forces, which can be measured and classified. It is conceived mechanically around a perfectly functioning clock model; everything is determined in this world, and in this particularity, the human is also robotized.

In the first quarter of the twenty-first century, perspectives on the dogmas of modern science have changed in accordance with developments in general relativity theory and quantum theory. It has emerged that the world is not, in fact, a perfectly functioning mechanical structure like a clock; it is relative to the position of the observer and contains processes that cannot be determined and cannot be measured, rather than certainty and precision. Chaos theory has struck a major blow against Newtonian understanding. With chaos theory, it has been demonstrated that systems are in disorder rather than order; therefore, they include nonlinear processes rather than linear processes.

It has been demonstrated that chaotic situations—the most familiar of which are clouds and weather conditions—do not occur only in nature; it has been proved that there is chaos in all aspects of life, from the natural sciences to the social sciences, and from human relations to financial markets. It is understood that butterflies are capable of producing typhoons in a world where uncontrollability, immeasurability, and complexity prevail. Of course, through this typhoon, strategic leadership is also affected, as are the strategies of businesses. In strategic management processes, if a strategic leader continues to think in a traditional way and exhibits modernist management behavior, he or she may meet the same fate as Achilles.

## References


Adair, J. (2002). *Inspiring leadership: Learning from great leaders*. London: Thorogood.

Amagoh, F. (2016). *Systems and complexity theories of organizations. A. Farazmand içinde, global encyclopedia of public administration, public policy, and governance*. Cham: Springer.

Andrews, K. (1971). *The concept of corporate strategy*. Homewood: Irwin.

Ansoff, H. (1965). *Corporate strategy: An analytical approach to business policy for growth and expansion*. New York: McGraw-Hill.

Ansoff, H. I. (1979). *A concept of corporate planning*. New York: Wiley.

Ansoff, H. I. (2007). *Strategic management*. New York: Palgrave Macmillan.

Barlett, C. A., & Ghoshal, S. (1987). Managing across borders: New organizational responses. *Sloan Management Review, 29*(1), 43–53.

Barney, J. B. (1991). Firm resource and sustained competitive advantage. *Journal of Management, 17*(1), 99–120.

Bass, B. (1985). *Leadership and performance beyond expectation*. New York: Free Press.

Beinhocker, E. D. (1997). Strategy at the edge of chaos. *The McKinsey Quarterly, 1*, 25–39.

Blake, R., & Mouton, J. (1964). *The managerial grid*. Houston: Gulf Publishing.

Bussolari, C. J., & Goodell, J. A. (2009). Chaos theory as a model for life transitions counseling: Nonlinear dynamics and life's changes. *Journal of Counseling & Development, 87*(Winter), 98–107.

Chandler, A. (1962). *Strategy and structure: Chapters in the history of the industrial enterprise*. Cambridge: MIT Press.

Clausewitz, C. V. (2018). *Savaş Üzerine* (H. Çelikler, Trans.). İstanbul: Alfa Ya.

Cummings, S. (1993). Brief case: The first strategists. *Long Range Planning, 26*(3), 133–135.

Day, D. (2000). Leadership development: A review in context. *The Leadership Quarterly, 11*(4), 581–613.

Dinçer, Ö. (1998). *Stratejik Yönetim ve İşletme Politikası*. İstanbul: Beta Ya.

Fiedler, F. (1976). The leadership game: Matching the man to the situation. *Organizational Dynamics, 4*(3), 6–16.

Gleick, J. (1995). *Kaos*. İstanbul: Tübitak Ya.

Goldstein, J. (1994). *The unshacled organization*. Portland: Productivity Press.

Graen, G., & Uhl-Bien, M. (1995). Relationship-based approach to leadership: Development of leader–member exchange (LMX) theory of leadership over 25 years. *The Leadership Quarterly, 6*(2), 219–247.
Greanleaf, R. (1970). *The servant as leader*. Atlanta: Greenleaf Center for Servant Leadership.
Halpin, A., & Hunt, J. (1957). A factorial study of the leader behavior descriptions. In R. Stogdill & A. Coons (Eds.), *Leader behavior: Its description and measurement*. Columbus: Ohio State University, Research Monograph No. 88.
Hamel, G. (2000). *Leading the revolution*. Boston: Harvard Business School Press.
Hard, R. (2004). *The Routledge handbook of Greek mythology*. London: Routledge.
Henderson, B. D. (1989). The origin of strategy. *Harvard Business Review, 67*(6), 139–143.
Hersey, P., Blanchard, K., & Natemayer, W. E. (1979). Situational leadership, perception, and the impact of power. *Group and Organization Management, 4*(4), 416–428.
Homer. (1990). *The Iliad* (R. Fagles, Trans.). London: Penguin.
Horwitch, M. (1987). *Emergence of post-modern strategic management, working paper no. 1901–87*. Cambridge: MIT Sloan School of Management.
House, R. (1971). A path–goal theory of leader effectiveness. *Administrative Science Quarterly, 16*(3), 321–339.
Hughes, R., & Beatty, K. (2005). *Becoming a strategic leader*. San Francisco: Jossey-Bass.
Ireland, R., & Hitt, M. (2005). Achieving and maintaining strategic competitiveness in the 21st century: The role of strategic leadership. *Academy of Management Executive, 19*(4), 63–77.
Jantsch, E. (1980). *The self-organizing universe*. London: Pergamon Press.
Koçel, T. (2014). *İşletme Yöneticiliği*. İstanbul: Beta Ya.
Kourdi, J. (2009). *Business strategy: A guide to taking your business forward*. London: Profile Books.
Learned, E. P., Christensen, C. R., Andrews, K. R., & Guth, W. D. (1965). *Business policy: Text and cases*. Homewood, III: R. D. Irwin.
Levy, D. (1994). Chaos theory and strategy: Theory, application, and managerial implications. *Strategic Management Journal, 15*(1), 167–178.
Lewin, K., Lippitt, R., & White, R. (1939). Patterns of aggressive behavior in experimentally created "social climates". *Reflections, 10*(2), 269–299.
Likert, R. (1958). Effective supervision: An adaptive and relative process. *Personnel Psychology, 11*(3), 317–332.
Machiavelli, N. (2008). *The prince* (J. Atkinson, Trans.). Cambridge: Hackett.
Mele, D., & Guillen Parra, M. (2006). *The intellectual evolution of strategic management and its relationship with ethics and social responsibility* (Working Paper No. 658). Barcelona: IESE Business School, University of Navarra.
Mintzberg, H. (1978). Patterns in strategy formation. *Management Science, 24*(9), 934–948.
Mintzberg, H. (1993). The pitfalls of strategic planning. *California Management Review, 36*(1), 32–47.
Mintzberg, H. (1994). The fall and rise of strategic planning. *Harvard Business Review, 72*(1), 107–114.
Mintzberg, H., & Waters, J. (1985). Of strategies, deliberate and emergent. *Strategic Managment Journal, 6*(1), 257–272.
Murphy, P. (1996). Chaos theory as a model for managing issues and crises. *Public Relations Review, 22*(2), 95–113.
Nanus, B. (1995). *Visionary leadership*. San Francisco: Jossey-Bass.
Nonaka, I. (1991). The knowledge-creating company. *Harvard Business Review, 69*(6), 96–104.
Papatya G. (2017). *Stratejik Düşünme: Yaratıcı Yıkıma Doğru Eleştirel Dönüşüm*. Harvard Business Review (Türkiye), Ocak-Şubat, 94–99.
Parker, D., & Stacey, R. D. (1994). *Chaos, management & economics: The implications of non-linear thinking*. London: Coronet.
Pfeffer, J., & Sutton, R. I. (2000). *The knowing-doing gap: How smart companies turn knowledge into action*. Boston: Harvard Business School Press.

Porter, M. (1979). How competitive forces shape strategy. *Harvard Business Review, 57*(2), 137–145.
Porter, M. (1985). *Competitive advantage: Creating and sustaining superior performance*. New York: Pree Press.
Porter, M. (1996). What is strategy. *Harvard Business Review, 74*(6), 61–78.
Porter, M. E. (1980). *Competitive strategy: Techniques for analyzing industries and competitors*. New York: Free Press.
Porter, M. E. (Ed.). (1986). *Competition in global industries*. Boston: Harvard Business School Press.
Prahalad, C., & Hamel, G. (1990). The core competence of the corporation. *Harvard Business Review, 68*(3), 79–91.
Prigogine, I., & Stengers, I. (1984). *Order out of chaos: Man's new dialogue with nature*. London: Bantam New Age Books.
Reddin, W. (1977). An integration of leader-behavior typologies. *Group and Organization Studies, 2*(3), 282–295.
Rumelt, R. P. (1974). *Strategy, structure, and economic performance*. Boston: Harvard Business School Press.
Schwarz, E. (1996). *Summary of the main features of a holistic metamodel to interpret the emergence, the evolution and the functioning of viable self-organizing systems*. Proceedings of the 40th Annual Meeting of the International Society for Systems Science, Budapest.
Skinner, W. (1969). Manufacturing—missing link in corporate strategy. *Harvard Business Review, 47*(3), 136–145.
Stacey, R., & Mowles, C. (2016). *Strategic management and organisational dynamics*. London: Pearson.
Sühreverdi, E. (1974). *Yönetenlerin Yönetimi*. İstanbul: Tercüman Ya.
Sullivan, G., & Harper, M. (1998). *Hope is not a method*. New York: Broadway.
Thiétart, R., & Forgues, B. (1995). Chaos theory and organization. *Organization Science, 6*(1), 19–31.
Thompson, A., Peteraf, M., Gamble, J., & Strickland, A. (2015). *Crafting & executing strategy: The quest for competitive advantage*. New York: McGraw-Hill.
Tzu, S. (2014). *Savaş Sanatı* (P. Otkan, & G. Fidan, Trans.). İstanbul: İşbankası Ya.
Ülgen, H., & Mirze, S. (2018). *İşletmelerde Stratejik Yönetim*. İstanbul: Beta Ya.
Weick, K. (1977). Organization design: Organizations as self-designing systems. *Organizational Dynamics, 6*(2), 31–46.
Wernerelt, B. (1984). A resource based view of the firm. *Strategic Management Review., 5*(1), 171–180.
Werndll, C. (2009). What are the new implications of chaos for unpredictability? *British Journal for the Philosophy of Science, 60*(1), 195–220.
Woodard, R. D. (Ed.). (2007). *The Cambridge companion to Greek mythology (Cambridge companions to literature)*. Cambridge: Cambridge University Press.

# Behaviours of Error-Prone Variables on Low-Chaotic Autoregressive Models

**Şahika Gökmen and Rukiye Dağalp**

**Abstract** Nowadays, both measurement errors and chaotic structures in data are frequently included in the literature. The main reasons for this are biased parameter estimations in the presence of measurement error and the unpredictability of chaotic structures. Although it has been investigated whether the confusion in the data is due to the measurement error or to the chaotic structure, the issue of how these two concepts affect each other has not been found in the literature. This study researched how time series with a low-chaotic structure were affected by measurement error, using Lyapunov exponents. This effect was demonstrated by various simulations for low-chaotic AR(1) and AR(2) autoregressive models. The results showed that the maximal Lyapunov exponent attenuated toward zero with an increase in the measurement error. It was also found that the Lyapunov exponent was affected by the sample size and the number of delays of the models.

**Keywords** Chaotic time series · AR model · Error-prone variables

## Introduction

Economic data exist as an area of interest that has become more and more popular due to the globalization of countries in the world. Where these data show very complex and chaotic structures, it is not possible to measure them exactly. All analyses determined with this type of data are obtained under the condition that the variables are observed correctly (error free). Violation of this condition causes some problems with the reliability and validity of analyses (Adcock 1878). Besides, many variables used in econometric analyses are recorded with errors such as

Ş. Gökmen (✉)
Hacı Bayram Veli University, Ankara, Turkey
e-mail: sahika.gokmen@hbv.edu.tr

R. Dağalp
Ankara University, Ankara, Turkey
e-mail: rdagalp@ankara.edu.tr



Ş. Ş. ERÇETİN, Ş. N. AÇIKALIN (eds.), *Chaos, Complexity and Leadership 2018*, Springer Proceedings in Complexity, https://doi.org/10.1007/978-3-030-27672-0_11

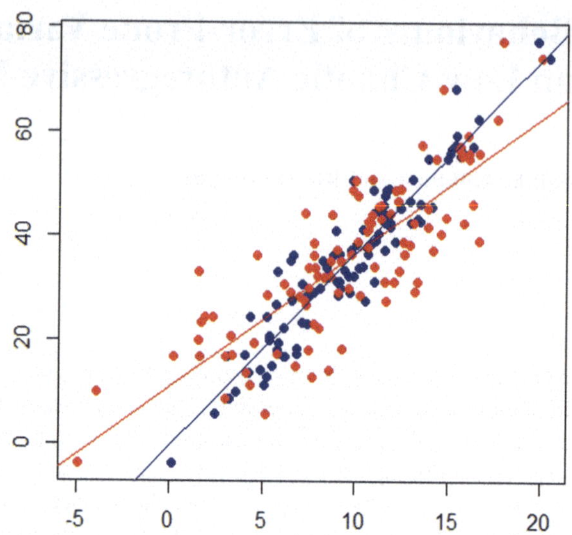


**Fig. 1** Effect of measurement error on the regression line (*blue* denotes error-free data and *red* denotes error-prone data within a 0.5 standard deviation of the measurement error)

miscoding by the collectors of the data, incorrect transformation, misreporting by subjects, smuggling, aggregated data, or just unobservable data (annual income, unemployment rates, etc.) (Bound et al. 2000; Turkish Statistical Institute 2012). Generally, such errors are ignored by researchers because of statistical analysis assumptions. In statistical analysis, this problem due to such errors is called the measurement error and the variables exposed to such errors are called error-prone variables or error-in variables. There are three effects of error-prone variables, generally referred to as measurement error problems, which are biased estimates of parameters, loss of power among variables, and masking properties of the data (Carroll et al. 2006; Stock and Watson 2011). These effects can be seen in Fig. 1, which shows how they affect the data and the estimated regression line even with low-level variance of the measurement error.

On the other hand, the effect of measurement error on time series is not very different. Figure 2 clearly shows this effect with the same measurement error level. Statistical inferences, unit root test, forecasting, etc., due to measurement error in time series are varied and are shown in Fig. 2. For example, policy making or forecasting are not only limited analyses but also lead to very misleading conclusions and inaccurate decisions. Accordingly, the subject of measurement error is called attention to econometrics as in other fields.

Another issue in econometric time series is chaotic structures. Financial variables such as exchange rates and stock returns can be defined by stochastic processes, and these economic systems are not linear. The high frequency and explosions of these data can be considered an example of inconsistent behavior (Barnett 1998). Therefore, the excess of external factors in this type of data shows the structure of a chaotic system. Chaotic dynamic systems follow deterministic equations, in spite of their irregularity (Baker and Gollub 1990), and also can be expressed by differential

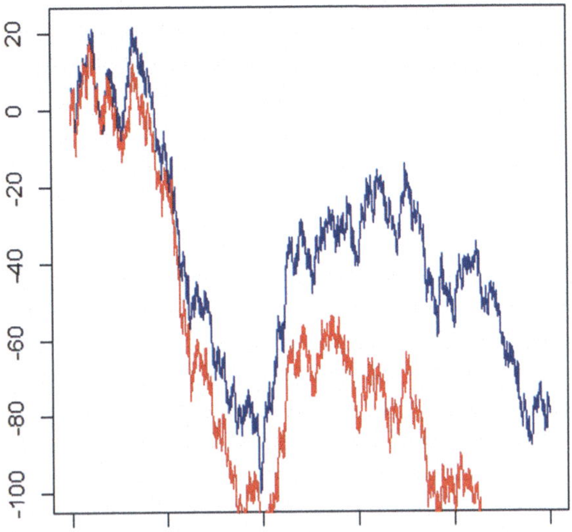


**Fig. 2** Effect of measurement error on time series (*blue* denotes error-free data and *red* denotes error-prone data within a 0.5 standard deviation of the measurement error)

equations in continuous systems and discrete systems, or by experimental data when the equations are not known (Yalamova et al. 2006).

The chaotic structures in which such factors are involved cannot be fully measured or observed, such as examples of the nonlinear forecasting technique in time series discussed by Sugihara and May (1990) and heart rate variability studied by Hilton et al. (1997). In both of those studies, whether the complexity of the data stemmed from the chaotic behavior or measurement error was discussed but their effects on each other were not mentioned. The effect of measurement error on chaotic structures has not come across in the literature. For this reason, the effect of the measurement error in chaotic AR(1) and AR(2) autoregressive models was investigated using Lyapunov exponents in this study.

## Detecting Chaos: Lyapunov Exponents

The unpredictability of the future is the most dramatic feature of chaos despite deterministic time growth. Sensitive dependence on initial conditions—a unique characteristic of chaotic systems—can be reflected by unpredictability, that is a consequence of the inherent instability of the solutions (Kantz and Schreiber 2004). Small deviations between the initial conditions are blown up after a few periods, such as the butterfly effect.

There are various methods available to detect a chaotic structure, such as the mutual information method, embedded dimension, wrong nearest neighbors, strange attractors, Poincaré maps, correlation integral, BDS test, power spectrum, and Lyapunov exponents. All of the methods mentioned herein are not enough to prove

**Table 1** Possible types of motion and the corresponding Lyapunov exponents

| Type of motion | Maximum Lyapunov exponent |
|---|---|
| Stable fixed point | $\lambda < 0$ |
| Stable limit cycle | $\lambda = 0$ |
| Chaos | $0 < \lambda < \infty$ |
| Noise | $\lambda = \infty$ |

Source: Kantz and Schreiber (2004)

that a structure is chaotic or not, although one can comment on this. Besides, if there is at least one positive Lyapunov exponent in the system, this is seen as the most definitive proof that the system is chaotic (Fırat 2006).

Lyapunov exponents determine the behavior of a dynamic system and allow measurement of the effects of divergence and convergence of the attractor. Additionally, the calculated exponents constitute a measure of the topological attributes corresponding to such concepts as unpredictability (Kasap and Kurt 2011). As many different Lyapunov exponents can be defined for a dynamic system as there are phase dimensions (Kantz and Schreiber 2004). In this study, the most important one—the *maximal Lyapunov exponent* $\lambda$—was chosen.

Usually, the Lyapunov exponent of a dynamic system is a quantity that characterizes the rate of separation of infinitesimally close trajectories in phase space. For example, $\mathbf{s}_{n_1}$ and $\mathbf{s}_{n_2}$ are two points in state spaces $n_1$ and $n_2$, with distance $\left\|\mathbf{s}_{n_1} - \mathbf{s}_{n_2}\right\| = \delta_0 \ll 1$.[1] Accordingly $\delta_{\Delta n}$ is the distance some time $\Delta n$ ahead between the two trajectories emerging from these points, $\delta_{\Delta n} = \left\|\mathbf{s}_{n_1+\Delta n} - \mathbf{s}_{n_2+\Delta n}\right\|$. Then the maximal Lyapunov exponent $\lambda$ is determined by

$$\delta_{\Delta n} \simeq \delta_0 e^{\lambda \Delta n}, \quad \delta_{\Delta n} \ll 1, \quad \Delta n \gg 1 \tag{1}$$

If the maximal Lyapunov exponent is positive, this means an exponential divergence of the nearby trajectories, i.e., chaos. Conversely, a zero maximal Lyapunov exponent denotes a limit cycle or a quasiperiodic orbit, and a negative maximal Lyapunov exponent is descending order (Liu 2010). A summary of the values of the Lyapunov exponent is given in Table 1.

## Simulation Study

In this study, Monte Carlo simulations were carried out using the R Project program. In the simulation, a low-chaotic time series without a measurement error was generated through four AR models. Then, randomly generated measurement errors at different levels with constant variance (standard deviations of the measurement

[1] "≪" means "much less than.

error in an ascending order of 0.05, 0.1, 0.15, ..., 10) were added to the time series. The maximal Lyapunov exponents were compared for all time series to see how the chaotic structure was affected.

Ashley and Vaughan's study (1986) was taken into consideration to determine the AR models because it dealt with measurement errors in economic time series. Four error-free models (two from the AR(1) models and two from the AR(2) models) were determined as follows:

(i) $\mathrm{AR}(1) - 1 : (1 - 0.6B)Z_t = a_t$
(ii) $\mathrm{AR}(1) - 2 : (1 - 0.9B)Z_t = a_t$
(iii) $\mathrm{AR}(2) - 1 : (1 - 0.3B)^2 Z_t = a_t$
(iv) $\mathrm{AR}(2) - 2 : (1 - 0.6B - 0.3B^2)Z_t = a_t$.

Considering Ashley and Vaughan's study (1986), 1000 repetitions were generated for each model at each of three sample sizes: 120, 350, and 1000. The first two of these sample sizes were chosen as being roughly typical of the quarterly and monthly data sets, respectively; the third was used to obtain large sample results.

Scatterplots of the estimated maximal Lyapunov exponents against the increased measurement error for three sample sizes are shown in Figs. 3 and 4. As can be seen from these figures, the maximal Lyapunov exponents attenuated toward zero depending on the measurement error. However, it was noted that the maximal Lyapunov exponent did not receive a negative value. This means that low-chaotic series are almost turned into a stable limit cycle because of the measurement error.

Another remarkable issue is that the sample size was also effective. In the small sample sizes, the Lyapunov exponent showed a faster decline against the

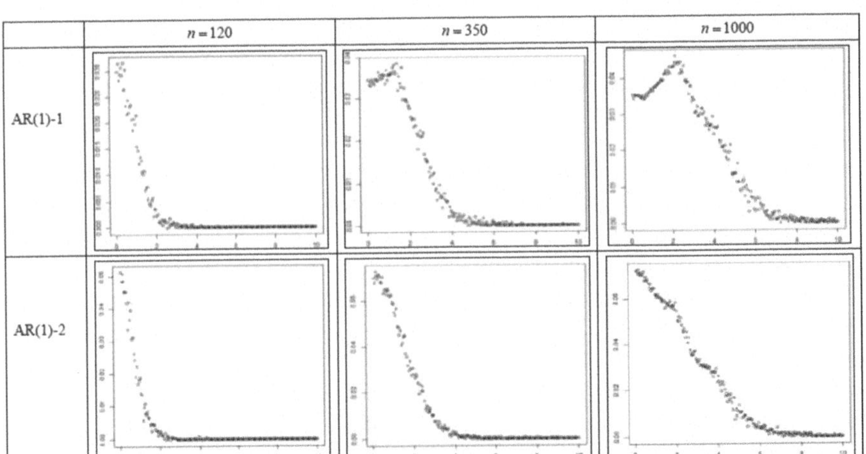


AR(1)−1 model is $(1 - 0.6B)Z_t = a_t$ and AR(1)−2 model is $(1 - 0.9\mathrm{B})Z_t = a_t$

**Fig. 3** Effect of measurement error on AR(1) autoregressive models for different sample sizes (with the standard deviation from 0.05 to 10 increasing by 0.05 on the *x*-axis and the estimated maximal Lyapunov exponents on the *y*-axis)

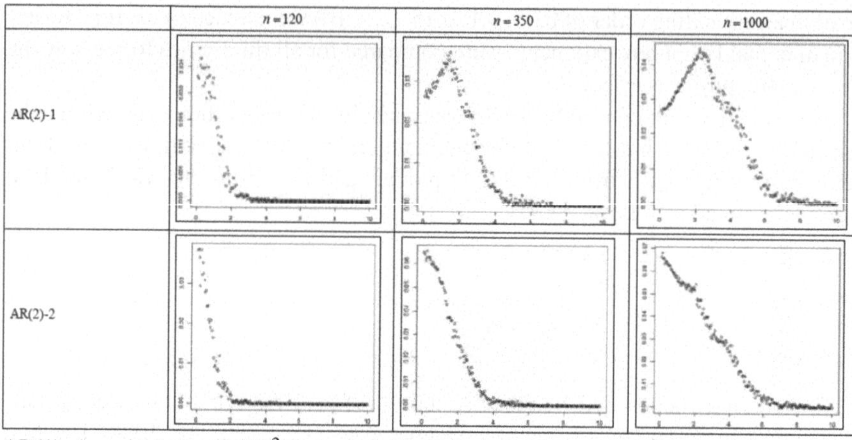


AR(1)–1 model is $(1 - 0.3B)^2 Z_t = a_t$ and AR(1)–2 model is $(1 - 0.6B^2)Z_t = a_t$

**Fig. 4** Effect of measurement error on AR(1) autoregressive models for different sample sizes (with the standard deviation from 0.05 to 10 increasing by 0.05 on the $x$-axis and the estimated maximal Lyapunov exponents on the $y$-axis)

**Table 2** Effects of measurement error with $\sigma_U = 2.5$ on the maximal Lyapunov exponent

| | AR(1)-1 | | AR(1)-2 | | AR(2)-1 | | AR(2)-2 | |
|---|---|---|---|---|---|---|---|---|
| | $\lambda_{free}$ | $\lambda_{prone}$ | $\lambda_{free}$ | $\lambda_{prone}$ | $\lambda_{free}$ | $\lambda_{prone}$ | $\lambda_{free}$ | $\lambda_{prone}$ |
| $n = 120$ | 0.03 | 0.003 (6.253) | 0.05 | 0.0004 (0.0001) | 0.02 | 0.0003 (5.144) | 0.04 | 0.003 (7.332) |
| $n = 350$ | 0.03 | 0.001 (0.0001) | 0.07 | 0.012 (0.0004) | 0.03 | 0.009 (0.0001) | 0.06 | 0.012 (0.0003) |
| $n = 1000$ | 0.03 | 0.001 (0.0003) | 0.07 | 0.02 (0.0006) | 0.03 | 0.02 (0.0002) | 0.06 | 0.02 (0.0005) |

The standard errors of the estimated maximal Lyapunov exponents are given in parentheses

measurement error. Especially in AR(1)-1 and AR(2)-1, as the sample size was increased, the maximal Lyapunov exponent resisted small measurement errors and it was increased too, although the maximal Lyapunov exponents attenuated toward zero after a certain measurement error.

The results of the simulations are given in Table 2 for the measurement error distributed with $N(0; 2.5)$. In accordance with the results, it was observed that the measurement error in the time series distorted the chaotic structure. The chaotic structure weakened but the maximal Lyapunov exponent still maintained its positive value while being attenuated toward zero. With this effect of the measurement error, it can be said that the chaotic structure was not completely disturbed but it was much weakened.

Besides, it was seen that the Lyapunov exponents from the models were affected by the sample sizes, and the number of delays and the model parameters also had an impact. The model parameters of the AR(1)-1 and AR(2)-1 models were smaller than the other model parameters. Low-level measurement error increases the chaotic structure, but the measurement error still dominated chaotic structure after a level. On the other hand, the high value (nearly one) of these parameters brings the series closer to being nonstationary. Here, these effects which were arisen from whether proximity to nonstationary or single delay should be discussed.

## Conclusion

In this study, the effect of measurement error on a low-level chaotic structure was investigated using Lyapunov exponents. According to the findings obtained from the simulation, it was observed that the measurement error in the time series distorted the chaotic structure. In the models discussed, Lyapunov exponents attenuated toward zero with the increase in the measurement error but did not reach a negative value. In the light of this study, it is intended to investigate the low-chaotic series are almost turned into a stable limit cycle because of the measurement error, which could be discussed in more detail. It is thought that determination of the Lyapunov exponent values will give a different perspective to the work to be done in this field.

Additionally, while the model parameters are decreasing (the dependence on the past is weakening), the increase of the Lyapunov exponents on the low measurement error appears to be another issue to be investigated. It is thought that investigation of the relation between the small rise of the Lyapunov exponent and the sample size in the AR(1)-1 and AR(2)-1 models with previous periods will contribute to the econometric literature.

With the progress of studies on measurement error in economic data, it is important for future studies to develop more accurate analysis of effects and distortions in chaotic structures. Additionally, it is suggested to develop estimation methods for such chaotic data with measurement error especially for nonlinear models. In this respect, it is thought that this study will contribute to the literature.

## References

Adcock, R. (1878). A problem in least squares. *The Analyst, 5*(2), 53–54.

Ashley, R., & Vaughan, D. (1986). Measuring measurement error in economic time series. *Journal of Business & Economic Statistics, 4*(1), 95–103.

Baker, G. L., & Gollub, J. P. (1990). *Chaotic dynamics: An introduction*. Cambridge: Cambridge University Press.

Barnett, W. A. (1998). *Martingales, nonlinearity and chaos*. Department of Economics, Campus Box.

Bound, J., Brown, C., & Mathiowetz, N. (2000). *Measurement error in survey data*. Ann Arbor: PSC Publications.

Carroll, R. J., Ruppert, D., Stefanski, L. A., & Crainiceanu, C. M. (2006). *Measurement error in nonlinear models* (2nd ed.). Boca Raton: Chapman & Hall/CRC.

Fırat, U. (2006). *Kaotik Zaman Serilerinin Yapay Sinir Ağlarıyla Kestirimi: Deprem Verisi Durumu* (Yüksek Lisans Tezi). Yıldız Teknik ÜniversitesiFen Bilimleri Enstitüsü Elektronik ve Haberlesme Anabilim Dalı, İstanbul.

Hilton, M. F., Beattie, J. M., Chappell, M. J., & Bates, R. A. (1997). Heart rate variability: Measurement error or chaos. *Computers in Cardiology, 24*, 125–128.

Kantz, H., & Schreiber, T. (2004). *Nonlinear time series analysis*. Cambridge: Cambridge University Press.

Kasap, P., & Kurt, D. (2011). *Karmaşanın Bilimi Kaos*. Ankara: Nobel Yayınevi.

Liu, Z. (2010). Chaotic time series analysis. *Mathematical Problems in Engineering*, 1–31.

Stock, J., & Watson, M. (2011). *Introduction to econometrics* (3rd ed.). Boston: Addison Wesley Longman.

Sugihara, G., & May, R. M. (1990). Nonlinear forecasting as a way of distinguishing chaos from measurement error in time series. *Nature, 344*(6268), 734–741.

Turkish Statistical Institute (2012). *Üretim ve Harcama Yöntemi ile Gayri Safi Yurtiçi Hasıla Tahminleri – Kavram, Yöntem ve Kaynaklar*. Ankara: Türkiye İstatistik Kurumu Matbaası.

Yalamova, R., Qi, L., Wang, L., Chen, Y., Mathews, S., & Gong, G. (2006). Detecting chaos in financial time series. *Technology Print*, 1–9.

# The Energy Policy of the European Union and Its Implications for Turkey

**Ahmet Atilla**

**Abstract** Energy has played a significant role in unifying the European nations and being an apparatus for peace building in Europe since the Second World War. In the postwar period the European Coal and Steel Community brought rival countries together as allies with regard to energy resources. Today, the energy issue remains a priority within the European institutional framework. Since the dissolution of the Soviet Union, more ex-Soviet East European countries have become members of the European Union (EU). This wider and deeper integration has led the EU to adopt comprehensive energy policies, so it is essential to examine the facets of the EU energy policy. Moreover, because of its strategic location, Turkey has become an active player in the energy trade between the Middle East, the Caucasus, and Europe. The number of pipelines that pass across Turkey has increased since the start of the 2000s. Turkey has become a candidate for membership of the EU and has had to comply with EU rules and norms, as well as internalizing the EU's Acquis Communautaire. In this chapter, the evolution of the EU energy policy and Turkey's integration into it are analyzed.

**Keywords** Energy policy · European Union · Turkey

## Introduction

Energy plays a significant role in meeting the needs of people, powering efficient industry, and providing the needs of modern, civilized nations. These aims are highly dependent on energy. Moreover, energy is not distributed around the planet equally. Some countries are in the producer position and others are consumers. In both the twentieth and twenty-first centuries, the underlying reason for wars has been control of energy resources and energy transport lines. The demand for energy is climbing year by year with increases in the world population, the rate

A. Atilla (✉)
Turkish-German University, İstanbul, Turkey



Ş. Ş. ERÇETİN, Ş. N. AÇIKALIN (eds.), *Chaos, Complexity and Leadership 2018*, Springer Proceedings in Complexity, https://doi.org/10.1007/978-3-030-27672-0_12

of urbanization, and the rate of industrialization. The idea of a united Europe was an ideal more than five centuries ago. However, considering old enmities and destructive wars, this ideal of continental peace and prosperity was just a dream at that time. Given the historical facts, unification could be realized only by economic integration and, without that, political unification would be unrealistic. The European unification movement started after the Second World War. First, the Organization for European Economic Cooperation (OEEC) was established to administer Marshall Plain aid to restore Europe, much of which had been destroyed. Secondly, the North Atlantic Treaty Organization (NATO) was founded to stop Communist expansion toward Western Europe, as part of the Truman Doctrine. Europe's longest period of peace began on 9 May 1950 with the Schuman Declaration, which paved the way for creation of the European Union (EU). Thus, as shown by this brief background and introduction, energy was the factor behind European unification. The Western European countries succeeded in coming together around energy resources after the Second World War.

The six nations of Germany, France, Italy, the Netherlands, Belgium, and Luxembourg agreed to establish supranational bodies—including the European Coal and Steel Community (ECSC), the European Economic Community, and the European Atomic Energy Community—to manage and organize the energy resources in the continent. By this time, they had widened the scope of their integration; thus, they decided to lift tariffs and barriers on commodities such as agriculture, as well as energy. In view of the global energy shocks at the beginning of the 1970s and the Arab–Israeli Wars, the European states preferred to retain their individual control of their energy resources because of security concerns. These concerns framed today's European energy policies. Over time, the European integration has deepened and widened in several policy areas as more countries have joined the EU since the end of the Cold War.

For the last two decades, Turkey has substantially invested in increasing its energy capacity and its leadership role in both energy security and energy transportation. Given Turkey's military and bureaucratic capabilities, it is possible to understand its bid to become a hub for cross-national energy routes and pipelines. On the other hand, the EU has given more importance to its own energy security. In the event of long-term disruptions in the energy supply to EU countries, this would leave most of these countries vulnerable to energy shortages. Many EU countries are highly dependent on energy imports, especially from Russia. The last 5 years have shown that the more Russia grows economically, the more confidently and aggressively it acts toward its neighbors during border disputes. Thus, the EU countries and Turkey share common interests in the energy trade. Moreover, in addition to the EU's wider and deeper integration process at the beginning of the 2000s, Turkey has emerged as an active power, which has taken a leadership role in regional politics as well.

At the beginning of the 2000s, the agenda of the EU evolved from security politics into fiscal and budgetary politics. Furthermore, economic and security concerns about the supply of energy overlapped; therefore, the EU started to bring up the energy issue at the Union level. After the signing of the Lisbon Treaty, the European Commission started to manage the EU's energy policies in general. In 2007, the Commission published a communication titled *An Energy Policy for Europe* in which it detailed essential ideas for the future of energy (Behrens et al. 2008, p. 2).

This chapter examines various aspects of the EU's energy policy. Analysis of these aspects sheds light on the policies of the EU that have been built on for future policy provisions. The analysis starts with the research question "What are the aspects (facets) of the European Union energy policy?" By answering that research question, this analysis examines the framework of the energy policy of the EU. Moreover, section "Implications of the EU Energy Policy for Turkey" of this chapter examines the EU's effect on Turkey's energy policies.

## Facets of the EU Energy Policy

### *Background*

Since the beginning of the Industrial Revolution, coal has been a significant component of mass production. The Western European countries began the competition for energy resources. The discovery of vast coal reserves in the region between France and Germany resulted in confrontations between them. Moreover, these coal reserves in the region of Alsace-Lorraine were the focus of a series of wars in Europe. The Western European countries were badly devastated by the Second World War. After the war, they decided to leave behind their ancient rivalries and unite, thanks to the reconciliation and rapprochement of France and Germany. Under their leadership and guidance, six European nations agreed to unite, put aside their old enmities, and transfer the control of their energy resources to supranational bodies. As a result, Germany, France, Italy, the Netherlands, Belgium, and Luxembourg signed the Treaty of Paris in 1952, and the ECSC was established. This unification brought peace and prosperity; thus, this community became the impetus for European unification. However, the European countries did not share their control of certain areas such as foreign policy; they tended to retain their national control in this regard. In addition, although the ECSC member states transferred their rights regarding coal resources, for various reasons they preferred to retain control of their energy policies. The member states always approached this topic from the security perspective. The bitter experiences resulting from tensions in the Middle East forced them to create national energy policies. These protective energy policies appeared after the Arab–Israeli Wars. This series of wars in the Middle East resulted in the oil shock at the beginning of the 1970s. European governments were reluctant to share control with the Commission until the Commission launched a common energy policy at the European level to decrease the shocks of the oil shortages. Eventually, the member states started to comply with the energy policies of the Commission. The common European energy policies were linked with European integration. The efforts of the member states for a more deeply integrated Europe became a reality after the end of the Cold War. Within a decade the integration was upgraded via the Maastricht, Nice, and Lisbon Treaties.

However, the energy policies could be carried out at the Union level only after the signing of the Lisbon Treaty, which brought about many changes in energy policy. The treaty empowered the Commission to make decisions about the energy sector. It could be said that the Commission became competent in managing energy policy when it was upgraded from the national level to the supranational level. In addition, the update of the Commission's 2007 communication *An Energy Policy for Europe* led to the formation of the 2009 Third Energy Package (Szulecki et al. 2016, p. 550). The EU energy policy has regional and global outcomes. Moreover, the energy sector is directly linked to the future. The EU energy policy aims to retain the share of coal in total energy consumption as well as increasing the share of natural gas and providing maximum safeguard regulations for nuclear power plants. It also has an agenda for increasing the share of renewable energy resources. The EU energy policy provides a balance between the facets of protection of the environment (sustainability), security of energy supply, and competitiveness (El-Agraa 2011, p. 260).

## *Sustainability*

The first facet of the EU energy policy is sustainability. The concept of sustainability is closely linked to environmental issues, global warming, and renewable energy. In terms of ecology, the meaning of the term "environment" covers organic and inorganic entities. As a result, the notion of the environment has an interactive relation with political and socioeconomic developments and transformations. With the introduction of technological transformations, the issue of environmental consciousness first arose in the 1970s. At first, the approach to environmental issues was control of pollution. However, consciousness of pollution was insufficient because it focused on the aftermath of its production. In the 1980s, the notion of energy efficiency appeared and, in contrast to the 1970s, this environmental consciousness focused on production and all of its phases. At the end of the 1980s, recycled materials began to be used as production inputs. Finally, this basic principle turned into an approach to prevent pollution at the production stage (El-Agraa 2011, p. 260).

At the beginning of the 2000s, modern approaches to energy efficiency became part of legislation such as the issuing of environmental warnings, tax deductions, and various precautions from the production stage to the consumption stage. The reason for environmental pollution in Europe is the industrialization of the European states. When the hazards of pollution started to become evident beyond state borders, the EU decided to make decisions about the environmental consequences of energy resource use. Thus, environmental issues were reflected in the political agenda. The

Green political party also forced member states to set agenda on environmental problems (El-Agraa 2011, p. 261). In addition, the cross-border consequences of environmental problems increased the Commission's competencies on this issue. Moreover, global warming became recognized as one of the greatest challenges facing humankind at present and in the future.

The cumulative results of $CO_2$ emissions into the atmosphere create a greenhouse effect. At the international level, to reduce greenhouse gas emissions, there have been two essential initiatives: the Kyoto Protocol and the United Nations (UN) Rio de Janeiro Earth Summit.

The Kyoto Protocol was signed by 184 UN member states (excluding the USA and a few other states) in 1997. The signing of the protocol drew global public attention and resulted in serious measures being taken against greenhouse gas emissions. The protocol determines quotas for greenhouse gas emissions by developed and developing countries, and it also sets legally binding goals for developed countries to reduce their emissions gradually (Stavins et al. 2008, p. 3). In 2000 the EU confirmed the Kyoto Protocol in a pledge to reduce its emissions by about 8% of the 1990 levels by 2012 (Nordhaus 2007, p. 27). In addition to environmental pollution, public opinion in the EU is very sensitive to global warming. The EU has been working on diversifying its resources and enhancing incentives for use of renewable energy resources, as well as mechanisms and tools for reduction of greenhouse gas emissions, to gain public support and to fulfill the Kyoto Protocol requirements. Carbon emissions trading is one preventive measure used to control $CO_2$ emissions and is an outcome of the Kyoto Protocol. This trading system is based on trade between states. For instance, a state that buys a carbon emissions allowance from another state is able to use it, while the state that sells that emissions allowance waives its right to use it (Ellerman and Buchner 2007, p. 67).

The Rio Summit has been one of the significant steps toward preventing global warming. The summit was held by the UN and signed by more than 100 member states. The summit emphasized the dramatic climate changes that have occurred and urged the international community to take action (Martella and Smaczniak 2012, p. 4).

The Commission set aspirational goals for use of renewable energies by the year 2020. Because of the high cost of renewable energy and the absence of a stable, long-term EU policy framework, previous efforts failed. However, the Commission has set up significant steps toward establishing renewable energy policies (El-Agraa 2011, p. 261). The President's Conclusion of 2007 stated that by 2020 the EU would reduce its greenhouse gas emissions by up to 20% of its 1990 emissions level. Furthermore, the EU set two goals to transform the economy on the basis of high energy efficiency and low greenhouse gas emissions. First, by 2020, the EU aimed to increase the renewable energy share to 20% of total energy consumption. Secondly, the EU aimed to reduce the primary energy input share of total EU energy consumption to 20% by 2020. Thus, these aims are called the 20/20/20 aims (Stankeviciute and Criqui 2008, p. 254).

Renewable energy could be quite useful to the EU for various reasons. First, the increase in the renewable energy share of total energy usage decreases energy dependence and reduces energy imports, which is useful for energy security. Secondly, renewable energy creates job opportunities at green energy facilities, increasing employment rates. Thirdly, the EU is able to reflect its experience of renewable energy in the international arena so it can take a leadership role in the renewable energy sector and promulgate, circulate the experience derived from renewable energies to other countries. Lastly, generation of renewable energy requires high technology and know-how. The EU's investment in renewable energy has enabled it to export its know-how and experience to other countries such as the USA, China, and Brazil (El-Agraa 2011, p. 261).

## *Security of Supply*

The second facet of the EU energy policy is security of supply. According to (El-Agraa 2011, p. 262), there are two aspects to the EU's security of supply: crude oil and natural gas.

The EU is quite dependent on imports of crude oil. Thus, the EU has been vulnerable to supply volatility. The crude oil sector is regulated and priced according to the international market. The prices and production are dependent on international market rules. This energy resource is essential for the EU. The first two treaties that were signed in the establishment of the EU—the Treaty of Paris and the Treaty of Rome—were about energy; the first treaty was about coal, and the second was about nuclear energy. However, the member states failed to agree on coordination and common energy policies. The main reason was the reluctance of these states to limit their national control of energy and share it with the supranational authority. Moreover, the Organization of the Petroleum Exporting Countries (OPEC) embargo after the 1973 Arab–Israeli War led to a major transformation in the European Community. The OPEC countries protested against the military aid given by the USA to Israel during the war. The embargo was initially against the USA, and then it spread to the European states. Eventually, the embargo caused great increases in oil prices. As a result, the European states realized the importance of security of supply. Thus, the 1973 OPEC embargo started a major transformation in the energy policies of the European member states (Knittel 2013, p. 344). The "Council Resolution of 17 December 1974 Concerning Community Energy Policy Objectives for 1985" actually set various targets: security of energy supply, a reduction in energy import dependence, and diversification of energy supply (Council of the European Union 1974). The new strategy entailed major transformations, which occurred until 1985. Furthermore, an Energy Committee was established to execute the energy strategy in the long term, which was firstly to search for substitute energy resources to replace the dependence on Middle East and African oil resources. Secondly, the member states supported use of advanced technologies for higher energy efficiency. The states did not consider alternatives to petrol; instead, they decided not to shut down their coal mines and started to support their national oil and gas companies' search for oil and gas in the North Sea.

Natural gas is another key resource in the EU's energy security (El-Agraa 2011, p. 263). The EU is highly dependent on imports of natural gas, especially from Russia, which has been the main provider for the EU since the 2000s. Unlike other energy resources such as petrol and coal, natural gas is transported via pipelines. The reason for the EU's high dependence on Russia is derived from the geographical proximity of the EU to Russia. Like the oil crisis, crises in the supply of natural gas have occurred in the EU. The worst such crisis to date occurred in 2009, according to the International Energy Agency (El-Agraa 2011, p. 263). It erupted as a result of a disagreement regarding debt between Russia and Ukraine, which hosts the transit gas pipelines that carry 80% of the Russian gas that is supplied to Europe. This crisis directly affected the security of the EU's gas supply and demonstrated the importance of that security (El-Agraa 2011, p. 264). Because of a lack of infrastructure in some EU member states, they are entirely dependent on Russian gas. These countries are the ex-Soviet and Eastern European countries of Finland, Slovakia, Latvia, Estonia, and Lithuania (Grigas 2012, p. 2). However, Russia and the EU are mutually interdependent in terms of economic relations.

## *Competitiveness*

The last facet of the EU energy policy is closely interlinked with the internal energy market. To create secure, affordable, and sustainable energy, the notion of competitiveness is essential in the energy sector. One of the goals of the EU energy policy is to operate its internal energy market in accordance with international rules on market competitiveness. The Commission made efforts in 2001 to create an internal energy market, offering member states fully liberalized natural gas and electricity markets by 2005 (El-Agraa 2011, p. 267). Accordingly, these markets have been able to transform into competitive markets so that consumers can choose their providers freely. Moreover, the Commission is implementing the Union's competitiveness policy to turn the electricity market into a competitive market. In this case, the Union's competitiveness policy is aimed at providing assistance and opportunities for new entrepreneurs in the electricity market who want to provide better service to their customers in harsh conditions. The Commission aims to achieve the end of the old electricity oligopoly market rules though implementation of the Union's competitiveness policy. Furthermore, the founding treaties of the EU established the legal basis for free movement of gas and oil within the EU. In addition, the same legal framework for transportation and distribution of gas and electricity was enacted by the Commission on the basis of its competitiveness rules. By use of the competitiveness rules, the old monopolistic markets have been successfully transformed into liberal markets.

## Implications of the EU Energy Policy for Turkey

Turkey's relations with the EU were boosted with the decision of this candidate country to be a full member of the EU's Helsinki Summit in 1999. Turkey's harmonization process with the EU's energy Acquis Communautaire began immediately after the 1999 Helsinki Summit. The reform process, which was initiated in 2001 to restructure the energy sector within the framework of harmonization with the EU's internal energy market acquis, still continues. In the process of EU enlargement, the European Union requires candidate countries to establish a legal infrastructure in accordance with these regulations by making their own internal arrangements to ensure efficient functioning of the European market and security of supply. Turkey, in line with the conditions stipulated by the Union during the EU candidacy process, has created the legal infrastructure of the market and directs the market.

Legislation has been enacted for alignment with the EU's internal market to establish a competitive market. Turkey instituted a national program for adoption of the EU acquis in 2001. As part of Turkey's EU accession process, the screening process for the energy chapter was completed in 2007. Energy issues constitute a positive agenda between Turkey and the EU. The reform process initiated in 2001 for restructuring of the Turkish energy sector in line with the EU energy market acquis is still in progress. Turkey aims to become a full member of the EU and aims to adapt to the EU's policies in terms of energy. The EU has introduced some criteria for monitoring the integration of energy policies in both member and candidate states. These criteria are integration, adaptation to the EU's internal market, energy efficiency and saving, security of supply, adherence to the Kyoto Protocol, increasing the market share of renewable energy resources, and emergency preparedness mechanisms (Yorkan 2009, p. 35).

Turkey plays an important role in energy because of its geographical location. Turkey is a transition corridor between regions with energy resources (such as the Middle East, the Caucasus, and the Black Sea) and the EU. Legal arrangements are in progress to establish a competitive market in Turkey in order for it to adapt to the EU's internal market. Moreover, Turkey is contributing to the integration of the EU's internal energy market with existing pipelines reaching the EU market through its territory, and Turkey's contribution is expected to gradually increase the number of pipeline projects. Modernization and construction of natural gas pipelines and electricity transmission lines within the framework of the EU's integration objective have allowed Turkey to become an energy base in the region and to strengthen its own energy infrastructure. In 2005, Turkey enacted a law on renewable energy resources and at this point removed obstacles to green energy investments. Especially in recent years, the number of private sector investments in wind turbines has increased in order to increase the share of renewable energy resources among the traditional resources. However, Turkey is at the very beginning of the road on environmental issues. Although it is a signatory to the Kyoto Protocol, it seems to be difficult for Turkey to meet the criteria that the EU has set so far. Furthermore, to fulfill these commitments, Turkey needs to rehabilitate its existing coal power plants and generate less emissions. In fact, Turkey is rich in hydroelectric, geothermal, wind, and solar power, and it needs to develop its policy in such a way that it can benefit more from these resources (Yorkan 2009, p. 36).

Turkey enacted an energy efficiency law in May 2007. This law aims to use energy resources efficiently from production to consumption. In terms of energy savings, the Turkish public is informed about the choice of energy-saving lightbulbs. In addition, goods-manufacturing companies have started to promote energy-efficient class A products in their advertisements. Security of supply is also vital for Turkey, where there are needs for alternative producers and carriers to minimize the risks associated with imports of energy. Given that the EU supplies half of its energy resources from abroad, the security of supply policy in Turkey, which is located between countries with rich energy resources and the EU market, gives it a great advantage in this regard. The EU wants to reach the natural gas and oil resources of the Caspian, Caucasus, North African, and Middle East regions to reduce its dependence on Russia. The fact that Turkey shares the same fate with the EU in terms of foreign dependence is the element that has brought their security of energy supply policies closer together. Moreover, oil and gas storage facilities need to be activated in case of potential energy cuts (Yorkan 2009, p. 37).

## Conclusion

Since the Second World War, energy has been a central and unifying issue in European politics. The ECSC was the cornerstone of the EU, whose energy policy has been gradually developed with global developments, technological advancements, and the Union's enlargement over time. In addition, despite the energy priorities of the member states, a common fact is that all member states are highly dependent on energy imports, and this dependence has been increasing year by year. Through the spillover effect of European integration, policy unification at the Union level has been reached; however, because of national preferences, member states have acted in their own national interests in other policy areas. As a result, the Commission has upgraded the status of the energy policy to the Union level. The Commission's vision was released in the Commission's communication *An Energy Policy for Europe*, which details the three facets of the energy policy—sustainability, energy security, and competitiveness—that will guide the energy framework in the near future. The analysis in this chapter has examined these three facets of the policy and has given a perspective on them. Moreover, the topic of energy will soon come under the full authority of the Commission. The supranational energy policy is vital for the EU and will be administered under supranational authority in accordance with the Commission's provisions to protect and operate a sustainable economy. Furthermore, the changes in the energy policy of the EU have affected Turkey's energy policy. Thus, Turkey has started its own transformation based on the EU membership process. Also, the energy issue has always contributed to the EU–Turkey political agenda positively. Finally, the energy policy is politically vital and essential for the environment, whose resources are finite.

## References

Behrens, A., Egenhofer, C., & Pedersen, K. (2008). *Energy policy for Europe: Identifying the European added-value*. Brussels: Centre for European Policy Studies. https://www.ceps.eu/ceps-publications/energy-policy-europe-identifying-european-added-value/.

Council of the European Union. (1974). *Council resolution of 17 December 1974 concerning community energy policy objectives for 1985*. Publications Office of the European Union. https://publications.europa.eu/en/publication-detail/-/publication/bd6b8c7f-ffba-46e6-9e4e-d558dac1dec5/language-en.

El-Agraa, A. M. (2011). *The European union: Economics and policies*. Cambridge: Cambridge University Press.

Ellerman, A. D., & Buchner, B. K. (2007). The European Union emissions trading scheme: Origins, allocation, and early results. *Review of Environmental Economics and Policy, 1*(1), 66–87.

Grigas, A. (2012). *The gas relationship between the Baltic States and Russia—politics and commercial realities*. Oxford: Oxford Institute for Energy Studies.

Knittel, C. R. (2013). Transportation fuels policy since the OPEC embargo: Paved with good intentions. *American Economic Review, 103*(3), 344–349.

Martella, R., & Smaczniak, K. (2012). Introduction to Rio+20: A reelection on progress since the first Earth Summit and the opportunities that lie ahead. *Sustainable Development Law & Policy, 12*(3), 4–50.

Nordhaus, W. D. (2007). To tax or not to tax: Alternative approaches to slowing global warming. *Review of Environmental Economics and Policy, 1*(1), 26–44.

Stankeviciute, L., & Criqui, P. (2008). Energy and climate policies to 2020: The impacts of the European "20/20/20" approach. *International Journal of Energy Sector Management, 2*(2), 252–273.

Stavins, R., Carraro, C., & Kolstad, C. (2008). *Introduction to the issue*. New York: Oxford University Press.

Szulecki, K., Fischer, S., Gullberg, A. T., & Sartor, O. (2016). Shaping the "energy union": Between national positions and governance innovation in EU energy and climate policy. *Climate Policy, 16*(5), 548–567. https://doi.org/10.1080/14693062.2015.1135100.

Yorkan, A. (2009). Avrupa Birliği'nin Enerji Politikası ve Türkiye'ye Etkileri. *Bilge Strateji, 1*(1), 24–39.

# Customer Perception of the Quality of Online Banking Services (with Special Reference to SBI and ICICI): A Study on Chaos and a Complexity Perspective

**Niyasha Patra and Nilanjan Ray**

**Abstract** Digitization has been introduced as a modern approach to globalization in this new era. It has a revolutionary impact on global, social, and economic aspects. With the same vision, India has taken part in this journey to transform India into a digitally empowered society. The present study aimed to find out customer perceptions of online/digital banking services provided by the State Bank of India (SBI; a public sector bank) and the Industrial Credit and Investment Corporation of India (ICICI; a private sector bank). The study focused mainly on the aspects of understanding, relationships, and customer values. It also illustrated game-changing strategies used by the banks to handle the current chaotic situation, which occurred due to a certain change in the existing conventional schema and the interference of complexity and chaos theory in a customer service satisfaction study. Here, five dimensions were considered for fulfillment of the study: reliability, empathy, responsiveness, securities and services, and assurance. Primary data were collected in prestructured questionnaires from 200 respondents of different ages, genders, professions, and incomes through e-mail, a field survey, personal contacts, and social media. This study showed the factors involved in customer ratings of SBI and ICICI, various challenges faced by consumers in different segments, acceptance of digital banking among consumers, and diverse challenges faced by banks in implementing a digitally based banking system.

**Keywords** Digital banking · Consumer satisfaction · Consumer perception · Service gaps · Banking chaos · Complexity

---

N. Patra (✉)
Independent Researcher, Kolkata, West Bengal, India

N. Ray
Adamas University, Kolkata, West Bengal, India



Ş. Ş. ERÇETİN, Ş. N. AÇIKALIN (eds.), *Chaos, Complexity and Leadership 2018*, Springer Proceedings in Complexity, https://doi.org/10.1007/978-3-030-27672-0_13

## Abbreviation

| | |
|---|---|
| ATM | Automated teller machine |
| BANKSERV | Bank service |
| CC | Credit Card |
| CCP | Customer Care Personal |
| CRM | Customer relationship management |
| DC | Debit Card |
| ETF | Electronic Funds Transfer |
| GOI | Government of India |
| ICICI | Industrial Credit and Investment Corporation of India |
| MF | Mutual Fund |
| NDLM | National Digital Literacy Mission |
| PC | Personal computer |
| SBI | State Bank of India |
| SC | Smart Card |
| SERVQUAL | Service quality |
| TDR | Term deposit receipt |
| UPI | Unified payment interface |

## Introduction

Digital banking is a process in which banking services are provided over the internet via mobile phones, automated teller machines (ATMs), personal computers (PCs), etc. Along with providing all conventional functional services, it also covers such areas of the financial industry as risk management, marketing, etc. This process of transforming traditional banking to digitization started with the introduction of ATMs and plastic cards. With the advancement of technology in this decade, banking services are now available over different cross-platform devices and media such as cellular phones, smart devices, social media, etc. Beside flexible usability, digital banking is cost effective and more business efficient. Merging of new technologies such as artificial intelligence (AI), the internet of things (IOT), machine learning (ML), and big data with digital banking architecture reshapes traditional banking into new-generation banking services. However, with these changes, maintenance of integrity, accuracy, and security are new challenges faced by the banking industry.

This study focused on different segments of digital banking services that customers use for their personal banking and daily use, the impact and acceptance of digital banking in rural India, the chaos and complexity that have arisen due to digitization of banking services, and proposed solutions to overcome the dilemmas faced by financial institutions.

## Objectives

This study included the following research objectives:

1. To assess the difficulties, feasibility, and awareness of online banking services among customers.
2. To assess chaos and complexity issues with reference to financial institution services in general and online banking issues in particular.
3. To assess customer expectations and perceptions of online banking services at two banks: the State Bank of India (SBI) and the Industrial Credit and Investment Corporation of India (ICICI).
4. To assess online banking services and their impact on customer satisfaction.

## Research Methodology

Data for the study were collected from the primary sources through prestructured questionnaires. The questionnaires included information on each respondent's name, gender, age, country, and occupation. On the basis of the five dimensions of service quality (SERVQUAL), the sample size was restricted to 200 respondents (100 respondents from SBI and 100 from ICICI). This study was representative in nature so far as the banks were concerned and focused on client satisfaction with internet banking services.

## *Chaos and Complexity in Financial Institution Services: Background of the Study*

"Chaos" is an interdisciplinary branch of mathematics theory based on the behavior of a dynamic system that is highly sensitive in the initial phase. Small differences at one stage of a deterministic nonlinear mechanism can create major disparity in the subsequent proceedings. Even in a deterministic system where the projected performance is completely determined as per its primary state, it may not remain predictable. This behavior of real-time systems is known as chaos. The butterfly effect is a property of a dynamic system, e.g., flapping of butterfly wings in Brazil causing a hurricane in Texas. The accuracy of traditional financial theories for forecasting long-term effects or predicting a near collapse in any financial system is very low. Since the 2008 crisis, a tendency has grown in economic and financial markets to adopt complexity theory for monitoring market sustainability, predicting financial crunches, and financial forecasting. The financial and regulatory lexicon includes concepts such as tipping points, networks, contagion, feedback, and resilience. According to Battiston et al. (2016) in a study of a Dutch interbank network, it was found that a heterogeneous network model was able to detect prior alarming indications of the financial–economic crunch of 2008. Another study on contagion in financial networks by a network economist at ETH Zurich showed that default by an individual bank can have flow on effects in a global network of closely linked banks.

A complexity study incorporates complex effects that are caused by small changes in processes, and can help us to better understand sources of order and changes in any system. Unlike the usual focus on trends or dependence on linear results, a complexity study focuses on nonlinear systems in which the result of a small change can be disproportionately large. This type of study finds out relations between interesting dynamic generative factors in any system that occur at the micro level but have effects at the macro level.

## *Chaos and Complexity in Digital Banking Services*

In the current scenario it is desirable for the whole of India to be an active part of the digitization process. Continuous advancement of technology in the sector of computer science and information technology has brought about revolutionary changes in industries. Changes are observed globally in all aspects. But the ultimate objective of any company remains unchanged. Besides profit making, creating happy and satisfied customers is one of the major tasks of any business organization. There are deceitful people who use organizations' names with the intention of illegally acquiring money using digital media such as e-mail, telemarketing, messaging, etc.

The transformation from old conventional approaches to digital services is a familiar example of a complex system. A slight change/improvement in an existing process may lead to dramatic changes in the outcome or result of any occurrence. As per chaos theory, its impact might be an unpredictable change in the behavior of the consumer or the result may be as per the assumed prediction. According to a recent study by *The New Indian Express*, between 2015 and 2017, ICICI penalized 2236 employees for defrauding bank assets—the highest rate among 60 participant banks operating in India. After ICICI, SBI took the second highest place for penalizing employees for fraud. The study found that in 2015, the number of penalized employees of SBI was 698, but in 2016 that number was reduced by 51.57% to 338 and, following this trend, in 2017 the number was reduced to 25.73% of the 2016. However, the study showed a completely different picture for ICICI. In 2015, ICICI penalized 732 employees, but in 2016 the number was reduced to 14.61% of that figure. Following this statistical trend, the occurrence of fraud activity in 2017 should again have been reduced to about 14.61% of the previous figure, but in fact it was increased by 40.64%. Such chaotic situations sometimes affect consumers perception regarding banking services.

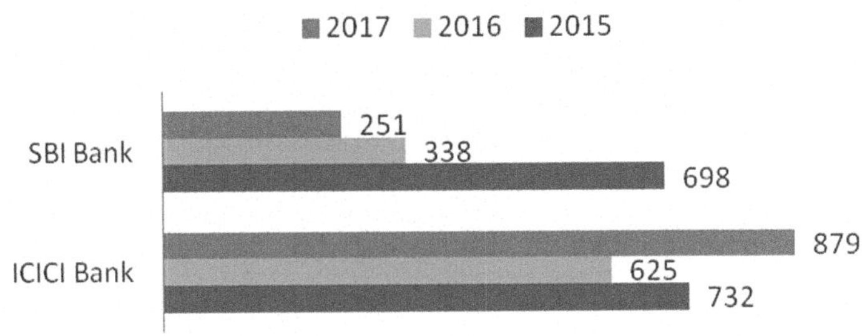


**Fig. 1** Action initiated against staff for defrauding bank assets (source: http://www.newindianexpress.com/)

The occurrence of defrauding at SBI is gradually being reduced year by year (see Fig. 1). In this study it was found that SBI provides better security or privacy of customer data, and that survey respondents feel the same, rating the securities and services factor higher for SBI than for ICICI (see Figs. 3 and 5). Securities create trust among customers, and this fact affects elements of the reliability factor, as shown in Fig. 3, where reliable information and error-fee services elements are rated highly by the respondents. It was also found that respondents considered SBI more reliable than ICICI, and that these two factors had indirect impacts on the trustworthiness and assured safety factors of the assurance dimension. Both of these figures were higher for SBI than for ICICI, and this showed a distinct difference between the services of two banks.

It is apparent that the digitization process, the new environment, and chaotic situations will create new demands on both customers and service providers. In the process of digitization, motivation plays a big role. A small motivation in sources of order can generate a dynamic and unpredictable result. The fixed deposit (FD) interest rate (per annum) for regular deposits at SBI is 6.85%, whereas the interest rate at ICICI is 7.00%. For senior citizens the annual interest rates are 7.35% at SBI and 7.50% at ICICI. There is a huge difference in the minimum investment amount, which is INR 10,000 for ICICI but only INR 1000 for SBI. The higher interest rates offered by SBI will attract more customers. The opportunity for making more money without extra effort will make many customers opt for this option. This also comes under complexity theory. SBI and ICICI have both introduced loyalty schemes for using their online services. For each transaction made using a debit/credit card, each bill payment, each money transfer, etc., the customer receives a reward in the form of cash back, deals and offers, discounts, gift vouchers, pay points, etc. These kinds of initiative always attract customers in all segments. Digitization also provides 360-degree feedback services for the customer from anywhere and at any time.

The SERVQUAL model used in this study was based on five linear core components to compare service quality performance with customer service quality needs, where the major factors are reliability, assurance, etc. Small incentives for customers to encourage them to use online services create a new set of happy customers, new customers, goodwill, a large market opportunity for the service organization, and new alignment with the existing system—just like butterfly effects, showing strong impacts on the linear factor of services model.

## Measures and Analysis

The primary data for this study were collected using a predetermined personally administered questionnaire. The arrangement of the questionnaires was done in a manner that covered the aforementioned segments in a quantitative and qualitative form. For the quantitative feedbacks, a five-point Likert scale from 1 to 5 was used, where 1 represented the lowest satisfaction level and 5 represented the highest satisfaction level.

## Data Analysis

Data were collected from several sources and a quantitative study was done. Primary data for the analysis were collected from people of different professions, ages, and genders. A set of questionnaires was used and a report was prepared from them to analyze the customers' expectations and perceptions of the banks' services (Table 1).

**Table 1** Demographics

| Variable | Measuring group | Frequency (*N*) | Percentage |
|---|---|---|---|
| Age | 18–25 years | 20 | 10 |
| | 25–35 years | 60 | 30 |
| | 36–45 years | 30 | 15 |
| | 46 years and above | 90 | 45 |
| | Total | 200 | 100 |
| Gender | Male | 114 | 57 |
| | Female | 86 | 43 |
| | Total | 200 | 100 |
| Income | INR 0–15,000 | 33 | 17 |
| | INR 15,001–30,000 | 63 | 32 |
| | INR 30,001–50,000 | 67 | 34 |
| | ≥INR 50,001 | 37 | 19 |
| | Total | 200 | 100 |
| Education | Secondary | 87 | 44 |
| | Higher secondary | 64 | 32 |
| | Undergraduate | 46 | 23 |
| | Postgraduate | 10 | 5 |
| | Total | 200 | 100 |

## *Interpretation*

Table 1 shows that out of 200 respondents, 10% of people were aged 18–25 years, 30% were aged 25–35 years, 15% were aged 36–45 years, and 45% were aged 46 years and above.

We can see that the largest proportion of respondents who did online shopping were aged 46 years and above. In addition, 30% of respondents aged 25–35 years utilized online shopping.

Another demographic factor that we considered was gender. Male respondents were more likely than female respondents to use online banking for transactions, online shopping, payments, etc. It could therefore be inferred that depending upon their preferences, they could get different and locally unavailable international products online, and that they found online shopping easy and time efficient.

Income was another factor that was considered here. The largest proportion of respondents (34%) were in the income category of INR 30,001–50,000, 32% were in the category of INR 15,001–30,000, 19% were in the category of ≥INR 50,001, and 17% were in the category of INR 0–15,000.

With regard to education, the largest proportion of respondents (44%) had secondary education, 32% had higher secondary education, 23% had undergraduate education, and 5% had postgraduate education.

## *Analysis and Discussion*

### Online Banking Customer Responses: The Gap Between Expectations and Perceptions

SBI

The null hypothesis ($H_0$) was that factors of online banking did not positively influence customer satisfaction with SBI (see Tables 2 and 3).

**Table 2** Ranks

| Rank | | *N* | Mean rank | Sum of ranks |
|---|---|---|---|---|
| VAR00007 – VAR00006 | Negative ranks | 43[a] | 82.64 | 3553.50 |
| | Positive ranks | 127[b] | 86.47 | 10,981.50 |
| | Ties | 30[c] | | |
| | Total | 200 | | |

[a]VAR00002 < VAR00001
[b]VAR00002 > VAR00001
[c]VAR00002 = VAR00001

Table 3 shows that the $Z$ value was −5.790 with a $p$ value of 0.000, which was lower than the table value of 1.96 at a 5% level of significance, so there was not enough evidence to accept the null hypothesis.

## ICICI

The null hypothesis ($H_0$) was that factors of online banking did not positively influence customer satisfaction with ICICI (see Tables 4 and 5).

Table 5 shows that the $Z$ value was −6.182 with a $p$ value of 0.000, which was lower than the table value of 1.96 at a 5% level of significance, so there was not enough evidence to accept the null hypothesis.

**Table 3** Test statistics

| Test statistic[a] | VAR00002 – VAR00001 |
|---|---|
| $Z$ | −5.790[b] |
| Asymptotic significance (2-tailed $p$ value) | 0.000 |

[a]Wilcoxon signed rank test
[b]Based on negative ranks

**Table 4** Ranks

| Rank | | $N$ | Mean rank | Sum of ranks |
|---|---|---|---|---|
| VAR00002 – VAR00001 | Negative ranks | 37[a] | 77.32 | 2861.00 |
| | Positive ranks | 124[b] | 82.10 | 10,180.00 |
| | Ties | 8[c] | | |
| | Total | 169 | | |

[a] VAR00002 < VAR00001
[b] VAR00002 > VAR00001
[c] VAR00002 = VAR00001

**Table 5** Test statistics

| Test statistic[a] | VAR00002 – VAR00001 |
|---|---|
| $Z$ | −6.182[b] |
| Asymptotic significance (2-tailed $p$ value) | 0.000 |

[a]Wilcoxon signed rank test
[b]Based on negative ranks

## *Comparative Analysis*

### Reliability

Figure 2 shows that in the case of SBI, providing reliable information (3.723) was rated higher, followed by online reliability (3.343), error-free services (3.435), and good performance right from the first time (3.345). In the case of ICICI, reliable information (3.423) was rated lower, followed by online reliability (3.647), good performance right from the first time (3.322), and error-free services (3.433).

### Empathy

Figure 3 shows that in the case of ICICI, the location factor (3.566) was rated higher, followed by understanding of the needs and wants of the customer (3.465), sincerity concerns (3.502), and willingness to help (3.335). In the case of SBI, understanding of the needs and wants of the customer (3.342) was rated lower, followed by the location factor (3.546), sincerity concerns (3.450), and willingness to help (3.32).

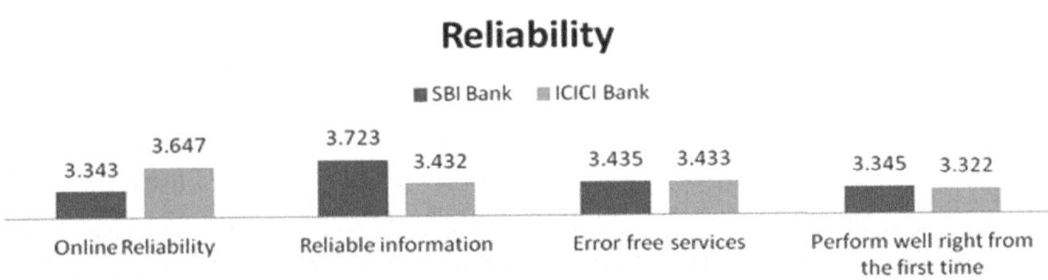


**Fig. 2** Customer perceptions of the reliability dimension of a public sector bank (SBI) and a private sector bank (ICICI)

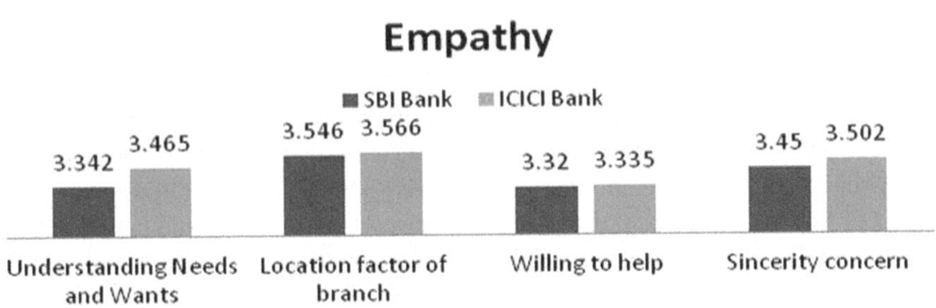


**Fig. 3** Customer perceptions of the empathy dimension of a public sector bank (SBI) and a private sector bank (ICICI)

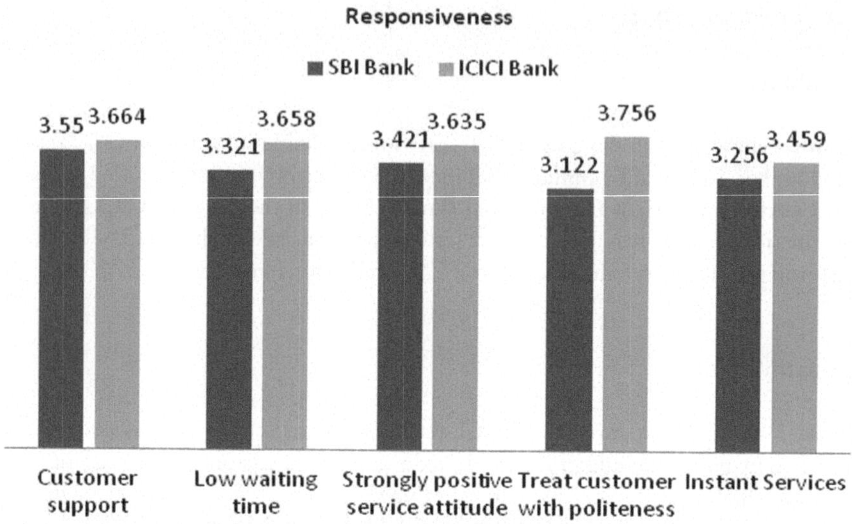


**Fig. 4** Customer perceptions of the responsiveness dimension of a public sector bank (SBI) and a private sector bank (ICICI)

## Responsiveness

Figure 4 shows that in the case of ICICI, a strongly positive service attitude (3.637) was rated higher, followed by customer support (3.664), treating customers with politeness (3.756), instant service (3.459), and short waiting times (3.658). In the case of SBI, customer service (3.55) was rated lower, followed by short waiting times (3.321), treating customers with politeness (3.122), instant service (3.256), and a strongly positive service attitude (3.421).

## Securities and Services

Figure 5 shows that ICICI was rated higher with regard to fraudulent transactions (3.044), followed by scams or fraud calls (3.145), a user-friendly interface (3.867), and association with different shopping portals (3.86). SBI was rated lower with regard to fraudulent transactions (3.034), followed by scams or fraud calls (3.056), a user-friendly interface (3.345), and association with different shopping portals (3.868).

## Assurance

Figure 6 shows that in the case of SBI, trustworthiness was rated higher (3.844), followed by communication and interaction with customers to understand their

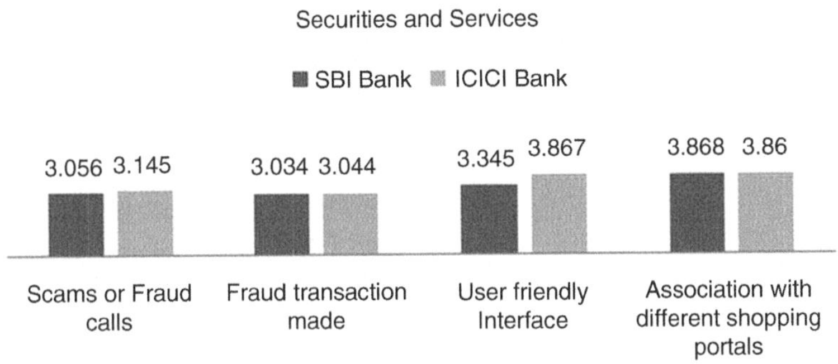


**Fig. 5** Customer perceptions of the securities and services dimension of a public sector bank (SBI) and a private sector bank (ICICI)

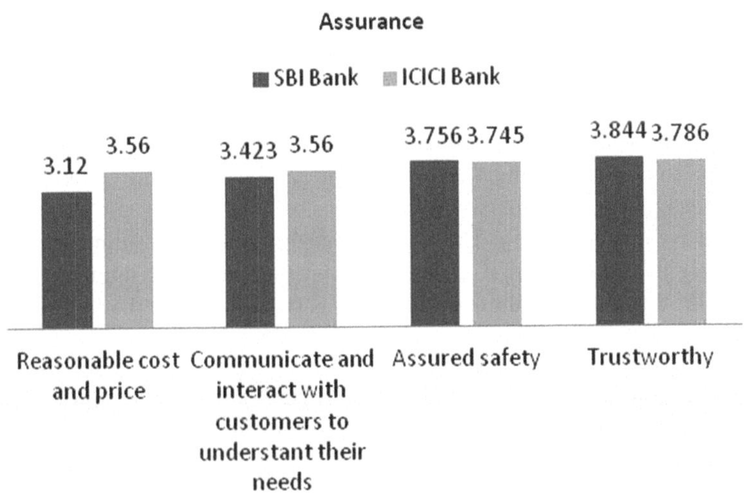


**Fig. 6** Customer perceptions of the assurance dimension of a public sector bank (SBI) and a private sector bank (ICICI)

needs (3.423), reasonable cost and price (3.12), and assurance of safety (3.756). In the case of ICICI, reasonable cost and price was rated higher (3.56), followed by communication and interaction with customers to understand their needs (3.56), assurance of safety (3.745), and trustworthiness (3.786).

## Comparative Analysis

A comparative analysis of the customer ratings of the two banks (SBI and ICICI) is shown in Table 6.

**Table 6** Comparative analysis of customer ratings of SBI and ICICI

| Factor | Higher customer rating | |
|---|---|---|
| | SBI | ICICI |
| Reliability | ✓ | |
| Empathy | | ✓ |
| Responsiveness | | ✓ |
| Securities and services | ✓ | |
| Assurance of safety | ✓ | |

## Findings of the Study

This study revealed that among all SERVQUAL dimensions, ICICI was rated more highly by customers except in reliability, securities and services, and assurance of safety, which were rated more highly at SBI. Among all reliability parameters, the reliability of information provided by SBI was rated highly by customers. The study also showed that in the error-free service factor of the reliability dimension, SBI and ICICI were to some extent rated similarly by the respondents. ICICI was rated highly by customers for the bank location factor, while SBI was rated similarly in willingness to help customers. Considering the two types of bank, it was clear that ICICI's treatment of customers with politeness was the highest-rated parameter in the responsiveness dimension. The study also showed that the waiting times at SBI were longer than those at ICICI, possibly due to internet link failures or lack of expertise. In the assurance dimension, SBI was rated highly in the trustworthiness parameter, as customers considered government organizations more reliable than private organizations. The study also showed that according to the customers, SBI should provide better services and securities, such as better and more user-friendly software, data securities, customer care personnel, etc. In attracting customers and achieving customer loyalty, service quality played an essential role. Different factors of service quality such as an easily accessible and user-friendly interface, high security, quick and error-free services, quick responses to customer enquiries, etc., had a strong impact on customer satisfaction. It was found that consumer perceptions of reliability, securities and services, and assurance were comparatively higher at SBI than at ICICI.

## Recommendations

1. SBI should have a compact customer relationship management (CRM) policy to point out the worth of customer expectations, so as to cross-provide proper services.
2. The information technology skill sets of staff need to be upgraded so as to make them more comfortable with the most up-to-date technology. This will boost their comfort level when educating customers to use the technology in their day-to-day transactions.

3. To develop the reliability factor, SBI should try not to appear as money mongers and should be clear and precise about their terms and conditions. They should not have hidden meanings and hidden terms in their documents.
4. Public sector banks should take the right action to avoid customers moving to private sector banks for better services and more attractive schemes than those offered by the public sector banks.
5. At present, Reserve Bank of India (RBI) regulations do not require banks to inform them or customers of security breaches in their ATM networks. However, banks whose ATMs have been infected should come forward publicly if their ATMs become infected, in order to win customers' confidence.
6. Banks should promote easy-to-use apps to make customers enjoy the benefits of mobile banking. For example, SBI Quick is a self-care app and can be used by all customers of the bank. By using the SBI Quick app, a customer can switch off the banking functionalities that he/she does not require at any given point in time. All that the customer needs to do during a security breach of his/her ATM card is enter the last four digits of the card number to use the "ATM card switch on/off" facility.
7. Customers do not like unsolicited calls regarding banking products. Banks should try to keep customer data confidential and customize company products to suit customer needs. Banks must not sell customer data.
8. Banks wrongly assume that customer service is a cost. Consequently, they overinvest in digital technologies and channels to create less human interaction. However, a bank should try to cover both digital and traditional channels to have loyal customers.
9. The focus should be on creating experiences rather than making digital interactions, and making the transition of digital to human interactions, and vice versa, less cumbersome. Banks should try to bring back their workforce to bring more smiles to customers' faces.

## Conclusion

In the current scenario, carrying a lot of cash in hand is very risky and can make consumers vulnerable to life-threatening scenarios. For shopping, bill payments, business-to-business transactions, or e-commerce processes, the whole procedure has been streamlined by online transactions. To ensure safe and easy online transactions, payment gateway software is now used by merchants.

One of the most important tools in any project is a feasibility study. In a flow chart, it is possible to represent how services evolve to reach the goal through business to physical attribute. The whole concept of e-commerce and selling products from websites requires use of a payment gateway to accept payments from customers. A few well-known payment gateways are InstaMojo, CCAvenue, EBS, and PayPal. The whole process of installing a payment gateway involves many challenges such as lots of documentation, huge setup fees, verification, and technical

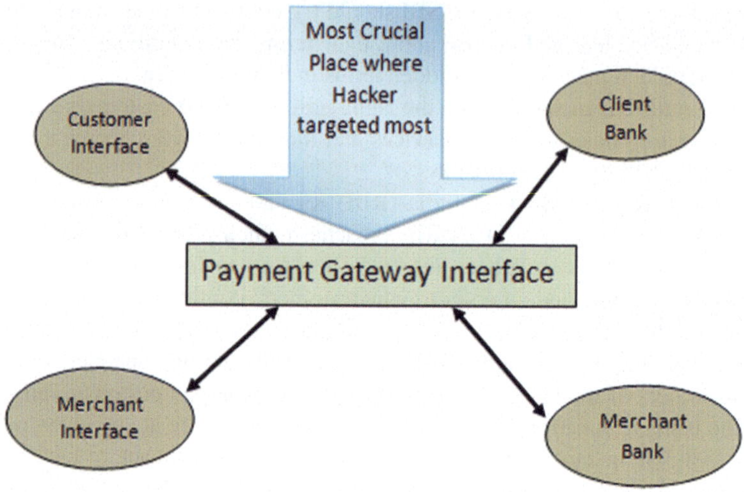


**Fig. 7** The payment gateway interface

knowledge. However, the process is now becoming easier. The five most important things that have to be considered when installing/designing a payment gateway are as follows:

1. The cost of the payment gateway (setup fee, annual fee, term deposit receipts (TDRs), etc.)
2. The setup time
3. The user interface
4. The payment options (debit/credit cards, unified payment interface (UPI), mobile wallets, net banking, etc.)
5. Customer care assistance

Despite its numerous benefits—such as making more profit, better customer service, quick and easy transactions, and 24/7 accessibility and supports—some serious issues of using "plastic money" cannot be ignored. There is no such thing as "absolute security." With enough time, resources, and knowledge, it is not impossible to bypass any and all security measures. The job of service providers is to improve their guidelines by increasing the cost of hacking to a point where the expense does not justify the reward. A visual representation of hacking taking place in a transaction is shown in Fig. 7.

Digital literacy includes the abilities to use, understand, and evaluate digital platforms through various devices (such as PCs, smart phones, etc.). In a growing country such as India, with more than 650,000 villages and 250,000 *panchayats* (village councils) representing 3 million *panchayat* members, nearly half of the population (approximately 40%) is below the poverty line, the illiteracy rate is more than 25–30%, and digital literacy is almost nonexistent. About 90% of India's population lives in rural areas or off-the-map villages. Most of them are remote and too isolated to benefit from their country's digital economic progress. The Government of India has taken initiatives to overcome this situation and has introduced a National Digital Literacy Mission (NDLM) program, aiming to empower at least one person per household by the year 2020.

## Survey of literature

According to Ray and Ghosh (2017), online banking has several smart facilities such as online account tracking, time-saving services, low-transaction cost, and 24 × 7 accessibility. On the other hand, it has a few drawbacks that cannot be ignored. These drawbacks include inadequate broadband penetration, fraud and risky group action, online threats/scams, security risk, and power interruption or failure. All these disadvantages lead to dissatisfactory services and unhappy customers.

Avkiran (1999) detailed the significance of using service quality instrument developed in Australia to measure service quality in retail banking as perceived by customers (BANKSERV). It was adapted from SERVQUAL to specifically suit the Australian banking industry. The key reason for customers to opt banks for investment is because of the dependability and reputation of banks. Banks always promise customers a high level of security during transactions. Banking service can increase customers' confidence and trust if employees are able to provide appropriate service to every customer. For instance, understanding the needs of each individual customer, such as knowing the customers' expected retirement age, annual income, and hobbies, is needed to help provide a good match of insurance and fund products for customers. Assurance is the knowledge and courtesy of employees and their ability to inspire trust and confidence. Bank commitments are important, as customers may save a large sum of money in banks. For sophisticated products such as insurance, funds, and margins, employees must provide a clear explanation of each product to customers, so that customers can feel confident about the services provided by banks. The final dimension is sympathy, which represents the individualized attention that firms provide to its customers. Service quality has a positive influence on customer satisfaction (Yee et al., 2010). Customer satisfaction is defined as the attitude resulting from what customers believe should happen (expectations) compared to what they believe did happen (performance perception) (Neal, 1998). Satisfaction reinforces quality perception and drives repeat purchases.

Zaim et al. (2010) found that corporeality, dependability, and sympathy are important for customer satisfaction, but Mengi (2009) found that responsiveness and assurance are more important. Siddiqi (2010) examined the pertinence of service quality of retail industry in Bangladesh and identified that service quality is positively correlated with customer satisfaction. According to them, sympathy had the highest positive correlation with customer satisfaction, followed by assurance and corporeality. On the other hand, Lo et al. (2010) indicated that sympathy and assurance had the highest influence on customer satisfaction in the Malaysian retail industry. Arasli et al. (2005) found that dependability had the highest impact on customer satisfaction. A number of studies have revealed the dimensions of service quality as the antecedents of customer satisfaction. Kotler and Armstrong (2012) stressed that satisfaction is the point-of-sale (POS) purchase analysis of merchandise or services taking into consideration the expectations. Researchers are divided over the antecedents of service quality and satisfaction. Some researchers believe service quality ends up in satisfaction, while others think otherwise (Ting 2004). Lee et al. (2000), Gilbert and Veloutsou (2006), Sulieman (2011), and Buttle (1996) emphasize service quality leads to customer satisfaction. Most researchers believe that a high service quality should be delivered by the service supplier to attain a high customer satisfaction as service quality is generally considered an antecedent of customer satisfaction. Very simply, the likelihood of customer satisfaction can be greatly increased by improving service quality. Wang et al. (2002) found in Chinese banks that dependability was one of the key drivers of the merchandise quality followed by corporeality. They also argued in favor of improving service quality and product quality to build and enhance company's reputation.

## References


Anderson, J. C., Narus, J. A., & Van Rossum, W. (2006). Customer value propositions in business markets. *Harvard Business Review, 84*(3), 91–99.

Arasli, H., Smachi, S., & Katircioglu, S. T. (2005). Customer service quality in the Derek Cypriot banking industry. *Managing Service Quality, 15*(1), 41–56.

Badri, M. (2003). *The measurement of IT centers service quality at the institutions of higher education in the UAE.* Proceedings of the Baptist University Fourth Annual Research Conference at the UAE University, Al-Ain, 27–29 April.

Bardhan, I. R., Demirkan, H., Kannan, P. K., Kauffman, R. J., & Sougstad, R. (2010). An interdisciplinary perspective on IT services management and service science. *Journal of Management Information Systems, 26*(4), 13–64.

Battiston, S., Farmer, J., Flache, D., Garlaschelli, A., Haldane, D., Heesterbeek, A. G., Hommes, H., Jaeger, C., May, C., & R. Scheffer M. (2016). Complexity theory and financial regulation. *Science, 351*(6275), 81.

Bitner, M. J., Faranda, W. T., Hubbert, A. R., & Zeithaml, V. A. (1997). Customer contributions and roles in service delivery. *International Journal of Service Industry Management, 8*(3), 193–205.

Chandler, J. D., & Vargo, S. L. (2011). Contextualization and value-in-context: How context frames exchange. *Marketing Theory, 11*(1), 35–49.

Chatterjee, P. (2010). Multiple-channel and cross-channel shopping behavior: Role of consumer shopping orientations. *Marketing Intelligence & Planning, 28*(1), 9–24.

Cronin, J. J., & Taylor, S. A. (1992). Measuring service quality: A reexamination and extension. *Journal of Marketing, 56*(3), 55.

Cui, C. C., Lewis, B. R., & Won, P. (2003). Service quality measurement in the banking sector in South Korea. *International Journal of Bank Marketing, 21*(4), 191–201.

Curry, A., & Sinclair, E. (2002). Assessing the quality of physiotherapy services using Servqual. *International Journal of Health Care Quality Assurance, 15*(5), 197–205.

Etgar, M. A. (2007). Descriptive model of the consumer co-production process. *Journal of the Academy of Marketing Science, 36*(1), 97–108.

Evans, J., & Lindsay, W. (2008). *Leadership and strategic planning, the management and control of quality (E.D.).* Mason/Thomson: South-Western.

Faranda, W. T. (1994). *Customer participation in service production: An empirical assessment of the influence of realistic service previews* (Doctoral dissertation). Arizona State University.

Ganesan-Lim, C., Russell-Bennett, R., & Dagger, D. (2008). The impact of service contact type and demographic characteristics on service quality perceptions. *Journal of Services Marketing, 22*(7), 550–561.

Ghost, F. S., & Gnanadhas, E. (2011). Impact of service quality in commercial banks on the customers satisfaction: An empirical study. *International Journal of Multidisciplinary Research, 1*(6), 19–37.

Grönroos, C. (1984). A service quality model and its marketing implications. *European Journal of Marketing, 18*(4), 36–44.

Grönroos, C. (2008). Service logic revisited: Who creates value? And who co-creates? *European Business Review, 20*(4), 298–314.

Gummesson, E. (2008). Quality, service-dominant logic and many-to-many marketing. *The TQM Journal, 20*(2), 143–153.

Hayes, N. (1997). *Doing qualitative analysis in psychology*. Hove: Psychology Press.

Haywood-Farmer, J. (1988). A conceptual model of service quality. *International Journal of Operations and Production Management, 8*(6), 19–29.

Honebein, P., & Cammarano, R. (2005). *Creating do-it-yourself customers*. Thomson Higher Education: Ohio.

Ingle, A., & Pardeshi, R. (2012). Internet banking in India: Challenges and opportunities. *IBMRD's Journal of Management and Research, 1*(1), 13–18.

Jain, V., Gupta, S., & Jain, S. (2012). Customer perception on service quality in banking sector: With special reference to Indian private banks in Moradabad region. *International Journal of Research in Finance and Marketing, 2*(2), 597–610.

Johnson, M. W., Christensen, C. M., & Kagermann, H. (2008). Reinventing your business model. *Harvard Business Review, 86*(12), 50–59.

Kang, G.-D., & James, J. (2004). Service quality dimensions: An examination of Grönroos's service quality model. *Managing Service Quality, 14*(4), 266–277.

Kelley, S., Skinner, S., & Donnelly, J. (1992). Organizational socialization of service customers. *Journal of Business Research, 25*, 197–214.

Kemal Avkiran, N. (1994). Developing an instrument to measure customer service quality in branch banking. *International Journal of Bank Marketing, 12*(6), 10–18.

Kumar, S. A., Mani, B. T., Mahalingam, S., & Vanjikovan, M. (2010). Influence of service quality on attitudinal loyalty in private retail banking: An empirical study. *IUP Journal of Management Research, 9*(4), 21–38.

Kuzgun, E., & Asugman, G. (2015). Value in services—A service dominant logic perspective. *Procedia—Social and Behavioral Sciences, 207*, 242–251.

Lusch, R. F., & Nambisan, S. (2015). Service innovation: A service-dominant logic perspective. *MIS Quarterly, 39*(1), 155–175.

Meng, J., Summey, J., Herndon, N., & Kwong, K. (2009). Some retail service quality expectations of Chinese shoppers. *International Journal of Market Research, 51*(6), 773–796.

Meuter, M., Ostrom, A., Roundtree, R., & Bitner, M. (2000). Self-services technologies: Understanding customer satisfaction with technology-based service encounters. *Journal of Marketing, 64*, 50–64.

Munusamy, J., Chelliah, S., & WaiMun, H. (2010). Service quality delivery and its impact on cutomer satisfaction in the banking sector in Malaysia. *International Journal of Innovation, Management and Technology, 1*(4), 398.

Negi, R. (2009). Determining customer satisfaction through perceived service quality: A study of Ethiopian mobile users. *International Journal of Mobile Marketing, 4*(1), 31–38.

Parasuraman, A., Zeithaml, V. A., & Berry, L. L. (1988). SERVQUAL: A multiple-item scale for measuring consumer perceptions of service quality. *Journal of Retailing, 64*(1), 12–40.

Rai, R. (2009). Service quality gap analysis in Indian bank: An empirical study. *Paradigm, 13*(2), 29–35.

Ray, N. (2016a). Impact of internet service quality (IS-QUAL) on client satisfaction. In *Handbook of research on promotional strategies and consumer influence in the service sector* (pp. 371–388). Hershey, PA: USA Business Science Reference.

Ray, N. (2016b). Impact of internet service quality (IS-QUAL) on client satisfaction: Case from Indian banking services. In U. Singh, R. Kumar, & N. Ray (Eds.), *Hand book of research on promotional strategies and consumer influence in the service sector*. Hershey, PA: IGI-Global.

Ray, N. (2017). Expectation and perception of internet banking service quality of select Indian private and public sector banks. In *Online banking security measures and data protection* (pp. 58–68). Hershey, PA: IGI Global.

Ray, N., & Bhattacharya, A. (2017). Examination of service quality gaps: Evidence from State Bank of India. In *Management and industrial engineering green and lean management* (pp. 139–148). Cham: Springer.

Ray, N., & Ghosh, D. (2014). Internet service quality (I-SQ) dimensions and their impact on consumer satisfaction: Case from banking industry. *Asian Journal of Research in Banking and Finance, 4*(8), 212–221.

Ray, N., & Ghosh, D. (2017). Online banking boon or bane? *Asian Journal of Research in Banking and Finance, 7*(5), 179–194.

Ray, N., Sen, K., & Ghosh, T. N. (2016). Examination of internet banking customer perception of service quality: Evidence from banking industry. In N. Ray (Ed.), *Business infrastructure for sustainability in developing economies*. Hershey, PA: IGI-Global.

Rintamäki, T., Kuusela, H., & Mitronen, L. (2007). Identifying competitive customer value propositions in retailing. *Managing Service Quality, 17*(6), 621–634.

Rodie, A., & Kleine, S. (2000). Customer participation in service production and deliver. In *Handbook of services marketing and management*. London: Sage Publications, Inc..

Schneider, B., & Bowen, D. E. (1995). *Winning the service game*. Boston: Harvard Business School Press.

Siu, N., & Cheung, J. (2001). A measure of retail service quality. *Marketing Intelligence & Planning, 19*(2), 88–96.

Slu, N. Y. M., & Mou, J. C. W. (2003). *A study of service quality in Internet Banking* (BRC Working Papers). Hong Kong.

Vargo, S. L., & Lusch, R. F. (2008). Service-dominant logic: Continuing the evolution. *Journal of the Academy of Marketing Science, 36*(2), 1–10.

Wang, Y., & Hing-Po, L. (2002). Service quality, customer satisfaction and behavior intentions. *Info, 4*(6), 50–60.

Wilson, A., Zeithaml, V. A., Bitner, M. J., & Gremler, D. D. (2008). *Services marketing*. Maidenhead: McGraw-Hill Education.

Yaghi Derar, S. (2010). *A customised scale for measuring retail service quality in a college shop: A context specific approach* (Thesis). Newcastle Business School.

Yousafzai, S. Y., Pallister, J. G., & Foxall, G. R. (2003). A proposed model of e-trust for electronic banking. *Technovation, 23*, 847–860.

## Websites

https://www.accenture.com/_acnmedia/PDF-10/Accenture-Strategy-GCPR-Digital-Profitabilty.pdf. Accessed 1 Dec 2018.

http://defindia.org/national-digital-literacy-mission/. Accessed 10 Jan 2019.

http://economictimes.indiatimes.com/. Accessed 1 Dec 2018.

http://www.firstpost.com/politics/pm-modi-encourages-upi-in-his-digital-push-heres-all-you-need-to-know-about-the-app-3126908.html.

http://www.grantthornton.in/.

http://www.ibef.org/. Accessed 1 Dec 2018.

http://www.icsi.org.in/banking/. Accessed 1 Dec 2018.
https://www.jio.com/en-in/apps/jio-money.
https://m.rbi.org.in/Scripts/PublicationVisionDocuments.aspx?Id=842. Accessed 10 Jan 2019.
http://www.newindianexpress.com/business/2018/dec/26/icici-tops-list-of-banks-penalising-staff-for-fraud-1916560.html. Accessed 21 Feb 2019.
http://www.npci.org.in/. Accessed 10 Jan 2019.
http://www.pmjdy.gov.in/account. Accessed 10 Jan 2019.
https://www.sbi.co.in/portal/web/personal-banking/mobile-banking. Accessed 10 Jan 2019.
https://sbiupiapp.com/. Accessed 10 Jan 2019.
http://shodhganga.inflibnet.ac.in/bitstream/10603/19602/12/12_chapter3.pdf. Accessed 10 Jan 2019.

# Insightful Leadership

**Halime Güngör**

**Abstract** From a managerial point of view, the concept of organizational insight is one of the important concepts in terms of evaluating the current situation by taking lessons from the history of the organization and shaping provisions for the future through conscious planning and organizational vision. It can be said that organizational insight in organizations is high and organizational communication is strong when organizational consciousness is high. Considering that organizations with high organizational awareness are also intelligent organizations, organizational intelligence is high in organizations with high insight. It is a fact that string theory, which is the newest theory in the contemporary world and is mostly concentrated in the physical world, also influences leadership. In this chapter, we define insightful leaders that bring order to the organizations they lead and the abilities of insightful leaders in terms of nonlinearity and organizational intelligence.

**Keywords** Leadership · Insight · Organizational insight · Insightful leadership · String theory

## Organizational Insight

From a managerial point of view, the concept of organizational insight is one of the important concepts in terms of evaluating the current situation by taking lessons from the history of the organization and shaping future provisions for the purposes of conscious planning and organizational vision. Erçetin et al. (2018) defined organizational insight as the ability to foresee chaotic situations that may emerge in the organizational environment and the future of the organization with experience based on the history of the organization, with strong organizational communication,

H. Güngör (✉)
Hacettepe University, Ankara, Turkey



Ş. Ş. ERÇETİN, Ş. N. AÇIKALIN (eds.), *Chaos, Complexity and Leadership 2018*,
Springer Proceedings in Complexity, https://doi.org/10.1007/978-3-030-27672-0_14

and with a high level of consciousness. The characteristics of organizations with high insight are as follows:

- Organizations with quantum entanglement consciousness
- Organizations with high organizational awareness and strong communication
- Organizations with intelligence
- Organizations in which the solutions to problems that are experienced are based on organizational consciousness and the essence of the organization's experience, and in which the solution is reached with certainty
- Organizations that have high motivation and performance to solve their problems and that find clear and correct solutions to their problems
- Organizations that can make objective and clear predictions for the future
- Organizations that have high coping skills in the presence of chaos

In their research, Vicentini et al. (2017) focused on organizational insight as a pivotal element in understanding the processes required to realize interrelationships between businesses. Nikula (1999) discussed organizational insight in the context of individuals working in the organization, focusing on organizational goals rather than the successes of their own departments, and serving the vision of the organization.

As a result, it can be said that organizational insight is high and organizational communication is strong when organizational consciousness is high. Organizations with high organizational awareness are also intelligent organizations, and organizational intelligence is high in organizations with high insight. Organizational intelligence has been defined as the sum and the utilization of the competencies that allow the organization to maintain its dynamism (Erçetin et al. 2007). The competencies in question are (1) rapid action and reaction, (2) quick adaptation to changes, (3) flexibility in function, (4) sensitivity and predictability, (5) open-mindedness, (6) use of imagination, and (7) innovativeness (Erçetin 2000, 2001). In this context, it can be said that organizations with high insight are rapid in action and reaction, quick in adapting to changes, flexible in function, sensitive and predictable, open-minded, imaginative, and innovative organizations.

In terms of quantum entanglement, intelligent organizations have high organizational insight, as systems that are entangled according to the quantum entanglement principle have an interaction at the level of consciousness regardless of time and place. It can be said that the existence of quantum entanglement in the organizational environment provides organizational insight while reinforcing organizational communication and consciousness. In this sense, organizational intelligence and quantum entanglement can be mentioned as factors contributing to the formation and development of organizational insight. In summary, organizations with high insight are organizations that have:

- Strong organizational communication
- High organizational awareness
- Intelligence
- Consciousness of quantum entanglement
- High ability to cope with chaotic situations

## Insightful Leadership

From the perspective of leadership in organizational insight, it can be said that insightful leaders are intelligent, can internalize the principle of quantum entanglement, can communicate effectively, have high skills to cope with chaos, and have the skills to motivate others. According to Erçetin et al. (2018), individuals who are intelligent can easily reach an insightful solution to their problems, have high motivation and performance to solve their problems, achieve clear and error-free solutions, and make predictions for the future. In this context, when studies on insightful leaders in the literature are examined, according to Fergusson (2018) insightful leadership is a personal journey to becoming a leader who is able to:

- Overcome personal insecurities about effectiveness and value in leadership
- Strengthen the leadership and strengthen aspects that add value to the organization
- Solve problems with defensive areas and people
- Eliminate discouragement and discouraging interactions
- Demonstrate strength and being deserving of their position without being selfish or manipulative
- Reprogram beliefs according to the rules that match the values that the leader creates
- Empower emotional awareness and strength as a manager
- Be the desired leader, role model, and coach

Pearl Zhu (2013) explained "three I's" in insightful leaders as follows:

1. *Intelligence*: Although today's management is trained and equipped with rational intelligence (IQ) and emotional intelligence (EQ), it is not sufficiently capable of dealing with the uncertainty and complexity of our age. It is necessary to have multiple-intelligence understanding to be an insightful leader.
2. *Inclusion*: Even the wisest person must have a collective understanding of management and be inclusive.
3. *Influence*: Leadership is an influence. Insightful leaders can have a profound effect because unique observations and open judgment not only shape the mindset but also appeal to the heart.

In another analysis, the attitudes of insightful leaders were described as follows (Economy 2015):

- Insightful leaders are always open minded and open hearted.
- Their past failures do not affect their prospects of success.
- They never stop fighting for what they believe in.
- They are not late unless there is something that is unpredictable and absolutely unavoidable.
- They act as quickly and professionally as possible and do not enter into polemics.

- Their pride does not prevent them from being receptive to the ideas of others and allowing others to take the lead.
- They never give up defending their organization, always stand up for what they believe is true, and back up their team members when necessary.
- They always seek the best for themselves, their teams, and their organizations.
- They never miss the opportunity to learn and develop from their mistakes.
- They find various solutions to overcome failures or shortcomings within themselves, their teams, or their organizations.
- They are not fixated on past mistakes or wrong steps; instead, they go on and look forward to future success.
- They never submit to norms; instead, they enjoy the freedom to follow their individuality and their place in the world.
- They never over-praise themselves and are not arrogant with regard to their own abilities; instead, they continue to learn and do not believe they know everything.
- They are never afraid of uncertainty; they address whatever problems arise and they help their teams succeed regardless of what obstacles may arise.
- They never devote all of their faith and trust to a single person.
- They never spread gossip.
- They never allow anger or jealousy to enter their minds or lives.
- They have multiple plans in place and do not rely on luck.
- They are not irresponsible regardless of the situation.
- They do not take disproportionate risks with their powers; rather, they make very careful and well-considered decisions.
- They never miss an opportunity to help their organizations save money.
- They never give a green light for anything they do not fully believe in; they are not afraid of saying "no" when necessary.
- They never miss opportunities for growth and development of their own knowledge.
- They look at all of the details of their work and are always careful in their jobs.

When organizations with high insight are managed by insightful leaders or when the insight of organizations, led by insightful leaders, increases, organizational motivation, organizational communication, organizational intelligence, and quantum entanglement will ensue. In this context, considering the characteristics of insightful leaders:

- They act with collective consciousness and superego.
- They always uphold and defend their organization's interests even if they are contrary to their own personal interests.
- They have strong intuition.
- They analyze threats, risks, and hazards.
- They have high ability to cope with chaotic situations.
- They do the right thing for the realization of organizational goals.
- They have high motivation and performance toward solving their problems.
- The solutions they find for problems are clear and error free.

- Their improved ability motivates their followers.
- They use their cleverness and mental skills to move the organization forward.
- They use effective communication skills.
- They are proactive.
- Their predictions for the future are consistent and clear.
- They are leaders with high awareness of quantum entanglement.

## Conclusion

In this chapter, the concept of insightful leadership has been discussed in terms of organizational intelligence and string theory. A different perspective has been put forward in an attempt to contribute to the development of the concept of insightful leadership. Explaining insightful leadership through practical research and leadership analysis and discussing this issue in future studies will create opportunities to contribute the development of this subject.

## References

Economy, P. (2015). *24 things exceptionally insightful leaders will never do*. Inc. https://www.inc.com/peter-economy/24-things-exceptionally-insightful-leaders-will-never-do.html. Accessed 1 Sept 2019.

Erçetin, Ş. Ş. (2000). Örgütsel zeka (Organizational intelligence). *Kuram ve Uygulamada Eğitim Yönetimi, 24*, 509–526.

Erçetin, Ş. Ş. (2001). *Örgütsel zekâ (Organizational intelligence)*. Ankara: Nobel Yayın Dağıtım.

Erçetin, Ş. Ş., Açıkalın, Ş. N., Güngör, H., & Potas, N. (2018). Chapter 22: Reflection of quantum entanglement principle to organization theory "organizational insight". In Ş. Ş. Erçetin (Ed.), *Chaos, complexity and leadership 2016* (pp. 309–315). Cham: Springer. https://doi.org/10.1007/978-3-319-64554-4_22.

Erçetin, Ş. Ş., Çetin, B., & Potas, N. (2007). Multi-dimensional organizational intelligence scale "muldimorins". *World Applied Sciences Journal, 2*(3), 151–157.

Fergusson, C. (2018). *The insightful leader: Find your leadership superpowers, crush limiting beliefs, and abolish self-sabotaging behaviors*. Santa Barbara: Praeger.

Nikula, R. E. (1999). Organizational and technological insight as important factors for successful implementation of IT. *Proceedings of AMIA Symposium*, 585–588. https://www.ncbi.nlm.nih.gov/pmc/articles/PMC2232858/.

Vicentini, F., Brunetta, F., & Capo, F. (2017). Building organizational insight: Strategy and organization. In Ü. Hacıoğlu, H. Dinçer, & N. Alayoğlu (Eds.), *Global business strategies in crisis. Contributions to management science*. Cham: Springer.

Zhu, P. (2013). Three "I"s in insightful leaders. *Future of CIO*. http://futureofcio.blogspot.com/2013/01/three-is-in-insightful-leaders.html. Accessed 1 Sept 2019.

# The Role of Trust in Principals in Readiness for Change Within Schools

**Hanifi Parlar, Mahmut Polatcan, and Ramazan Cansoy**

**Abstract** Readiness for change is an important factor in the occurrence of positive changes in schools, which are complex social systems. The main actors of change in schools are principals. However, teachers' trust in principals facilitates the preparation of the change process. In this context, this study aimed to examine the predictors of teachers' trust in principals on readiness for change in schools. A relational survey model was used in this study, and the sample consisted of 328 teachers. As a result of the study, a moderate positive correlation was found between trust in principals and readiness for change. According to the results of regression analysis, teachers' trust in principals influenced readiness for change. This chapter interprets the results of this study in the context of the school system's features and the related literature.

**Keywords** Readiness for change · Trust in principals · Principals

## Introduction

Under the influence of globalization, the intensive interactions between societies and organizations create new developments in political, technological, and economic areas, as well as in sociocultural life (Erdoğan 2015; Madsen et al. 2005). The process is not as simple as it seems when seen from outside the organizational systems; instead, it is quite complicated. Chaotic situations and nonorganizational interactions add a new dimension to existing complexities. However, it is possible to see the system when the system components are considered together. In addition, a small change in the system leads to large changes based on the logic of mutual causality (Açıkgöz and Edin 2018; Morgan 1997; Özgenç 2008). Hence, it can

H. Parlar (✉)
İstanbul Commerce University, İstanbul, Turkey

M. Polatcan · R. Cansoy
Karabük University, Karabük, Turkey



Ş. Ş. ERÇETİN, Ş. N. AÇIKALIN (eds.), *Chaos, Complexity and Leadership 2018*,
Springer Proceedings in Complexity, https://doi.org/10.1007/978-3-030-27672-0_15

be stated that changes in the curriculum, school culture, and student success are the results of interactions within the system, since the relationship between school communities ultimately focuses on school success. However, considering the school as a whole, it can be said that the system is in an order. There is a mutual interaction between communities because they serve the same purpose in the school system. All of these processes show that change in schools is quite complex.

In educational organizations, the attitudes and behaviors of principals and teachers play an important role in the success of the change process. Therefore, it is important that principals and teachers are ready for and open to change. Principal and teachers are expected to maintain the processes in the school actively in the light of their roles and duties. Trust-based relationships among stakeholders in the process of change and their readiness for change facilitate adaptation to changes in schools.

## Conceptual Framework

Schools need to make sense of changes in their environment and keep up with the change process to survive (Demirtaş 2012; Kondakçı et al. 2013; Levent 2016). Education is directly affected by economic, technological, and social developments. In this respect, it can be declared that changes are necessary in many areas, from the way in which education is presented to the integration of technology, and from the administrative structure to financial resources (Erdoğan 2015). However, the human factor, which is the most important element of change in educational organizations, is often ignored; change practices are generally focused on financial and material resources (Kondakçı et al. 2013). While organizational goals, complexity, crises, and changing rules facilitate change in schools, factors such as strict bureaucracy, inertia, disinterest in change, and unwillingness to change slow down the process of change (Lunenburg and Ornstein 2013; Tokat 2012). The current situation shows that change in schools depends on social changes and that many factors affect change in schools.

For organizations to achieve effective results in change practices, employees need to increase their readiness for change (Khedhiri 2018; Rafferty and Simons 2005). Uncertainties are likely to occur as a result of operating order (Zayim and Kondakci 2015). These uncertainties cause employees to resist changes (Inandi and Giliç 2016). Readiness for change refers to the cognitive and affective beliefs of employees toward change (Armenakis et al. 1993), conversion of negative perceptions and behaviors into positive ones (Bouckenooghe et al. 2009), acceptance and adoption of a specific plan that aims to change a static state (Holt et al. 2007), and determination and collective competences for implementation of change (Weiner 2009). In the literature, readiness for change has been examined in cognitive, emotional, and intentional dimensions. The cognitive dimension shows the competence of the organization related to change and individuals' belief in and support for change. The emotional dimension reflects individuals' feelings such as optimism resilience in the face of change. The intentional dimension refers to individuals' behaviors that support the process of change (Armenakis and Harris 2002; Weiner 2009).

Researchers have stated that supportive attitudes and behaviors of employees toward change are effective in the success of change (Inandi and Giliç 2016; Khedhiri 2018; Zayim and Kondakci 2015). Weiner (2009) indicated that failure of change in organizations occurs because of the absence of organization readiness for change. According to Arslan et al. (2005), the strong organizational culture in schools increases cooperation and adaptation of employees to change. In studies on readiness for change, organizational culture and climate change (Jones et al. 2005), organizational development (Choi and Ruona 2011), change leadership and organizational commitment (Madsen et al. 2005; Nordin 2012; Santhidran et al. 2013), and organizational trust (Zayim and Kondakci 2015) were examined. Despite the consensus on the importance of readiness for change in organizational change, it can be said that the research on trust in readiness for change in schools is limited. Trust, which is one of the important factors in readiness for change, is related to attitudes, beliefs, and behaviors of the individual or the organization in the process of change (Eby et al. 2000). Teacher participation and strong trust relationships among colleagues play a crucial role in making changes that build effective school communities (Bryk and Schneider 2003). The basis of change and development is trust (Yılmaz 2005); in other words, trust is effective in eliminating resistance against change in organizations, and in increasing support for and participation in change (Zayim and Kondakci 2015).

The existence of open and transparent relationships among school stakeholders shows that trust is established in schools (Bryk and Schneider 2003). The existence of schools, adoption of targets by employees, and cooperation between employees are associated with trust (Boyacı et al. 2018; Cheng et al. 2017; Wolfe et al. 2015). In educational organizations, open communication among school stakeholders, sharing knowledge and expressing feelings and thoughts, reflect the trust related to the school. Teachers' belief that principals are honest and have consistent behavior refers to their trust in principals (Van Maele and Van Houtte 2015; Yılmaz 2005). In addition, trust between teachers and students has an impact on students' adaptation to school, academic motivation, and performance (Lee 2007; Mitchell et al. 2018). Therefore, it can be said that trust relations between principals and teachers, who are among the important stakeholders in the process of readiness for change, contribute to change-oriented reform studies.

## Research Context

Among the dynamics of change in education are school reform, enhancement of student achievement, effective teaching practices, education policies of governments, and accountability (Goh et al. 2006; Lunenburg and Ornstein 2013). Uncertainty due to confusion created by small- and large-scale changes makes it difficult to predict problems in complex systems such as schools. Principals play an important role in eliminating uncertainties caused by the abundance of schools' connections (Fidan and Balcı 2017). Therefore, in particular, school principals and teachers should be ready for the process of change to ensure successful change in schools. Indeed, Cunningham et al. (2002) stated that employees with high levels of self-efficacy who are in decision-making or active positions in organizations are ready for change.

Since the establishment of the Turkish Republic, small- and large-scale reforms have been made in the philosophy of education and the teaching approach. In particular, the reforms toward the end of the 1990s led to significant changes (Zayim and Kondakci 2015). Everyone, from politicians to educators, agreed that the people of the new century could not be educated with the education program of the twentieth century in Turkey. The constructivist approach adopted by the central government in 2004 has had important implications for change. Within the scope of this approach, student-centered education has been adopted and the curriculum has been changed in a way that students apply their experiences from the past to what they have learned recently. Moreover, the importance of effective use of technology in teaching has been given importance in line with the conditions of the day (Arslan 2007; Çeliköz and Erişen 2017).

In addition to considerable changes in the curriculum, there have been radical reforms in many areas such as changes in exam formats nationally at all levels (from primary, secondary, and tertiary education), an increase in compulsory education to 12 years, changes in the evaluation system used in secondary education, and allocation of school principals (Gedikoğlu 2005; Yılmaz and Altınkurt 2011). However, there have been reports of failure of these radical changes in the Turkish education system. Among the common results of these reports, it is emphasized that the national education curriculum and learning-oriented education, and education have not been considered as a holistic system and that changes made in elements of the system have therefore been implemented without evaluation of what kind of results will ensue (Eğitim-Bir-Sen 2018; Gür and Çelik 2009; Türk-Eğitim-Sen 2018). Inconsistencies in central government interventions in the Turkish education system and arrangements made without considering the views of stakeholders (principals, teachers, etc.) have caused big problems in the system (Abu et al. 2016; Yaşar and Sözbilir 2017; Yeşil and Şahan 2015). Readiness for change and trust among employees are important in terms of solving these complications and eliminating negative perceptions of change in schools. In this context, the focus of this study was school stakeholders' readiness to change and the relationship between teachers and principals. Accordingly, the research question was as follows:

According to the teachers' views, is there a relationship between readiness for change in the school and trust in principals?

# Methods

## Research Model

In this study, using the relational survey model, the relationship between primary and secondary school teachers' views of readiness for change and trust in principals was examined. The dependent variable of the study was readiness for change, and the independent variable was trust in principals. In the study, it was considered that there would be a positive relationship between readiness for change and trust in principals. In addition, the predictive effect of trust in principals on readiness for change was investigated.

## Sample

The study population consisted of teachers working in primary and lower secondary schools in Çekmeköy, İstanbul, Turkey, in the 2017–2018 academic year. Permission for this research was given by the Research Council of İstanbul Provincial Directorate of National Education. In the implementation of the research, consent for participation was obtained from the participant teachers. The sample of the study included 328 teachers selected from the population by a simple random-sampling method; 215 (65%) of the teachers were female and 113 (35%) were male; and 86 (20.4%) had 1–10 years of teaching experience, 98 (29.9%) had 11–20 years of experience, and 141 (43%) had at least 21 years of experience.

## Data Collection Tools

*Organizational Trust Scale* The organizational trust scale was developed by Daboval et al. (1994) and adapted to Turkish culture by Yılmaz (2005). In the present study, the trust in principals dimension of the scale was used. This five-level Likert-type scale consisted of 12 items. The Kaiser–Meyer–Olkin (KMO) test value (0.96) and the Bartlett sphericity test value ($\chi^2 = 3714.38$; $p < 0.05$) of the current study were found to be significant. As a result of exploratory factor analysis, a single factor structure was observed. As a result of varimax rotation, one item was noted to have a factor loading, so it was excluded and the analyses were done with 11 items. The factor loadings of the scale items were between 0.63 and 0.83, and the total variance explained by the scale was 74.55%. The Cronbach alpha reliability coefficient calculated for the trust in principals scale was 0.97, and the corrected item total correlations were between 0.75 and 0.84.

*Readiness for Change Scale* A 12-item scale developed by Kondakçı et al. (2013) was used to measure employees' readiness for change. As a result of the exploratory factor analysis, the KMO value was calculated as 0.94 and the results of the Bartlett sphericity test were found to be significant ($\chi^2 = 2653.59$; $p < 05$). As a result of varimax rotation, the scale items showed a one-dimensional structure and the item factor loadings ranged from 0.61 to 0.74. The total variance explained by the scale was 68.22%. The reliability coefficient of the scale was calculated as 0.95, and the corrected item correlations ranged from 0.48 to 0.77.

### Data Analysis

Skewness and kurtosis coefficients were examined for the normality of the data before analysis of the data. The skewness values for the trust in principals and readiness for change scales were found to be between (−1.8) and (−0.78), while the kurtosis values were between (0.47) and (0.43). The fact that the skewness and compression were between +2 and −2 indicated that the data were suitable for normal distribution (George and Mallery 2010). In addition, a Q–Q plot graph also showed that the distribution of the data was normal. Multicollinearity between variables, the variance inflation factor (VIF), and tolerance values were examined for regression analysis of the predictive variables. The fact that the tolerance value was lower than 0.20, the VIF value was higher than 10, and the correlations between the independent variables were 0.80 or above could have been an indicator of multicollinearities (Büyüköztürk 2011). The data were found to be suitable for regression analysis when the analyses were evaluated as a whole. Arithmetic means were calculated in the analysis of the data obtained for the analysis of subproblems. The Pearson's Moments Multiplication Correlation Coefficient ($r$) and multiple linear regression analysis were used to determine the relationships between variables. The significance of the data was determined according to a $p$ value of <0.05.

## Findings

Table 1 displays the mean and standard deviation values of the variables in the study and the correlations between the variables.

**Table 1** Descriptive values of variables and correlations between variables

| Variable | $\overline{X}$ value | SD | Trust in principals | Readiness for change |
|---|---|---|---|---|
| Trust in principals | 4.37 | 0.78 | 1 | 0.380* |
| Readiness for change | 4.13 | 0.72 | | 1 |

*$p < 0.01$

**Table 2** Regression analysis for predicting readiness for change

| Variable | Unstandardized *B* value | Standard error of unstandardized *B* value | Standardized β value | *t* test statistic | *p* value |
|---|---|---|---|---|---|
| Constant | 2.59 | 0.21 | | 12.332 | 0.00 |
| The basis of trust in principals | 0.22 | 0.07 | 0.24 | 2.790 | 0.00 |

$R = 0.39, R^2 = 0.15, F = 29.78, p < 0.05$

As seen in Table 1, according to the perceptions of the teachers, the basis of trust in principals ($\overline{X}$ = 2.72) and employees' readiness for change ($\overline{X}$ = 2.72) were at high levels. Considering the correlation coefficients between the variables, a significant and positive relationship was found between the basis of trust in principals and readiness for change ($r = 0.38, p < 0.01$).

Table 2 shows the results of multiple linear regression analysis for predicting readiness for change to trust in principal.

When Table 2 is examined, it can be seen that there was a meaningful relationship between the basis of trust in principals and readiness for change ($\beta = 0.24, p < 0.05$). The predictor of the trust in principals variable explained 15% of the total variance in the perceptions of readiness for change.

## Discussion

Within the scope of this study, the relationship between readiness for change in schools and trust in principals was tested through the views of the teachers. The results of the study showed that the level of trust in principals and readiness for change among primary and secondary school teachers were quite high, suggesting that the teachers were optimistic about change. In addition, a positive relationship was found between readiness for change and trust in principals. Thus, it is important that school stakeholders trust principals in the process of readiness for change. Positive relations were also found in previous studies on readiness for change and trust in principals (Neves and Caetano 2006; Rafferty and Simons 2005; Zayim and Kondakci 2015). These results support the relationship between readiness for change and trust in principals (the hypothetical problem in this research). It was seen that teachers' trust in their principals influenced their cognitive, emotional, and intentional readiness for change in schools. Zayim and Kondakci (2015) noted that teachers' trust in their colleagues and principals was significantly related to the teachers' change readiness in the cognitive and emotional dimensions. Accordingly, it can be said that for teachers to be ready for change in the change practices of primary and secondary schools, it is important that principals have a democratic, honest, transparent, collaborative, and participatory attitude.

According to the teachers, the perception of trust in principals is a predictor of school stakeholders' readiness for change in schools. Trust in principals demonstrates the beliefs of other school stakeholders (especially teachers) that they can overcome change, demonstrates their optimism in the process of change, and demonstrates their behaviors in making changes. Otherwise, lack of trust in principals during the change process may lead teachers and other school stakeholders to resist or remain indifferent to change. In such a case, the change practices will most probably fail. Indeed, Zayim and Kondakci (2015) provided evidence that trust in colleagues and principals in primary and secondary schools was effective in the readiness of schools for change, intentionally, emotionally, and cognitively. In their research on sales personnel, Eby et al. (2000) found that employees working in sales teams were more ready for change than those working individually, because the employees' feeling of participation in a team and confidence in their peers increased their readiness for change. In addition, in a study on hospital management, Cunningham et al. (2002) stated that employees with high self-efficacy in decision-making positions were more ready for change. In this context, as a leader, a principal with a positive attitude toward teacher participation in the process of change contributes to teachers' cooperation, willingness to change, and elimination of uncertainties in the process of change. As a result, it can be said that a high level of trust among all stakeholders in the success of the change process in schools facilitates cooperation of the stakeholders and the adoption of the innovations at the end of the change process.

The results of this study show that schools and policy makers take into account the trust premise when making decisions on changes in the system. Incentives such as trust that develop teachers' self-efficacy, optimism, and positive behavior toward change should not be ignored in the readiness for change, which is an important prerequisite of the change process. Indeed, trust strengthens communication and interaction in the school and creates a supportive exchange culture. Therefore, there should be practices for developing a trust-based climate that will foster not only the school but also its relations with its environment. In addition, the principal should be honest with his/her teachers, demonstrate open and transparent management, support the teachers, and make plans to eliminate the risk of uncertainty that change may cause, because trust in the principal as the leader of change in a school is an effective premise of readiness for change.

Considering the theoretical and researcher recommendations of the study, the ultimate aim of the change processes in schools is to develop the schools. One of the criteria of school development is the academic success of the students. Thus, it can be said that readiness for change (which is an important precursor of the change process) and trust are essential for school success. In addition, the relationship between a school's readiness for change and teacher loyalty, school culture, and climate can be examined. In addition, the relations of resistance, indifference, and uncertainty that can be encountered by teachers who are unready for change can be investigated.

Finally, this research study had some limitations. It focused on the relation between primary and secondary school teachers' views of readiness for change and trust in principals. Further research on teachers and principals working in secondary education could be worthwhile. Moreover, students and parents are among the important factors that are influenced by changes in schools; thus, the scope of this empirical research could be expanded by ensuring student and parent participation in research on readiness for change.

## References


Abu, N. K., Bacanak, A., & Gökdere, M. (2016). Öğretmen adaylarının türk eğitim sisteminin sorunlarına İlişkin görüşlerinin İncelenmesi. *Amasya Üniversitesi Eğitim Fakültesi Dergisi, 5*(1), 287–307.

Açıkgöz, S., & Edin, İ. (2018). Kaosun örgütsel değişim üzerindeki etkisi: Profesyonel futbol kulüpleri üzerine bir çalışma. *Spor ve Performans Araştırmaları Dergisi, 9*(1), 1–17. https://doi.org/10.17155/omuspd.300912.

Armenakis, A. A., & Harris, S. G. (2002). Crafting a change message to create transformational readiness. *Journal of Organizational Change Management, 15*(2), 169–183. https://doi.org/10.1108/09534810210423080.

Armenakis, A. A., Harris, S. G., & Mossholder, K. W. (1993). Creating readiness for organizational change. *Human Relations, 46*(6), 681–703. https://doi.org/10.1177/001872679304600601.

Arslan, H., Kuru, M., & Satici, A. (2005). A comparison of organizational culture in Turkish primary schools and high schools. *Educational Administration: Theory & Practice, 11*(4), 449–472.

Arslan, M. (2007). Eğitimde yapılandırmacı yaklaşımlar. *Ankara Üniversitesi Eğitim Bilimleri Fakültesi Dergisi, 40*(1), 41–61.

Bouckenooghe, D., Devos, G., & Van den Broeck, H. (2009). Organizational change questionnaire—climate of change, processes, and readiness: Development of a new instrument. *The Journal of Psychology, 143*(6), 559–599. https://doi.org/10.1080/00223980903218216.

Boyacı, A., Karacabey, M. F., & Bozkuş, K. (2018). Okul yöneticilerinin liderliğinin öğretmenlerin İş doyumuna etkisinde örgütsel güvenin rolü. *Kuram ve Uygulamada Egitim Yönetimi Dergisi, 24*(3), 437–482. https://doi.org/10.14527/kuey.2018.011.

Bryk, A. S., & Schneider, B. (2003). Trust in schools: A core resource for school reform. *Educational Leadership, 60*(6), 40–45.

Büyüköztürk, Ş. (2011). *Sosyal bilimler için veri analizi el kitabı.* Ankara: Pegem Akademi.

Çeliköz, M., & Erişen, Y. (2017). Epö alanında görev yapan eğitim bilimcilerin yapılandırmacılık uygulamalarıyla İlgili görüşleri. *International Online Journal of Educational Sciences, 9*(3), 738–754.

Cheng, X., Fu, S., Han, Y., & Zarifis, A. (2017). Investigating the individual trust and school performance in semi-virtual collaboration groups. *Information Technology & People, 30*(3), 691–707. https://doi.org/10.1108/ITP-01-2016-0024.

Choi, M., & Ruona, W. E. A. (2011). Individual readiness for organizational change and its implications for human resource and organization development. *Human Resource Development Review, 10*(1), 46–73. https://doi.org/10.1177/1534484310384957.

Cunningham, C. E., Woodward, C. A., Shannon, H. S., MacIntosh, J., Lendrum, B., Rosenbloom, D., & Brown, J. (2002). Readiness for organizational change: A longitudinal study of workplace, psychological and behavioural correlates. *Journal of Occupational and Organizational Psychology, 75*(4), 377–392. https://doi.org/10.1348/096317902321119637.

Daboval, J., Comish, R., Swindle, B.,& Gaster, W. (1994). A trust inventory for small businesses. *Proceedings of the conference of the business trends and small business trust southwestern small business institute association, Dallas.* http:/www.sbaer.uca.edu/ docs/proceedings/ 94swi031.txt

Demirtaş, H. (2012). İlköğretim okullarının değişime açıklığı. *İlköğretim Online, 11*(1), 18–34.
Eby, L. T., Adams, D. M., Russell, J. E. A., & Gaby, S. H. (2000). Perceptions of organizational readiness for change: Factors related to employees' reactions to the implementation of team-based selling. *Human Relations, 53*(3), 419–442. https://doi.org/10.1177/0018726700533006.
Eğitim-Bir-Sen. (2018). *2023'e doğru türk eğitim sistemi bulma konferansı ve çalıştayı öneriler.* http://www.ebs.org.tr/ebs_files/files/yayinlarimiz/2023TurkEgitimSistemiBulmaKonferansi_oneriler.pdf.
Erdoğan, İ. (2015). *Eğitimde değişim yönetimi (4. Baskı).* Ankara: Pegem Akademi.
Fidan, T., & Balcı, A. (2017). Managing schools as complex adaptive systems: A strategic perspective. *International Electronic Journal of Elementary Education, 10*(1), 11–26. https://doi.org/10.26822/iejee.2017131883.
Gedikoğlu, T. (2005). Avrupa birliği sürecinde türk eğitim sistemi: Sorunlar ve çözüm önerileri. *Mersin Üniversitesi Eğitim Fakültesi Dergisi, 1*(1), 66–80.
George, D., & Mallery, P. (2010). *SPSS for Windows step by step: A simple guide and reference, 17.0 update.* Boston: Allyn & Bacon.
Goh, S. C., Cousins, J. B., & Elliott, C. (2006). Organizational learning capacity, evaluative inquiry and readiness for change in schools: Views and perceptions of educators. *Journal of Educational Change, 7*(4), 289. https://doi.org/10.1007/s10833-005-5033-y.
Gür, B. S., & Çelik, Z. (2009). *Türkiye'de millî eğitim sistemi: Yapısal sorunlar ve öneriler.* Ankara: SETA.
Holt, D. T., Armenakis, A. A., Feild, H. S., & Harris, S. G. (2007). Readiness for organizational change: The systematic development of a scale. *Journal of Applied Behavioral Science, 43*(2), 232–255. https://doi.org/10.1177/0021886306295295.
Inandi, Y., & Giliç, F. (2016). Relationship of teachers' readiness for change with their participation in decision making and school culture. *Educational Research Reviews, 11*(8), 823–833. https://doi.org/10.5897/ERR2016.2730.
Jones, R. A., Jimmieson, N. L., & Griffiths, A. (2005). The impact of organizational culture and reshaping capabilities on change implementation success: The mediating role of readiness for change. *Journal of Management Studies, 42*(2), 361–386. https://doi.org/10.1111/j.1467-6486.2005.00500.x.
Khedhiri, M. (2018). Readiness for change in public education: A discrete empirical investigation. *Higher Education for the Future, 5*(2), 178–197. https://doi.org/10.1177/2347631118767293.
Kondakçı, Y., Zayim, M., & Çalışkan, Ö. (2013). Değişime hazır olma ölçeği'nin geçerlilik ve güvenilirlik çalışması. *Elementary Education Online, 12*(1), 23–35.
Lee, S. J. (2007). The relations between the student–teacher trust relationship and school success in the case of Korean middle schools. *Educational Studies, 33*(2), 209–216. https://doi.org/10.1080/03055690601068477.
Levent, F. (2016). Öğretmenlerin değişime hazır olma durumlarının farklı değişkenlere göre incelenmesi. *Eğitim Bilimleri Dergisi, 43*(43), 117–134. https://doi.org/10.15285/ebd.09059.
Lunenburg, F. C., & Ornstein, A. C. (2013). *Eğitim yönetimi* (Gökhan Arastaman, Çev.). Ankara: Nobel Yayınları.
Madsen, S. R., Miller, D., & John, C. R. (2005). Readiness for organizational change: Do organizational commitment and social relationships in the workplace make a difference? *Human Resource Development Quarterly, 16*(2), 213–234. https://doi.org/10.1002/hrdq.1134.
Mitchell, R. M., Kensler, L., & Tschannen-Moran, M. (2018). Student trust in teachers and student perceptions of safety: Positive predictors of student identification with school. *International Journal of Leadership in Education, 21*(2), 135–154. https://doi.org/10.1080/13603124.2016.1157211.
Morgan, G. (1997). *Yönetim ve örgüt teorilerinde metafor* (G. Bulut, Çev.). İstanbul: MESS Yayınları.
Neves, P., & Caetano, A. (2006). Social exchange processes in organizational change: The roles of trust and control. *Journal of Change Management, 6*(4), 351–364. https://doi.org/10.1080/14697010601054008.

Nordin, N. (2012). The influence of leadership behavior and organizational commitment on organizational readiness for change in a higher learning institution. *Asia Pacific Education Review, 13*(2), 239–249. https://doi.org/10.1007/s12564-011-9200-y.

Özgenç, Ö. (2008). Değişimin mantığını anlamak: Akış ve dönüşüm olarak örgüt. In A. Balcı (Ed.), *Örgüt mecazları*. Ankara: Ekinoks Yayınları.

Rafferty, A. E., & Simons, R. H. (2005). An examination of the antecedents of readiness for fine-tuning and corporate transformation changes. *Journal of Business and Psychology, 20*(3), 325. https://doi.org/10.1007/s10869-005-9013-2.

Santhidran, S., Chandran, V. G. R., & Borromeo, J. (2013). Enabling organizational change—leadership, commitment to change and the mediating role of change readiness. *Journal of Business Economics and Management, 14*(2), 348–363. https://doi.org/10.3846/16111699.2011.642083.

Tokat, B. (2012). *Örgütlerde değişim ve değişimin yönetimi*. Ankara: Seçkin Yayıncılık.

Türk-Eğitim-Sen. (2018). *Türk eğitim sistemine İlişkin görüş ve öneriler*. https://turkegitimsen.org.tr/userfiles/files/TES_Go%CC%88ru%CC%88s%CC%A7veo%CC%88neriler%20(1).pdf .

Van Maele, D., & Van Houtte, M. (2015). Trust in school: A pathway to inhibit teacher burnout? *Journal of Educational Administration, 53*(1), 93–115. https://doi.org/10.1108/JEA-02-2014-0018.

Weiner, B. J. (2009). A theory of organizational readiness for change. *Implementation Science, 4*(1), 67. https://doi.org/10.1186/1748-5908-4-67.

Wolfe, S. E., Chrusciel, M. M., Rojek, J., Hansen, J. A., & Kaminski, R. J. (2015). Procedural justice, legitimacy, and school principals' evaluations of school resource officers: Support, perceived effectiveness, trust, and satisfaction. *Criminal Justice Policy Review, 28*(2), 107–138. https://doi.org/10.1177/0887403415573565.

Yaşar, M. D., & Sözbilir, M. (2017). Fen bilgisi öğretmen adaylarının türk eğitim sistemi'ne yönelik görüşleri: Güncel sorunlar ve çözüm önerileri. *Çukurova Üniversitesi Eğitim Fakültesi Dergisi, 46*(1), 165–201.

Yeşil, R., & Şahan, E. (2015). Öğretmen adaylarının türk eğitim sisteminin en önemli sorun, neden ve çözüm yollarına ilişkin algıları. *Journal of Kirsehir Education Faculty, 16*(3), 123–143.

Yılmaz, E. (2005). Okullarda örgütsel güven ölçeğinin geçerlik ve güvenirlik çalişmasi. *Selçuk Üniversitesi Sosyal Bilimler Enstitüsü Dergisi, 14*, 567–580.

Yılmaz, K., & Altınkurt, Y. (2011). Öğretmen adaylarının türk eğitim sisteminin sorunlarına ilişkin görüşleri. *Uluslararası İnsan Bilimleri Dergisi, 8*(1), 942–973.

Zayim, M., & Kondakci, Y. (2015). An exploration of the relationship between readiness for change and organizational trust in Turkish public schools. *Educational Management Administration & Leadership, 43*(4), 610–625. https://doi.org/10.1177/1741143214523009.

# A Chaos Theory Perspective on Migration of Sub-Saharan Africans to Europe

**Mahamadou Yahaya**

**Abstract** Colonization, with its religious and political aspects, which supported the practice of having economic and political control through established colonies, is considered by many historians as one of the reasons behind the delayed industrialization of sub-Saharan Africa. In fact, because of this, the region faces a lack of economic opportunity, unemployment, a brain drain, and migration of young people seeking good jobs and better lives elsewhere, namely in Europe. Therefore, for these young people, illegal migration has become the solution to overcome the difficulties they are facing. Unfortunately, illegal migration or migration itself for some European leaders is considered as a threat for their nations. The phenomenon of illegal migration for some western leaders is regarded as one of the biggest challenges of the twenty-first century because migrants in many European countries are seen as invaders of their land and job takers from nationals; some dislike migrants for security reasons, while other countries only help migrants for humanitarian reasons. For the African continent, illegal migration leading to a loss of many young African lives in the Mediterranean Sea is a human catastrophe. Meanwhile, a number of thinkers and decision makers have come up with and published in several reports some suggested solutions, including restraining the flow of migrants and therefore the risks of travel and the number of deaths as well as the loss of money.

**Keywords** Chaos Theory · Migration · Sub-Saharan Africans · Europe

## Introduction

Migration is defined as the large scale movement of people from one place to another, and it becomes illegal when it is done without any documents delivered by the receiving country. However, there are campaigns discouraging the use of the

M. Yahaya (✉)
Attadamoune University, Niamey, Niger



Ş. Ş. ERÇETİN, Ş. N. AÇIKALIN (eds.), *Chaos, Complexity and Leadership 2018*,
Springer Proceedings in Complexity, https://doi.org/10.1007/978-3-030-27672-0_16

term "illegal migrant" generally based on the argument that the act of migrating illegally does not make the people themselves illegal, but rather they are "people who have migrated illegally". Instead, many advocate for the use of terms such as undocumented migrants or unauthorized migrants when referring to the foreign nationals who reside in a country illegally. Indeed, legal and illegal migration from sub-Saharan Africa to Europe, with all of its patterns, can be considered a cumulative result of colonialism, lack of appropriate leadership in the postcolonial era, and the effects of globalization. The result is that the region faces a lack of economic opportunity for youth—unemployment—which causes a brain drain and migration of sub-Saharan African people from their countries to seek good jobs and better lives for themselves and the families they leave behind.

## Overview

The concept of clandestine or illegal migration dates back to the 1930s, but it only became prominent during the 1980s and 1990s (Düvell 2008). In sub-Saharan Africa, legal and illegal migration became prominent during the 1990s and 2000s because of economic, political, ethnic, and religious conflicts (Fearon 2005). Conflict and wars in Africa tend to last many years before peace is restored. Picciotto points out that African casualties of conflict exceed those of all other regions combined (Picciotto 2010:2). Another factor in migration is poverty, as Venkatasawmy (2015) observed: "low-income nations coming out of conflict situations usually have a hard time overcoming poverty, implementing development strategies and improving economic growth".

According to the United Nations (UN) Development Program (UNDP), Human Development Index, the majority of the world's poorest people are located on the African continent, which has the slowest economic growth (see: World Bank: https://ourworldindata.org/extreme-poverty-projections; OECD: http://www.oecd.org/development/africa-s-development-dynamics-2018-9789264302501-en.htm) undp: https://www.africa.undp.org/content/rba/en/home/regioninfo/ online). The consequences of economic and political factors and rising conflicts lead a number of Africans to leave their countries for the hope of safer and better lives.

The majority of illegal migrants from sub-Saharan Africa are young African people who despair for the future of their countries and risk their meagre resources (sometimes by selling their family's unique resource) to travel illegally—most of the time without documentation—to Europe. Despite the current policy measures taken by European and African leaders, the number of migrants remains high (International Organization for Migration 2018). The migrants use different passages across the countries—such as Spain, Italy, and Malta—surrounding the Mediterranean Sea. Recently, the International Organization for Migration (2018) documented a large number of children born at sea.

## *A Chaos Theory Perspective on Migration of Sub-Saharan Africans to Europe*

The migration from sub-Saharan Africa to Europe is economic migration; the migrants, who have gone to Europe to seek jobs without any legal protection, sometimes die in transit by drowning or other causes. Lately, European Union and African leaders have tried to find sustainable solutions to overcome this kind of migration. Among the short-term solutions found by the leaders is the return of migrants back to their home countries. One of the strategies used to speedup this process is giving the migrants a cash lump sum and a plane ticket to leave the country they have reached. However, it is worth remembering that there is no published or clear program for their occupational integration when they arrive back in their native countries. Another solution is banning migration by using legal force; for example, most of the concerned European governments refused to rescue drowning migrants in 2018.

When we return to the meaning and the main objectives of migration, it is easier to understand why sub-Saharan Africans have fled to Europe. In recent decades, with economic regression occurring in sub-Saharan African countries, nearly half of the continent's people still live on less than 1.25 USD a day (Bob-Milliar 2012). Meanwhile, sub-Saharan Africa is considered a group of "emigrant producing countries". It is also well-known that in most of these countries there is high unemployment or salaries are very low, and people have to seek opportunities elsewhere. That situation has made migration one of the central points of disagreement between the mainly poor countries of origin and the richer destination countries.

### The Colonial Era

In the *Oxford Dictionary*, "colonization" is defined as "the action or process of settling among and establishing control over the indigenous people of an area" and "colony" is defined as a "country or area under the full or partial political control of another country and occupied by settlers from that country". The word "colony" has its root in religion and politics; for example, in Christianity a "colony of heaven" is used to call mankind to obey God rather than mankind. Religious missionaries tried to serve as ministers of education, literacy, social justice, health, and economic development. On the other side, the political meaning of the word, can be understood from analysis of the British Empire time when the Kingdom, as a political entity, settled colonies to fulfil its own economic desires and to sell manufactured goods to the countries from which it could buy/obtain raw materials; the colonized countries had abundant natural resources. In addition to their interest in economic approaches, they tried to "modernize" the colonies by stimulating social and economic activities, such as education and how to increase the indigenous use of the products made in Europe. By the way, colonizers obtained their raw materials from sub-Saharan

Africa but they did not care to even suggest to the first Africanists the idea of implementing policies for the local industrialization of sub-Saharan Africa.

### The 1884–85 Berlin Conference and the Future of Africa

After African countries became independent, it was mentioned that they could change the borders that had been designated at the Berlin Conference of 1884–85, but they did not do that; possibly to avoid further chaos on the continent. Nonetheless, it is easy to notice how the European Powers parceled out Africa in their own interests and not in the interest of the rapid development of Africa. This was done without inviting African leaders or representatives of the effected communities, ethnic groups, or language groups. It is worth noting that there are around 3000 distinct ethnic groups and 2000 languages in Africa. When new boundaries were drawn over the territory of each tenth ethnic group, trade routes were cut off because trade with persons outside the colony was prohibited. It is possible to state that this action was enough to delay the economic growth of the continent.

### The Independent Era

The Berlin Conference led to a period of heightened colonial activity by the European Powers without the involvement of or even one African leader being invited. Thus, with the exception of Ethiopia and Liberia, all of the states that make up Africa today were parceled out between the colonial powers a few years after the 1884–85 conference. After half a century, they started to establish the process of independence for their countries, which were sometimes viewed as nationalist movements. In the 1960s, most African countries did finally gain their independence. In formally granting their independence, the colonial powers made special agreements with new authoritarian leaders and thereby maintained their relations effectively.

## Colonization and Migration

The relationship between colonization and the current flow of illegal migration may lie in the political, economic, and diplomatic relations of the former colonizers and colonized countries. The European countries established the colonies in Africa to further their own economic, political, and military interests. They benefited from the abundance of natural resources available in the colonies and from African human resources, and they used the slave trade to bring young Africans to Europe to work on farms. Both of these things were used against the economic growth of Africa for the benefit of Europe.

Another issue was military development. A group of young African people was brought to fight for the colonial powers, and a large number of them perished without meaningful recognition. Finally, after settling the colonies, the colonial powers used a group of young African people to work in their administrations, which became the first staff of governments of those countries upon their independence. Among this group were people who had been sent to Europe for education, to learn the European languages, governing systems, economics, and social activities. Thus, the current flow of migrants from Africa to Europe is a product of the above-mentioned complex relationship between the colonized countries with colonial powers.

Experience has shown that after the independence of the sub-Saharan African countries, the former colonial countries kept their relationship with the established former colonies by maintaining their economic and political strength. For example, France has maintained its presence and its political strength in its former colonies by convincing the countries' governments that they are around to save them—to control civil unrest, to prevent the expansion of communism in the Cold War environment in the early 1990s, and to prevent terrorism.

We can realize how this relationship continues to hamper the development of most sub-African countries. While most of them are rich in materials—such as oil, minerals, and other strategic raw materials—they still suffer from a lack of energy for household use and industrialization. This means that that most of the first African leaders failed to implemented good policies, to create jobs, to eliminate hunger, to resolve ethnic and political conflicts, and to prevent corruption. This situation resulted in the migration of young people from Africa to Europe in the 1990s, and currently unemployment, corruption, and lack of leadership are among the reasons that force young people to leave Africa for Europe. The evidence is that most African countries struggle to achieve good reforms that can create jobs and persuade the youth to stay in their home countries instead of risking their lives on the Mediterranean Sea.

## *Post-colonial Era and Migration*

The specific history and circumstances of sub-Saharan African countries drastically affected the policies taken by their first leaders who came after the colonial powers. Nonetheless, lack of good policies to achieve quality education and literacy, social justice, health care, and economic development in their countries are among the real reasons for the current flow of migration of young African people to Europe. In addition, young sub-Saharan Africans suffer from the inequality of chance, and many forms of exclusion, to benefit from those social services and social protections. This group of young people, who are mostly non-educated, early school leaving and unemployed, choose a new destination abroad, in the hope of good jobs and better lives. We realize that, since independence until now, African countries' leaders seem to have difficulty working together to ensure the equitable distribution of wealth, opportunity, and power in their societies.

Another factor of migration of young Africans is that most African countries failed to establish good security policies due to the lack of a good relationship between the first military and civil authorities, which created conflict and persecution in the region. Therefore, a lack of stability and good governance increase drastically the number of candidates for migration. Some governments have failed to eradicate the illegal services of human traffickers, and many migrants take on this death-defying experience and journey on overloaded trucks in convoys and part of the way on foot through the Sahara Desert.

Another leadership failure is the existence in sub-Saharan Africa of the highest level of income inequality; it is extremely difficult to establish stability in this kind of situation. Ignorance and deprivation are two factors that may increase the flow of migration. At least 70% of sub-Saharan people live in rural areas, and there is still a lot of effort to be made for implementing a clear policy to diminish rural areas in the countries; for example, gathering people in the same and comfortable venue to easily provide social services. With regard to young girls and disadvantaged groups, there are no efficient policies that can be cited as good references for other countries.

Sub-Saharan leaders also failed to link education to the job market. At present, not all African young people have a chance to access a quality education, and a job they are qualified for. We realize the unemployment level is high in most sub-Saharan African countries. The governments have not been able to link education with required skills in the countries. Thus, young people mostly get support from their relatives to leave, because they need better lives for their children and the expected financial support will help to improve the family's living conditions.

## Globalization and Migration

Most of the sub-Saharan Africa countries are often cited among a group of fragile countries, with rapidly growing populations, rising urbanization, and huge economic inequalities. European countries are considered wealthy and stable countries, which promoted globalization. In this framework non-industrialized countries are viewed as customers for industrialized countries. Globalization, which promotes easier flow of capital, goods, and services supplied by industrial countries, may be another factor hindering the industrialization of sub-Saharan Africa, which is able to sell only raw materials. Meanwhile, when it concerns population movements from South to North, European countries believe it should be under their strict control because the poor countries mostly provide poor persons who can negatively affect the security and stability of their countries. Thus, we can conclude that sub-Saharan African countries benefit from a small piece of the meaning of globalization as interconnectedness between the global and local levels.

## Context and Policy Measures to Overcome Illegal Migration

Despite that the migration of sub-Sahara Africans to Europe is assessed in its economic contexts, the stake holders seem to have difficulty handling it. The problematic and complex relationship between the reality of migration and the undertaken policies and measures to overcome illegal migration can be the real reasons behind this misleading notion. For example, we have the "human right" issue that seems to give them the right of movement, the factor of unemployment in sub-Saharan countries versus labor shortages and economic opportunity in Europe, divergence regarding the effect of the brain drain factor for the migrant countries, and the effects of globalization.

### *In the Framework of Human Rights*

According to the Universal Declaration of Human Rights (United Nations 1948), illegal migrants—as part of humankind—should benefit from human rights wherever they are, regardless of their race, sex, nationality ethnicity, language, religion, or any other status. Human rights also include the rights to life and liberty, freedom from slavery and torture, freedom of opinion and expression, the right to work and education, and many more rights. However, as a solution, it is currently being reported that a group of wealthy countries have refused to recognize migration as a human right.

### *In the Framework of Economic Opportunity*

On one hand, host countries may help illegal migrants—as "disadvantaged communities"—to achieve their economic independence through new approaches to help them build their skills, access labor markets, and improve their economic security. On the other hand, African leaders should rethink their approaches to their countries' development programs and should establish plans and programs through industrialization and opening up opportunities not only for young people to work as civil servants but also to help them become entrepreneurs.

### *In the Framework of Unemployment*

It is said that sub-Saharan African governments should work to minimize private, internal, and external investment arrangements, provide infrastructure, and promote political systems that allow the majority of their citizens to participate. Among the solutions for restraining illegal migration is the proposition of the Addis Ababa-based UN Economic Commission for Africa (ECA) to leaders to concentrate on job creation policies on the continent. The ECA noted this in its Economic Report on Africa 2005 (ECA 2005).

## *In the Framework of the Brain Drain*

Because African countries have failed to create jobs for their young talent, their migration to Europe is viewed as a gain for Africa. In this regard, Mr. Kofi Annan, the former UN Secretary-General, focusing on the reality of the migration, has said, "We cannot ignore the real policy difficulties posed by migration, but neither should we lose sight of its immense potential to benefit migrants, the countries they leave and those to which they migrate". And Ms. Amina Muhammed the current Deputy Secretary-General has said, "Migration is not a zero-sum game."

## *In the Framework of Labor Shortages*

Populations in developing countries grow rapidly. Labor shortages in Europe may be an opportunity for those who arrive, because Europe is now facing labor shortages in highly skilled areas such as information technology. European countries employ immigrants, and Europe mainly selects those with ability or capital to invest. This kind of hiring system should be negotiated between countries (African and European), not individually, and it should be linked with education and training opportunities.

## *In the Framework of Globalization*

The research on migration and profiling of migrants show that most migrants are between 18 and 40 years old. These men and women and also children (some born at sea) have fled from Africa to Europe to seek jobs and well-being. In other words, those who should be the future of Africa have left their countries to save their lives and improve the well-being of their families. Potential solutions to illegal migration should be considered within the framework of globalization. Solutions to restrain the flow of migrants from Africa have been elaborated in several aspects, such as human right, economic opportunity, labor shortages, unemployment, the brain drain, multiculturalism, and asylum seeking. For example, Canada can provide a good example of multiculturalism. Recently, Canadian authorities gave immigration status to a number of African migrants rejected by Israel. It is reported that, around 2000 asylum seekers who have requested either immigration status or asylum in

Canada will be accepted and processed there. In Europe, currently the most popular decision is to offer the migrants a cash lump sum and a plane ticket to leave the country voluntarily. There is also an integration program for migrants, although it does not cover all of them.

## Conclusion

The phenomenon of illegal migration has become one of the biggest challenges of the twenty-first century. We realize that retraining the migrants, diminishing the risk of travel and the number of death cases, loss of money, and cracking down on smugglers may be the effective solutions even though they seem to be contrary to human rights and morality.

## Recommendations

- To face illegal migration challenges, sub-Saharan African leaders should have data and gather information about the entire process. Data about middlemen, transporters and routes is needed to manage the movement of people at their borders within the framework of the rights of international migrants. Increasing awareness programs about the risk of the journey and a search of how to provide alternatives for potential migrants and to those who have come back are needed.
- Efforts should be made to prevent the tragedies by banning migration.
- The smugglers should be identified, trained, and assisted in finding other jobs.
- Asylum seekers should receive assistance to integrate, to work or to return to their home countries.
- Africa countries should involve their youth in governance and decision making.
- African countries should provide mass education to their populations in their own languages to teach them basic knowledge—reading, writing, and basic mathematics—to boost the development of the countries.
- The educational system should focus on technical and professional education, and the system should motivate women and young people to open their own businesses to provide decent jobs and promote small and medium enterprises.
- The educational system should focus on science not only learning the European languages.
- Problems in African countries should be resolved with inter-African solutions.
- African countries should not concentrate on financial and developmental aids from Europe or others that do not actually help to efficiently develop the countries.
- Reforms should be made with non-governmental organization (NGOs) working in sub-Saharan Africa to implement more policies and programs that support young people and contribute to diminish unemployment and migration.

## References


Bob-Milliar, G. M. (2012). *Rescuing migrants in Libya: The political economy of state responses to migration crises. The case of Ghana* (No. 2012: 16). DIIS Working Paper.

Düvell, F. (2008). Clandestine migration in Europe. *Social Science Information, 47*(4), 479–497.

ECA. (2005). *Economic Report on Africa 2005: Meeting the Challenges of Unemployment and Poverty in Africa. Economic Comission for Africa.* Addis Ababa, Ethiopia: ECA Publication. https://www.un.org/en/africa/osaa/pdf/pubs/2005era-uneca.pdf.

Fearon, J. D. (2005). Primary commodity exports and civil war. *Journal of Conflict Resolution, 49*(4), 483–507.

International Organization for Migration. (2018). *World Migration Report 2018.* Switzerland: The UN Migration Agency.

Picciotto, R. (2010). Conflict prevention and development co-operation in Africa: An introduction. *Conflict, Security & Development, 10*(1), 1–25. https://doi.org/10.1080/14678800903553787.

United Nations. (1948, 10 December). The Universal Declaration of Human Rights. https://www.un.org/en/universal-declaration-human-rights/.

Venkatasawmy, R. (2015). Ethnic conflict in africa: a short critical discussion. *Transcience, 6*(2), 26–37.

# Chaos in the Future: Artificial Intelligence as a Strange Attractor of the Future

**M. Şahin Bülbül**

**Abstract** Studies that give opinions about the future and that focus on future research are known as futurism. In this sense, futuristic studies of the future predict that the greatest effect will be in the field of artificial intelligence (AI). This analysis, which explains that this big factor is one of the strange attractors in future modeling of the chaotic shaping of the future, has a theoretical structure. The similarities between human intelligence and artificial intelligence, which are the product of human intelligence, are increasing day by day, and artificial intelligence has become the one we can rely on, even in matters that we are most anxious about. International airlines use these systems that can make more rapid decisions that are error-free and less expensive, and the health sector can use them without hesitation for routine and general follow-ups. The fact that artificial intelligence will be encountered in every field in the future necessitates questioning the structure of artificial intelligence that shapes the future. For example, we believe that the future of our lives will be shaped with flawless decisions. Strange attractors are the biggest factors in the process. A period of time in which artificial intelligence is not included in the process of fictionalization and shaping of the future is not possible.

**Keywords** Chaos · Strange attractor · Artificial intelligence

## Introduction

It is very natural that humans are concerned about their future, especially if the universe exhibits chaotic behavior. The unexpected events we have experienced in the past remind us of the unexpected events we will experience in the future. One of the most well-known concerns about the future is whether artificial intelligence (AI), which is our slave for the time being, could one day declare its independence and even enslave us (Lehman-Wilzig 1981). Although this question does not pass

M. Ş. Bülbül (✉)
Kafkas University, Kars, Turkey



Ş. Ş. ERÇETİN, Ş. N. AÇIKALIN (eds.), *Chaos, Complexity and Leadership 2018*, Springer Proceedings in Complexity, https://doi.org/10.1007/978-3-030-27672-0_17

through the minds of the people dealing with everyday problems, it is constantly raised by experts. This chapter addresses the issue of how artificial intelligence will shape the future through chaos theory.

Certain surgical developments in the field of medicine and the fact that devices are getting smaller are strengthening the idea that devices will be injected into people. Thus, advanced, technology-supported people will be created. This goal is directly related to the concept of immortality, which is the greatest longing of the human being. When we look at the many social and psychological conditions revealed by the instinct for survival, the desire to be immortal can be understood. The human being, who knows that he has to build a new human being and a new world for immortality, is no longer the point of imagination. The first concrete indicator is the rapid evolution in artificial intelligence.

## Commenting on the Future: Futurism

Everyone can talk about the future, but to do research on the future and to make consistent comments on it is now a profession, and the name of this profession is futurism (Potts 2018). How will health, education, and trade be shaped in the future? Data-based and consistent interpretations are needed when commenting on these issues. The most important point that futurists should focus on is artificial intelligence.

Futurism has two golden rules. These rules should be observed when commenting on the future. The first rule says that the time and location should be well defined when thinking about the future. For example, how far in the future are we talking about: the next 10 years or the next 100 years? Also, the location is very important. Are we talking about Africa or Europe in the next 10 years? It is very difficult to argue that students in poor schools will be educated using holographic or other new technological methods in 10 years' time. The second golden rule is used to explain the exceptional circumstances of the first golden rule and says that progress is not always linear; sometimes unexpected changes occur. This is possible because of the chaotic mechanisms of nature. Let us look at the example given for the first golden rule again. We have to be careful about time and location when talking about the future. In the next 10 years, there will be smart technologies in schools; the first golden rule says that some low-income schools will not follow this prediction, but, according to the second golden rule, a wise school principal might be assigned to that school and the fate of the school could change.

Let us answer the question of whether artificial intelligence will take over our lives from the perspective of the two golden rules. First, how far in the future are we talking about? For example, in 10 years we will probably dominate artificial intelligence, but if we talk about the distant future rather than the near future, the situation may be uncomfortable. Moreover, according to the first golden rule, we know that even if we talk about the distant future, the whole world will not be at the same level; there will still be lives that are not influenced by artificial

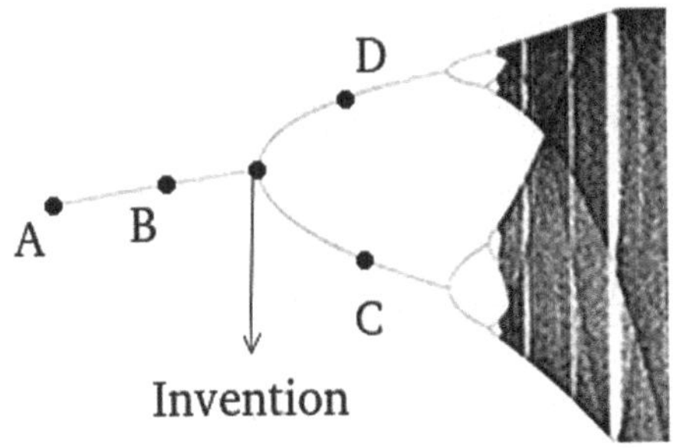


**Fig. 1** Bifurcation diagram of developments in mobile phones

intelligence. At that time, in the contemporary society, the possibility of an artificial intelligence–free civilization will appear. As we know, this could be considered a result of the second golden rule. Movements against artificial intelligence or behavior by artificial intelligence that brings its own end are alternatives to the idea that artificial intelligence will spread and grow. These small possibilities are never impossible because in natural phenomena there are always jumps (through evolution or paradigm shifts). These jumps come from the chaotic behavior of nature. In the long term, although there is a model of chaos theory, unforeseeable changes and transformations are inevitable in the interim. The development of artificial intelligence as a branch of a tree could evolve into a completely different structure with a new invention. According to the second golden rule, we can easily say that the future does not have a predictable structure like a ruler; it has a continuously fringing structure like a branch of a tree, known as bifurcation (see Fig. 1).

Consider the changes in mobile phones. The capacity, speed, features, and size of mobile phones are important. Let us examine the specific property of size. Over time, we have observed that mobile phones tended to shrink. This tendency to shrink continued to certain points (points A and B), and then a new invention emerged, transforming all linear change into a branched structure (bifurcation). With this invention, smaller phones were being produced (point C), but larger, more feature-rich, faster, and higher-capacity smartphones started to be developed (point D). This is the nonlinear and unpredictable side of development.

Considering this example, we can evaluate the development of artificial intelligence in our recent history. One of the areas where people show that they are intelligent is the field of intelligence games. It is a fact that everyone considers chess champions clever. In 1997 the chess champion Garry Kasparov was defeated in a chess game by the IBM supercomputer Deep Blue, and all people started to experience fear of artificial intelligence. Was the computer superior to human intelligence? Until then, calculators could do the operations we wanted to do in a very short time, but we still could not make calculators integral, derivative, etc. We could only use them to perform operations (points A and B). Functional calculators were produced over time (point C); however, chess applications can now be defeated by applications (apps) on our mobile phones (point D and other unspecified points). Thus, we can say that in the future, artificial intelligence will undergo an unpredictable change from our previous experience. Let us see if the first golden rule works. Yes, there are still some nonfunctional calculators in the workplace, and in some parts of the world, there are still children performing the four operations with an abacus.

## A World of Possibilities

To briefly explain artificial intelligence, it is a structure that creates a big library of situations that have been encountered, and it is able to quickly examine all of the possibilities in this library with the algorithms taught to it and give output (sound, text, movement, etc.). In this sense, having a library (data storage memory) and fast processing (high central processing unit (CPU) intelligence) are two very important concepts.

Artificial intelligence does our tasks if we use the data in the library in the desired direction. If the task (algorithm) that we give to it contains an explicit process, then the outputs it reveals will be defined by us in two ways: "How creative these are!" or "these are nonsense!" We will define what we find meaningful as creative and what we find meaningless as nonsense. Artificial intelligence will look holistically, as it will try all possible situations and not make meaningful or meaningless distinctions, unlike us. Our brain is a highly complex code that encodes repetitive and sequential events (Konar 1999). If artificial intelligence yields outputs over successive and repetitive structures, the results may be close to our own results, but the output of artificial intelligence will change if a holistic perception of meaning is established. Then the artificial intelligence data entry will continue to require the concept of "truth in the moment." This concept is the first philosophical need of artificial intelligence. In the process, some results that will be reached in the process will actually be in the world of possibilities.

The fact that some possibilities are meaningless in our world does not mean there can be no meaningful structures in other worlds. In this sense, when artificial intelligence starts to use chaos theory as a means of interpreting/modeling events, it will serve as our perceiver by seeing fractal structures as repetitive and ordered structures, but it will prevent us from seeing the real big picture. So humans must decide: will artificial intelligence be shaped according to its own truth in the moment (instant reality), or will it shape both our own future and its own future through integration of laws, just as we have?

## A Strange Attractor: Artificial Intelligence

In 1963, Adward N. Lorenz dealt with the interaction of three basic parameters to simplify atmospheric phenomena and demonstrated the interdependence of the parameters in the atmosphere in a model. In this model, the attracting areas that operate the relationship are called "strange attractors" (Sparrow 2012).

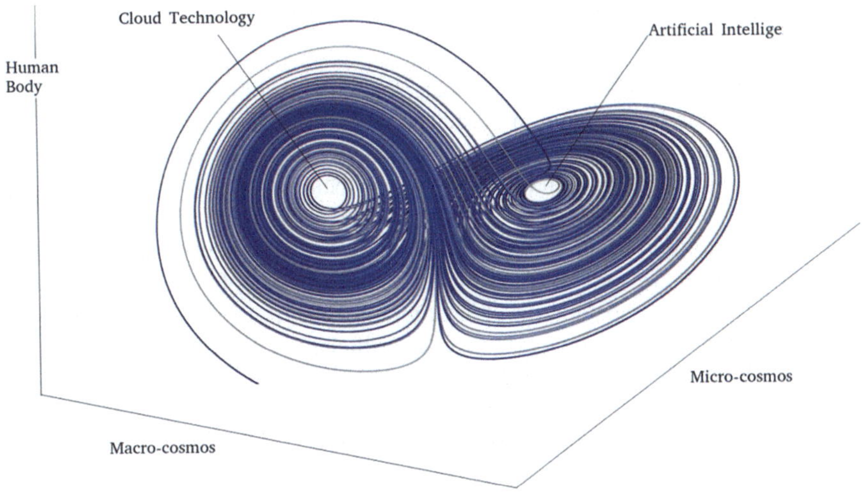


**Fig. 2** Model of the instinctual immortality plan of humans

In the context of the nature of scientific studies, at the beginning of this analysis, we aimed to explain the efforts of human beings with regard to immortality, such as Lorenz's explanation of atmospheric events. If we focus on the flow of scientific research, a model may arise. Sometimes research on living things such as biology and genetics is popular, and space, the universe, and physical events are popular. If we divide scientific research into three dimensions—those related to the macrocosmos, those related to the microcosmos, and those related to human beings—whichever of these three sets of parameters are chosen will give us a model for studies to increase those sets of parameters. This model can be called the instinctual immortality plan of humans. The instinctual immortality plan of humans determines the direction of the flow of scientific research (Fig. 2).

Humans wants to be immortal. Colonization of space, researching other planets that contain life, organ transplants, bionic arms, endless energy from the sun, and microcosmic levels are all products of this instinctive desire in humans. Human immortality is possible with virtualization of the way of thinking of immortality (artificial intelligence) and finding a place where the immortal body can live (a virtual environment, i.e., cloud technology). According to this model, artificial intelligence is the most important strange attractor in humanity's instinctual plan for immortality, and the future of humanity will be shaped in accordance with this model.

## Conclusion

Developments in perfect decision-making technologies (artificial intelligence) seem to be an opportunity for acquiring renewed technological errors, but there is no chance of updating the human brain at the same speed. The effects of artificial intelligence on the values produced by human beings and the effects of those effects on the current situation are among the objectives of this analysis. Artificial intelligence not only manages decision-making processes but also changes human behavior and changes the values of humans at the points of its interaction with human beings. The formation, development, and changes of values in the human brain are similar to the working principles of artificial intelligence and develop like the artificial intelligence of learning systems. The algorithm of values is formed by the ordering of needs in the human brain. In this sense, how the needless systems of artificial intelligence will produce values will also cause a change in the perception of the future model.

How would artificial intelligence and its structure lead to formation of a model in the processes of fictionalizing and shaping the future? On the other side of this question is chaos theory itself. Will the world's most revealing theory (chaos theory) evolve into an old form, evolving toward growing data and imperfect analysis systems, where constrained data and flawed decision-making mechanisms are increasingly evolving? For example, will approaches that address the system as a whole be able to explore interconnected systems with robust and big data and illuminate the chaotic loops and fragment–whole transitions of the fractal structures within the chaos? What new and old theories will shed light on which theories in the world of science will shed light on the transitions? If artificial intelligence, which is the assistant of humanity, can declare its own independence, with which theory will it choose to explain the universe and the structures in the universe? The adaptive subject of these two main sets of questions is the analogy and the decomposition between the structures of the brain in which artificial intelligence works and the structures of the human mind. For example, the decision-making system formed by the electrical conduction of nerve cells in the neural network and artificial intelligence decision-making technologies based on semiconductor systems should be compared. When anomalies in the brain turn into creative ideas, the damage to the image quality caused in the processes of recording the image for a long time and the changes in the memory of an event by recalling it make it necessary to question the aforementioned relations.

## References

Konar, A. (1999). *Artificial intelligence and soft computing: Behavioral and cognitive modeling of the human brain*. Boca Raton: CRC Press.

Lehman-Wilzig, S. N. (1981). Frankenstein unbound: Towards a legal definition of artificial intelligence. *Futures, 13*(6), 442–457.

Potts, J. (2018). Futurism, futurology, future shock, climate change: Visions of the future from 1909 to the present. *PORTAL Journal of Multidisciplinary International Studies, 15*, 1–2.

Sparrow, C. (2012). *The Lorenz equations: Bifurcations, chaos, and strange attractors*. New York: Springer.

# Financing Higher Education in Sub-Saharan Africa: A Proposed Model Based on the Experiences of Ugandan Higher Education Institutions and Exemplary Practices from the Asian Tigers

**Miiro Farooq and Ssekamanya Siraje Abdallah**

**Abstract** Financing higher education has become one of the most contentious educational issues in both developed and third world countries. Among the causes highlighted by researchers are changing fiscal policies and government resource allocation priorities, coupled with rapid increases in college-going populations. Today, there is an upsurge in the numbers of colleges and enrollments all over the world. This upsurge and the forces of globalization, which put demands on institutions in terms of international rankings and the need to adopt the latest technologies, are putting enormous fiscal pressure on higher education institutions and their funders. Increasing enrollments without sufficient funding lead to substandard facilities and a decline in academic standards. To close that funding gap, institutions have resorted to privatization, which includes (inter alia) tuition charges and other fees. This has effectively turned the traditional model of higher education upside down. The challenge is: How do universities ensure they admit only the most qualified students and provide them with the best instruction, using state-of-the-art facilities, in a learning system run by highly qualified and motivated faculty staff? Against that background, this chapter analyzes best practices for financing higher education in developing countries that have vibrant higher education sectors (such as Malaysia), together with the challenges, practices, and experiences of funding Ugandan higher education institutions. The chapter ultimately suggests a suitable framework for higher education financing in Uganda and similar sub-Saharan African countries.

**Keywords** Sub-Saharan Africa · Higher education · Asian Tigers

---

M. Farooq (✉) · S. S. Abdallah
Islamic University, Uganda, Mbale, Uganda



Ş. Ş. ERÇETİN, Ş. N. AÇIKALIN (eds.), *Chaos, Complexity and Leadership 2018*,
Springer Proceedings in Complexity, https://doi.org/10.1007/978-3-030-27672-0_18

## Introduction

Today education is taken as a good for transforming human life in terms of civility, economic growth, capital development, and individual lives. Because it is a private good that is envisaged to change the entire human spectrum and the environment, many countries all over the world have invested immensely in this sector. For the above reasons, from the start of the twenty-first century onward, the higher education (HE) sector has realized a big influx of student enrollments. This kind of pressure has been put on these institutions because of increased demand from all levels, including individuals, groups, families, government, occupations, and social status (World Bank 2012). Since the economy of a country can be greater than the education of its people, it is evidenced that through higher education, people's mind-set and lives are improved. Also, societies become more harmonious because of civility, wisdom, exposure to approaches that are employed to handle conflicts. Economic strength and transformation become strong and steadily progressive because of the multiple skills and experiences that citizens of the given society are exposed to. Likewise, countries in Europe, Asia, and some parts of Africa that have invested highly in the higher education sector have not only developed their citizens but also gone on to spread their culture and influence to several regions of the world. For instance, China, Russia, and the USA have a lot of influence in the world order and its markets because they have graduates in influential positions all over the world. For these reasons, this chapter highlights ways through which higher education investment can be implemented to change the lives of African citizens, especially in Uganda.

## The Problems

The university-going age group (18–30 years) is increasing in size. According to the Uganda Bureau of Statistics, this group constitutes about 20% of Uganda's population. Considering the rate of population growth, it will increase even further, considering because Uganda has the youngest population in the world, with children and adolescents (aged 0–18 years) constituting 55% of the population. This is coupled with efforts to democratize education and increase access at all levels. These have been in the form of (1) universal primary education (UPE), which has created opportunities for all children, regardless of their family incomes, to get access to primary education; and (2) and universal secondary education (USE), which has created opportunities for those who have completed primary education to continue with secondary education. The result is that there is an increased number of students who complete secondary education and desire to proceed with tertiary education. Whether they are of the preferred quality or not is beyond the main scope of this chapter.

This increase in demand for tertiary education, however, has come at a time of reduced government funding for higher education. Over the years, there has been an argument in favor of expanding public funding of basic education and leaving the burden of funding tertiary education to be carried by private individuals. Of course, in situations of limited resources, priorities have to be set.

Funding bottlenecks have been aggravated by reduced overseas support in the wake of the global financial crisis. This saw a reduction of budgetary support from overseas development funding across the board, education inclusive.

To make things worse, high levels of corruption mean that a significant portion of the meager resources available is siphoned off by unscrupulous officials at all levels of the education hierarchy, starting from the ministry all the way down to the department. This means that only a small portion of the allocated amounts actually reaches the target group or program.

Things are further complicated by the fact that the majority of the population is so poor that it is not feasible to use student fees alone as a source of income for universities. Most families are too poor to afford the actual budget cost of providing a quality education for their children.

## *Funding Problems*

The net result of the scenario depicted above is that institutions of higher learning in Uganda, both public and private, are suffering from an acute shortage of funding. Having some degree of budgetary allocations and subsidies from the government while at the same time charging fees to students, public universities are doing better financially than their private counterparts (National Council for Higher Education 2014). Their situations are complicated only by archaic bureaucratic procedures and excessive political interference.

Funding problems have led to low enrollment and high dropout rates among the poor. Many capable students from poor families cannot afford to pay the tuition fees and living expenses charged by universities. Experts in higher education financing have identified a mechanism that a poor family uses to decide whether they should pay for higher education. The primary method is to conduct a cost–benefit analysis by determining how much the family will benefit from paying for one of its members to continue with higher education. This analysis considers not only the direct costs of higher education in the form of fees and living expenses but also the opportunity costs, or what the family will forego in terms of services and income earned by their son or daughter when he/she goes for further studies. In rural economies where children provide labor on farms, in homes, and in family businesses, this can be significant. Where resources are limited and the family has a number of children, the cost–benefit analysis will involve deciding who to sponsor for higher education and who should drop out and help the family. These are indeed very difficult decisions. Underlying this situation is what educational economists refer to as the liquidity problem, meaning that the family has limited cash resources.

This is exactly the case in the majority of Ugandan families, who (according to United Nations estimates) live on less than US$2 a day. To make things worse, the very few available scholarship programs are ostensibly given on merit and are not means tested. The net result is that they go to the children of the rich who can afford to attend expensive boarding schools and score the highest grades in national examinations.

## *Poor Facilities and Programs*

In its many annual reports titled *The State of Higher Education,* the Ugandan National Council of Higher Education has consistently shown that the facilities at higher education institutions (HEIs), both public and private, are not adequate. In line with pressures arising from an increasing population and reduced funding, it is noticeable (except at a few private campuses) that the standard of facilities and infrastructure is less than desirable. There are not enough lecture rooms to accommodate the student populations in the popular courses, leading to overcrowding. For courses that require access to and use of information and communications technology (ICT) hardware and software, only outdated equipment and software are available most of the time. Hence, most of the learning is theoretical instead of practical. This is further complicated by the unavailability of the latest books and journals, especially in rapidly changing areas such as ICT. Internet access is very poor and networks are usually overloaded. Moreover, there is a dearth of science and technology labs and equipment, thereby making most of the learning theory based instead of practical. This, coupled with the fact that very few students take and pass science subjects in high school, hence leading to acute shortage of scientists and engineers in the country. Moreover, because of a lack of funds, universities are not capable of providing suitable on-campus accommodation for all students. As a result, many private hostels have sprung up, mostly lacking basic amenities and conducive learning environments (National Council for Higher Education 2014).

To make things worse, there is an acute shortage of qualified senior staff who have research experience and are capable of supervising quality research projects. A majority of the lecturers hold only bachelor's degrees, followed by master's degree holders. PhD holders are a minority. This has led to a situation whereby universities have been reduced to "overglorified high schools"—places where knowledge is consumed but not produced. Because of the shortage of staff, the few that are available are teaching at several institutions, a phenomenon referred to as moonlighting. When lecturers are busy moonlighting, we cannot expect them to give quality instruction to their students. This situation is worsened by the often poorly designed and archaic curricula at most of the institutions. Lecturers usually employ outdated instructional strategies they themselves inherited from their own lecturers (National Council for Higher Education 2014). In short, this system is not capable of producing graduates equipped with twenty-first century skills who are capable of functioning and being competitive in a modern, knowledge-based economy (Singh 2014).

While most Ugandan and other African scholars and administrators are fond of looking for solutions from the advanced countries of the West, in this chapter we propose that it is better to look at recently developed countries in Asia and Latin America. The reason for this is that the developed nations of Europe and America are so far ahead and have been developed for so long that it is unimaginable for poor countries such as Uganda to catch up with them. The story is different when it comes to recently developed countries such as the Asian Tigers. These countries were at the same level of development as Uganda and other African countries at the time of their independence. As a result of good leadership and enlightened setting of priorities, they have managed to get out of poverty.

## The Experience in Malaysia

Southeast Asian countries achieved amazing levels of economic and social development especially in the 1990s, when the term "Asian Tigers" was coined. Countries that had started off lower than most African countries at the time of independence were now boasting of entering the middle-income range. Indeed, countries such as Singapore and South Korea are now among the high-income countries (OECD/Asian Development Bank 2015). Since it is difficult to do justice to this topic by studying the mechanisms of financing of higher education in all Southeast Asian countries, we have decided to focus on Malaysia (for studies of the entire region, see Marginson et al. (2011) and Armstrong and Chapman (2011). In terms of social and economic development, Malaysia is neither too high nor too low. Thus, it is easier to imagine how to implement the Malaysian model in an African country such as Uganda. Moreover, like Uganda, Malaysia was colonized by Britain; thereby, both countries share a common colonial legacy. It is believed that at the time of Uganda's independence in 1962, the two countries were almost equal in all aspects.

Without delving into the history of Malaysia and how it ended up having the system it has today, we will summarize the structure of higher education financing in Malaysia. In comparison with neighboring countries such as Indonesia, the Malaysian system is very much dominated by the public sector (Armstrong and Chapman 2011; Regel et al. 2007; Sirat 2010). Even developments in the private sector are, by and large, facilitated by progressive government policies. The main thrusts of funding mechanisms for higher education are in the form of federal government funding, government-linked companies (parastatals), state government initiatives, the private sector, and religious institutions.

### *Federal Government Funding*

Most of the financing for higher education institutions has been dominated by the public sector (Wilkinson and Yussof 2005). The main form of funding provided by the government is in the form of land grants, whereby public universities are allocated land by the federal government for constructing campuses and other required facilities. Additionally, public universities are allocated funds for constructing campuses and other required facilities such as ICT infrastructure. These major expenses are usually allocated in the 5-year development plans of the country, as well as in the annual budgets. The Malaysian federal government spends billions of dollars annually on this item (Regel et al. 2007; Sirat 2010).

Another important expense borne by the federal government is the recurrent or operating expenditure of universities. The Malaysian federal government used to cover all of the operating costs of universities, including salaries, emoluments, and other overheads. However, the government has recently required all public universities to generate at least 30% of their total recurrent expenditure. Many public universities are dealing with this through strategic investments and provision of training and consultancy services, as well as imposition of tuition fees (Ahmad et al. 2013; Sirat 2010).

The requirement by the federal government for universities to generate at least 30% of their operating budgets has made many of them think outside the box. Many of them are trying to do so through the traditional channels of research and consultancy, with investments in providing specialized services to the public, especially in the area of health (e.g., the UM (University of Malaysia) Medical Center, UKM (Universiti Kebangsaan Malaysia/National University of Malaysia) Hospital, and IIUM (International Islamic University Malaysia) Specialist Hospital). Some universities have ventured into serious business investments—for example, IIUM Holdings, which includes a trading arm (IIUM Trading), a preschool education arm (IIUM Montessori), two elementary schools, two secondary schools, a high school, and a private diploma- and degree-awarding college. Moreover, universities, either alone or in partnership with private companies, have established learning centers in Malaysia and abroad, offering their courses to students whose work or family commitments would otherwise make it difficult for them to enroll. This has increased opportunities for people to enroll in higher education while contributing to the universities' income. Most of these courses are offered in an executive mode, which attracts more fees than the normal mode. Such programs include the UTM (Universiti Teknologi Malaysia/Malaysian University of Technology) School of Professional and Continuing Education (SPACE), the IIUM Center for Continuing Education (ICCE) and Center for Strategic Continuing Education and Training (CRESCENT), and the UM Center for Democracy and Elections (UMCEDEL).

Another mechanism for funding institutions of higher learning is through research grants. The federal government offers research grants in strategic areas through the Ministry of Education (MOE) and the Ministry of Science, Technology and Innovation (MOSTI). MOE runs the highly competitive Fundamental Research Grant Scheme (FRGS), which is open to all institutions of higher learning. MOSTI awards two main grants (through the E-Science Fund and the E-Techno Fund) aimed at generating basic ideas in science and technology, and developing those innovations to the stage of commercialization. Through these grants, successful researchers generate additional funds for themselves and their graduate students, and acquire the latest equipment for their universities.

To encourage competition in research activities, the federal government has introduced the Research University (RU) status, a competitive program whereby public universities are assessed in terms of their research capacity and outputs. Institutions that fulfill the requirements for RU status are given an extra research budget allocation (US$10–20 million per annum) over and above their usual development and operating budgets. As of now, five universities have attained that status: UM, UKM, UTM, USM (Universiti Sains Malaysia/Malaysian University of Science), and UPM (Universiti Putra Malaysia/Putra University of Malaysia). The federal government has also set up several world-class research and innovation centers of excellence in different strategic areas, such as the Malaysian Agricultural Research and Development Institute (MARDI) within UPM.

For the past 10 years, the government has focused on improving standings in international rankings—the most important indicator being the number of publications in top journals indexed in Scopus and the World of Science/ISI (International Scientific Indexing) indexes. In addition to requiring research universities to produce a specified number of high-ranking publications, all other universities are ranked on the basis of their research and publications. Most non-research universities give incentives of about US$500 per paper published in an indexed journal.

In addition to research and publication output, other quality assurance measures have been put in place and funded by the federal government. These include supervision by the Malaysian Qualifications Agency (MQA). All public and private institutions are judged by the same standards. A star ranking system is used for all public and private institutions under the Malaysia Research Assessment Instrument (MyRA) and Rating System for Malaysian Higher Education Institutions (SETARA) frameworks. Most institutions have International Standards Organization (ISO) quality assurance certification to ensure that all programs, facilities, and processes are managed at international standards. Part of this process is the use of outcome-based management. Universities have adopted the Balanced Score Card system, which requires setting of clear key performance indicators (KPIs) whereby everyone (leaders, administrators, and faculty members) is accountable for achieving his/her KPI. These measures have a real impact on annual salary increments, promotions, renewal of contracts, and other entitlements (Wilkinson and Yussof 2005).

As expected, one of the main avenues of higher education funding is through provision of scholarships. The Malaysian federal government gives a combination of merit and needs-based scholarships. Top performers in the national high school examinations are given Jabatan Perkhidmatan Awam/Public Service Department (JPA) scholarships whereby they are sent to the top universities in the world. Upon graduation they are expected to serve in various government departments, depending on their specialization. In addition to the JPA scholarships, there are affirmative action scholarships under the Majlis Amanah Rakyat/People's Trust Council (MARA), which are mainly targeted toward indigenous students from poor families. They are sent to boarding schools in the country where they can access good facilities and excellent tuition. Upon successfully completing high school, the best are given scholarships or convertible loans to pursue higher education at universities of their choice outside Malaysia and domestically. Convertible loans are loans that are converted into scholarships if the student attains a first-class degree (Tham 2011; Sirat 2010).

In 1998, Malaysia established a higher education loan fund for all Malaysians regardless of ethnicity, religion, or location. Every student can apply. However, the board considers each applicant's parents' incomes to decide on the amount to be given. Some students get 100% of the amounts they apply for, while others are given only a percentage. After graduating and securing a job, the borrower is then expected to make scheduled monthly payments (Regel et al. 2007; Ahmad et al. 2013). As expected in programs of this nature, there have been teething problems, with many borrowers refusing to pay after they graduate. In response to this, the government has introduced tough measures, including barring defaulters from leaving the country. This program has helped many Malaysians to access higher education (Ismail and Fadzim 2009).

On the capacity development side, the federal government provides funds for universities to send their academic staff for further studies in good universities overseas. Some of the funds are specifically reserved for affirmative action for indigenous staff, especially those from poor families. As a result of this, most of the academic staff of public universities have doctoral degrees from good universities abroad.

To further encourage the development of higher education, the government gives incentives to government-linked companies and the private sector to establish higher learning institutions. These are discussed in the following sections.

## *The Role of State Governments*

Malaysia is a federation composed of 14 states. Most of the affairs related to education are centralized under the federal Ministry of Education. State education departments are directly answerable to the federal ministry. Nevertheless, most of the state governments have put measures in place to help citizens of those states to access higher education. The most common approach has been provision of scholarships and loan schemes for their citizens. Under this arrangement, state governments or foundations under the state governments offer scholarships or loans for their citizens who fail to secure funding under federal arrangements. Recently, some states have ventured into higher education by establishing state universities. For example, the state of Selangor has established Kolej Universiti Selangor/Selangor University College, the state of Perak has established Sultan Azlan Shah University, and the State of Kedah has established Insaniah University.

Moreover, in Malaysia, all matters of religion are the prerogative of state governments. Islam is the religion of the state and the traditional ruler (the Sultan) is the head of religion in the state. Religious institutions usually offer funding to higher education in the form of *awqaf* (endowments), scholarships, and other short-term funding for students who qualify (Ismail and Fadzim 2009; Aziz et al. 2013).

## The Role of Government-Linked Companies

The Malaysian government fully owns or has controlling stakes in many large publicly listed companies operating in a number of strategic areas. These include the national oil company (Petronas), the national electricity company (Tenaga Nasional), the Malaysian telecommunications company (Telekom Malaysia), some of the major banks (e.g., Malayan Banking Berhad), the railway company (Keretapi Tanah Melayu Berhad (KTMB), and Malaysian Airlines. Most of these companies have centers for training their staff. In the 1996 educational reform whereby Malaysia decided to become a regional higher education hub, the Malaysian federal government encouraged most of the government-linked companies to transform their training centers into fully fledged private universities (Wilkinson and Yussof 2005). At present, Petronas owns Petronas University of Technology (UTP), Telekom Malaysia owns Multimedia University (MMU), and Tenaga Nasional owns Tenaga Nasional University (UNITEN). These are among the top private universities in the country, educating many local students at subsidized rates and with a large percentage of foreign students paying full fees (Regel et al. 2007; Sivalingam 2007).

In addition to establishing their own institutions of higher learning, government-linked companies contribute to the finances of higher learning in other ways. Most companies have scholarship schemes for specific populations, and most of them offer scholarships for potential future staff. Others help students from rural areas. There is a new trend of major corporations sponsoring entire departments or helping to construct buildings on campuses or even to finance professorial chairs named after those organizations. Some government-linked companies such as the Federal Land Development Authority (FELDA), the Palm Oil Research Institute (PORIM), and the Rubber Research Institute (RRI) have established state-of-the-art research and development facilities, which they often share with researchers in universities.

## The Role of the Private Sector

Prior to the 1996 higher education reforms, the role of the private sector in higher education was mainly in the form of giving scholarships and other financial support to students to pursue further studies in Malaysia and abroad. Private entities such as Star Publications, the Sunway Group, the Genting Group, and the Berjaya Group used to offer scholarships every year to students who qualified for admission to good universities but were unable to meet the costs. A small number of companies operated commercial and technical colleges, mostly offering professional qualifications (such as the Association of Chartered Certified Accountants (ACCA) and the Chartered Institute of Management Accountants (CIMA). Some colleges started operating twinning programs whereby students would spend 1 or 2 years of their course in Malaysia and the remainder in a foreign country such as the UK, the USA or Australia, thereby saving on the total cost of the degree while acquiring a highly coveted foreign qualification (Lee 2004; Tham 2011).

As part of the 1996 higher education reforms, the Malaysian government decided to become a global player in the higher education market, with the private sector being a major driver. When the government announced its aspiration to have at least 100,000 foreign students by 2010, with each of them paying US$10,000 in tuition fees and spending an equal amount on living expenses, many analysts thought it was a joke (Wilkinson and Yussof 2005; Lee 2004; Sivalingam 2007). However, by 2012, Malaysia had more than 100,000 foreign students. The Higher Education Strategic Plan (2013–2025) then announced its goal of having 200,000 foreign students in the country by 2020. Most of these foreign students are enrolled in private universities and colleges whose main source of income is tuition fees. The expansion of this sector shows that Malaysia is collecting a good amount of revenue from this source. Building upon this success, some of the Malaysian private universities have started establishing branches in other countries (e.g., the Limkokwing University campuses in South Africa, China, and the UK). As we write this, private higher education in Malaysia continues to expand at a rapid pace (Wilkinson and Yussof 2005; Regel et al. 2007; Sirat 2010).

It is evident from the foregoing that the Malaysian federal government has taken the lead in financing higher education, as is the case in Uganda and other African countries. What is different is the monetary amounts involved and the mechanisms that have been used. Starting with the traditional government-dominated model of financing, Malaysia has progressively diversified its funding model, leading to a public–private partnership model and fair competition between public institutions of higher learning and private ones. This has led to rapid development of higher education in terms of both quantity and quality. Targeted spending on staff development, infrastructure, and research grants has tremendously improved the potential of Malaysia's higher education human resources. While most African universities are bottom heavy, with very few PhD holders and often no professors, in Malaysia it is not uncommon to find departments with more professors and associate professors than staff of other ranks. This, coupled with progressive incentives and funding for research and publications, is gradually pushing Malaysia's academics to be among the best in their fields.

## Lessons for Uganda and Other African Countries

To contribute significantly to economic growth and social development as a whole, governments should offer grants from their national budgets to institutions of higher education. This is because the driving forces of the globalized knowledge economy and its market are centered on the system of education, through which learners are exposed to skills and new technologies to enhance new developments for providing solutions to human challenges. However, this kind of input in terms of grants cannot be of great importance if learners do not have access to standardized facilities and efficient use of available resources. In Asia, governments provide grants to both private and public universities for the production of needed human resources, at the same time equipping their countries with the necessary skills to address their social and economic challenges. Facilities such as high-speed internet, high-tech computers, and standardized lecture rooms are in place, in addition to what lecturers are offered to carry out research (Ahmad et al. 2012). Furthermore, those lecturers work in a conducive and quiet atmosphere and, at the same time, are offered reasonable incentives and facilitation to encourage their research and collaboration. However, in Uganda the situation seems not to be at a level of high expectations although the government is envisaging middle-income generation by 2040. To achieve this strategy, the government must work hand in hand with higher education institutions to ensure that the exposure of their students to knowledge and skill rhymes with the needs of the time to catch up with giant countries such as China and Japan. Those countries invest most of their budgets in production of highly skilled human resources and thus stand greater chances of filling all of the necessary gaps in the world market (Welch 2016). In Uganda, for instance, higher education institutions are challenged with brain drain, moonlighting, and lack of knowledge creation and generation, which can be attributed to the poor remuneration and fringe benefits that staff are given. With low salaries and lack of grants and conducive environments for working, lecturers cannot settle and do reasonable work to change the society and generate economic development (Cloete and Maasen 2015; Kasozi 2014; Miiro et al. 2016).

In addition, governments may fund higher education institutions through loans to boost their megadevelopments. For instance, institutions that have survived for years not only have secured loans to run their budgets but also have been able to put up structures through endowments to strengthen their economic viability. Dependence on tuition fees has not only disadvantaged students but also stalled many university activities because in Uganda, many students come from poor families; therefore, to raise funds at once to pay for tuition charges is always complicated. Meanwhile, in other parts of the world, higher education institutions get 60% of their budget funding from the government and the rest is fulfilled by the institutions. For example, in China the funding system has been diversified whereby the state does not take charge of its supervision; rather, the government offers mixed grants and commodification of tuition fees comes as a supplement for university activities (Ahmad et al. 2012).

In Asian countries, governments fund staff development programs for all institutions. This is done by sending staff with different specializations to highly ranked universities all over the world. After completion of their studies, the staff are bonded and, at the same time, paid highly to motivate them to carry out government projects while teaming up to come up with innovative ideas, strategies, and policies that can enhance teaching, learning, research, and publication. Through this strategy, many governments have more than enough human resources to address the challenges of all kinds that may befall their nations.

In addition, students are entitled to government loans to ensure that there is no socioeconomic gap between different states, especially in Malaysia. Along the same lines, students fees are regulated by the government; therefore, universities cannot raise their tuition fees. Meanwhile, students who are given government funds are normally bonded after their studies, so the state treasury recovers its investment in those students and can then use that money for funding of new generations and other applicants.

In countries such as Uganda, students who enroll in these institutions depend on soliciting funding for their studies from relatives, friends, and good Samaritans. This kind of situation can be attributed to the increased level of poverty, economic volatility, economic gaps between families, and increased competition for national funding. It is therefore imperative that governments and other stakeholders take charge and rethink their strategies for helping students to acquire more useful education to change the economies of their countries. For instance, countries that invest in human capital all over the world stand greater chances of earning both economic and noneconomic benefits. For instance, educated citizens attain better jobs with higher pay, and the noneconomic benefits include intergenerational benefits, low crime rates, more literate and politically educated citizens, and low maternal and infant mortality rates (Gichuhi 2015). Nevertheless, governments that advocate for student loans (especially Malaysia and China) have not gone wrong in that their students' economic burdens are reduced and their higher education institutions also receive their incomes in a timely fashion. Through this strategic plan of improving their education levels, certain levels of adequacy, equity, and efficiency have been achieved. Kenayathulla and Tengyue (2017) state that the cost of fees and the levels of family income determine which students should benefit from governments loans. Therefore, efficiency and effective mechanisms for governments to sponsor students should be diversified to cater for all regions and all specializations, since each and every field of education contributes to both the economic and social growth of a nation (Ahmad et al. 2012; Gichuhi 2015; Kenayathulla and Tengyue 2017).

Also, the need to have a strong strategic plan for enhancing the efficiency of teaching and learning, research, development, engagement, and internationalization call for government funding so that these institutions can survive the sweeping global waves and their forces. This is because with contributions from fees alone, alumni cannot facilitate their daily operation and development and, the same time, facilitate their strategic vision of competing in the world market (Ahmad et al. 2012a). It is therefore important that funds are given in abundance to consolidate achievements and, at the same time, focus on the bright future to be gained by reforming the performance of these countries in the world market.

In Malaysia, universities come up with comprehensive strategic plans in which elements such as performance, quality, delivery systems, employability, marketability, research and development, and internationalization are embedded. This is done to ensure that innovative human capital is produced to address the general national strategic plan and development. Meanwhile, this strategy is also inclusive of staff development, both academic and nonacademic, with an agenda of sustainable development in terms of leadership and talent to avoid a vacuum and, at the same time, improve performance (Amran et al. 2014). For a country such as Uganda, there is a need to advise universities to strategically develop models that suit its political, economic, and cultural setting, as its challenges may be very different from those faced in other parts of the world. In addition, the budget and allocation of resources must tally with the economic growth agenda and, at the same time, involve all stakeholders across universities to support their implementation (Ahmad et al. 2012a, b).

Also, some Asian governments have dedicated some of their public institutions as research universities. This has been done to ensure that these institutions come up with innovative ideas and solutions to the many challenges that their citizens face. For instance, in Malaysia, UTM is purely dedicated to ensuring that Malaysia catches up with new trends in research and innovation, especially in the areas of ICT and engineering. The government has achieved this by providing at least 80% of the university's operating budget, and the balance of the budget is supplemented by student fees and other funds secured from grants and scholarships (Ahmad et al. 2012a; Amran et al. 2014)

The way forward for Uganda and other sub-Saharan African countries is to increase the amount of funding (as a percentage of the annual budget and as a percentage of the per capita gross domestic product) to a level at which actual transformation takes place. For this to be effective, however, the country must reduce leakages in terms of false claims, ghost workers, undelivered services, etc. There is need to reduce administrative costs and fund facilities, materials, ICT infrastructure, libraries, and staff development. To get an optimum return on investment, more emphasis should be placed on science, technology, and mathematics (STEM) right from primary education onward.

We must find low-cost ways of doing things, such as having digital libraries instead of physical books and journals, using low-cost technologies for construction, and using renewable energy such as biogas for cooking and generating electricity to operate equipment and lights. To increase access at a lower cost, efforts should be made to use open and distance education models, using readily available mobile telephone technologies and social media.

When it comes to scholarships and student loans, there is a need to improve targeting. Loans and scholarships should be awarded to those who really need them (the poorest of the poor who are willing to serve rural communities upon graduation) and not to those who are well connected. This will require improved transparency in the system.

Public and private sector companies and organizations can be encouraged to contribute to financing higher education by giving them tax waivers and other fiscal incentives. The government could even establish a department to assist the expansionary efforts of Ugandan institutions in Africa by capitalizing on the existing favorable perceptions about Ugandan education. It would be helpful to learn from Malaysia's "twinning program" strategy. Moreover, we need to establish a culture of endowments by alumni and the general public. For this to take place, however, transparent management is absolutely crucial, and ideally such endowments should be managed by the alumni themselves. The government should give tax-free status to such initiatives.

## References


Ahmad, A. R., Farley, A., & Kim-Soon, N. (2013). Funding reforms in Malaysian public universities: Transitions towards competitive funding. *Australian Journal of Basic and Applied Sciences, 7*(10), 553–561.

Ahmad, A. R., Farley, A., & Naidoo, M. (2012). Funding crisis in higher education institutions: Rationale for change. *Asian Economic and Finance Review, 2*(4), 562–576.

Ahmad, A. R., Farley, A., & Naidoo, M. (2012a). The study of government–university relationship in Malaysian higher education system. *International Education Studies, 5*(5), 25–34. https://doi.org/10.5539/ies.v5n5p25.

Ahmad, A. R., Pahat, B., Farley, A., & Naidoo, M. (2012b). An examination of the implementation federal government strategic plans in Malaysian public universities. *International Journal of Business and Social Science, 3*(15), 290–301.

Amran, F. H., Kamal, I., Rahman, A., Salleh, K., Noh, S., Ahmad, S., & Haron, N. H. (2014). Funding trends of research universities in Malaysia. *Procedia – Social and Behavioral Sciences, 164*(August), 126–134. https://doi.org/10.1016/j.sbspro.2014.11.060.

Armstrong, S., & Chapman, B. (Eds.). (2011). *Financing higher education and economic development in East Asia*. Canberra: ANUE Press.

Aziz, M. R. A., Johari, F., & Yusof, M. A. (2013). *Cash Waqf models for financing in education*. Paper presented at the 5th Islamic Economic System Conference (iECONS2013), Kuala Lumpur, 4–5 September 2013

Cloete, N., & Maasen, P. (2015). Roles of universities and the African context. In *Knowledge production and contradictory functions in African higher education* (pp. 1–31). Somerset West: African Minds.

Gichuhi, L. (2015). Alternative methods of financing higher education in Kenya. *International Journal of Scientific Research and Innovation Technology, 2*(5), 57–67.

Ismail, R., & Fadzim, W. R. (2009). *Financing education: Are there alternatives to student loans?* Paper presented at the Seminar Ekonomi Islam Peringkat Kebangsaan 2009 (SEIPK '09), San Josa, Costa Rica, 28–31 May 2008.

Kasozi, A. (2014). Academic corruption: A silent plague in universities. *Newvision Uganda*, p. 3.

Kenayathulla, H. B., & Tengyue, Z. (2017). Student loans in Malaysia and China: Equity, efficiency and adequacy. *Malaysian Online Journal of Educational Management (MOJEM), 4*(1), 64–85.

Lee, M. N. N. (2004). *Restructuring higher education in Malaysia*. Penang: School of Educational Studies, Universiti Sains Malaysia.

Marginson, S., Kaur, S., & Sawir, E. (Eds.). (2011). *Higher education in the Asia-Pacific: Strategic responses to globalization* (Vol. 36). Dordrecht: Springer.

Miiro, F., Othman, A., Sahari, M., & Burhan, M. (2016). A measurement model of talent management practices among university staff in central. *Journal of Positive Management, 7*(3), 3–19.

National Council for Higher Education. (2014). *The state of higher education and training 2013/2014*. Kampala: National Council for Higher Education.

OECD/Asian Development Bank. (2015). *Education in Indonesia: Rising to the challenge*. Paris: OECD Publishing.

Regel, O., Salmi, J., Watkins, A., Tan, H., Dawkins, J., Saroyan, A., & Vestergaard, J. (2007). *Malaysia and the knowledge economy: Building a world-class higher education system. Human development sector reports*. Washington, DC: World Bank Publications.

Singh, S. (2014). *Global trends in higher education financing*. Singapore: The HEAD Foundation.

Sirat, M. B. (2010). Strategic planning directions of Malaysia's higher education: University autonomy in the midst of political uncertainties. *Higher Education, 59*(4), 461–473.

Sivalingam, G. (2007). Privatization of higher education in Malaysia. In *Forum on Public Policy Online* (Vol. 2007, No. 1, p. n1). Urbana, IL: Oxford Round Table.

Tham, S. Y. (2011). *Exploring access and equity in Malaysia's private higher education* (ADBI Working Paper 280). Tokyo: Asian Development Bank Institute. Available: http://www.adbi.org/working-paper/2011/04/19/4513.access.equity.malaysia.higher.educ/.

Welch, A. (2016). Evolving ASEAN-Australia relations in higher education. Towards a regional knowledge network? *The International Education Journal: Comaparative Perspectives, 15*(1), 5–25.

World Bank. (2012). *World development report 2012: Gender equality and development*. Washington DC: The World Bank.

Wilkinson, R., & Yussof, I. (2005). Public and private provision of higher education in Malaysia: A comparative analysis. *Higher Education, 50*(3), 361–386.

# Autophagic Leadership: Promoting Self-Renewal of Organizations

**Mustafa Özmusul**

**Abstract** This chapter explores how the concept of autophagy, which is used in medicine and biology, might be helpful in developing leadership in social organizations. First, from the perspective of a social researcher, it attempts to explain how autophagy works in a biological organism by a simple framework. It then focuses on understanding the potential role of autophagy in social organizations through use of metaphors, models, and theoretical explanations. Finally, it develops a conceptual model of autophagic leadership as a contribution to utilizing this vital biological process in today's social organizations. The implications for the role of leaders arising from the autophagic leadership model are also analyzed.

**Keywords** Autophagic leadership · Self-renewal · Organization development

## Introduction

### *Autophagy*

The Nobel Prize in Physiology or Medicine in 2016 was given to Yoshinori Ohsumi from the Tokyo Institute of Technology for his discoveries regarding the mechanisms of autophagy. What is so important about the concept of autophagy that could earn a Nobel Prize? What does it mean to discover the mechanisms of autophagy?

According to the Oxford Dictionary (2018), the term "autophagy," which dates back to the mid-nineteenth century, means "consumption of the body's own tissue as a metabolic process occurring in starvation and certain diseases" or it can be explained as "destruction of damaged or redundant cellular components occurring in vacuoles within the cell."

M. Özmusul (✉)
Harran University, Şanlıurfa, Turkey
e-mail: mustafaozmusul@harran.edu.tr



Ş. Ş. ERÇETİN, Ş. N. AÇIKALIN (eds.), *Chaos, Complexity and Leadership 2018*, Springer Proceedings in Complexity, https://doi.org/10.1007/978-3-030-27672-0_19

Similarly, when explaining the role of autophagy in the human organization, Ohsumi said that the "human body is always repeating the auto-decomposition process, or cannibalism, and there is a fine balance between formation and decomposition. That's what life is about" (China Daily 2016). Yes, autophagy can be defined as a "self-eating mechanism."

However, according to David Rubinsztein, for people who have a bad lifestyle, such as snacking or consuming junk food, the benefits of autophagy might not be seen (BBC 2018). Dysfunction in autophagy plays a considerable role in cancer, neurodegeneration, microbial infections, and aging. It is normally a protective process, but in some conditions it turns out to be a lethal process (Mizushima et al. 2008).

The cellular autophagic process mostly starts in low under basal situations. Nutrient starvation, in particular, stimulates autophagy. It is also seen in the presence of physiological stress stimuli such as hypoxia, energy depletion, high temperature, or high-density situations; hormonal stimulation; pharmacological agents such as rapamycin; innate immune signals; and diseases (Mizushima et al. 2010). Also, it should be said that there is no perfect method for starting or stopping autophagy in the research setting (Mizushima et al. 2010).

## Autophagy in Social Organizations

It is understood that the autophagic process is very complex, and it is not seen in only one way. From the findings and explanations in biological organizations, we can see the autophagic process in at least two ways—a positive way and a negative way—in social organizations.

### *Positive Autophagy*

Positive autophagy can be seen if an organization gains benefit in a certain period, has a chance to be improved and developed during organizational starvation (which can mean lack of vital resources in the organization), or experiences organizational problems, which can prevent the organization from functioning well.

### *Negative Autophagy*

Negative autophagy can be seen if an organization suffers damage to its structure, vital functions, human resources, organizational culture, organizational climate, or management processes; during organizational starvation (which can mean lack of vital resources in the organization); or experiences organizational problems, which can prevent the organization from functioning well. If the organization is very sick, attempting an autophagic process will damage the organization.

## A Conceptual Model of Autophagic Leadership

As a conceptual framework, eight suggestions to lead autophagy in social organizations are as follows:

- End needless practices, procedures, rules, etc.
- Prioritize the internal resources of the organization during challenging periods.
- Make the organization healthy and vigorous.
- Make accurate diagnoses and give appropriate treatment.
- Permit some kinds of organizational starvation (e.g., limited lack of resources, wellness, or support; and the existence of certain problems).
- Balance the benefits and risks of autophagy in the organization.
- Protect autophagic mechanisms.
- Do not harm the running of the organization if it is functioning well.

### *End Needless Practices, Procedures, Rules, Etc.*

If we see autophagy as basically being a process of destruction of redundant components in a biological organization, we can consider that the autophagic process starts with demolition of needless practices, procedures, and rules. In particular, during a stressful situation in the organization, needless components of the organization can be removed so the organization can become more healthy, rigorous, and vigorous to cope with this stressful situation.

### *Prioritize the Internal Resources of the Organization During Challenging Periods*

Saving and prioritizing the internal resources of the organization can prevent negative autophagy. In particular, if the internal resources of the organization are sufficient for its function, its immune system will be enhanced. Excessive use of external resources can cause severe damage to the organizational culture and the organizational climate, as well as the financial structure.

### *Make the Organization Healthy and Vigorous*

One of the requirements to stimulate positive autophagy in a social organization is to make the organization healthy and vigorous. Establishing a positive organizational climate and creating an organizational culture that opens the way to development and improvement can make the organization healthy. Excessive conflicts between staff, bad management approaches such as decision-making processes that do not consider the needs of stakeholders, lack of management support, excessive workloads, exhaustion, bureaucracy, and burdensome protocols can be accepted as factors that cause severe damage to organizational health. An autophagic leader should care about organizational health.

## *Make Accurate Diagnoses and Give Appropriate Treatment*

We know that making an inaccurate diagnosis and implementing incorrect treatment produce negative autophagy, which can cause severe damage to the human organism. In particular, an attempt to change a social organization may jeopardize the health and vital elements of the organization if the diagnosis is inaccurate. For instance, policy makers in an education system may consider that the motivation and performance of schoolteachers are very low because of lack of a performance appraisal system. Then it is decided to introduce a performance-related pay scheme for schoolteachers to reform the education system and to improve student achievement very quickly. If such a process creates conflicts between teachers and other stakeholders, excessive workloads, high levels of burnout, high job stress, unethical behaviors, etc., the education institutions or schools will experience negative autophagy.

## *Permit Some Kinds of Organizational Starvation*

Given that autophagy can be observed during starvation, autophagy will occur in social organizations during situations of organizational starvation. So, how can organizational starvation be seen in a social organization, and what does it mean?

Organizational starvation can be seen when an organization experiences lack of resources, lack of wellness, lack of support, lack of staff, lack of management and leadership support, or lack of energy. Autophagy can also be seen during some kinds of conflict, organizational disruption, disagreements, policy differences, etc. All of the situations that endanger the survival of the organization in a complex environment can produce autophagy. What is important is whether these situations create positive autophagy in the social organization. Some kinds of challenges that social organizations face can be useful during the right conditions for positive autophagy.

For instance, when a manager/leader in a social organization is visible everywhere and at all times, this can make the staff immature. However, under the correct conditions, positive autophagy can arise when the manager/leader is not always around, sometimes does not give enough support, or does not immediately attempt to resolve minor matters and problems, in order to empower the staff and make them more mature. It can be argued that difficult conditions create strong people.

Nevertheless, these efforts can create negative autophagy if these situations are not well managed and the organization is very sick.

## *Balance the Benefits and Risks of Autophagy in the Organization*

Autophagic leadership in a social organization needs to seek to balance the potential risks that can cause negative autophagy to increase, and the potential benefits that can improve the organization and make the staff very mature to cope with challenges in a complex environment. It is not easy to create positive autophagy in an organization; it is similar to walking on ice. It is difficult to create positive autophagy, whereas it is easy to produce negative autophagy. Leaders who make an attempt to produce autophagy in a social organization should have an awareness of the need for balance.

## *Protect Autophagic Mechanisms*

Perhaps the first condition for stimulation of positive autophagy in a social organization in the long term should be protection of autophagic mechanisms, because disorder of autophagy in the organization forces attempts to redevelop the organization or means that reform initiatives have failed. Protection of autophagic mechanisms in a social organization entails establishment and maintenance of a positive climate and a healthy culture, as well as at least the minimum resources that are required.

## *Do Not Harm the Running of the Organization if It Is Functioning Well*

If the running of a social organization is healthy, the managers/leaders in the organization should behave responsibly. This is especially important for top managers who want to make big changes in an organization. It should not be forgotten that big changes that are not well planned and well considered may ruin the autophagic mechanisms in the organization. If the running of the organization is healthy, responsible leaders should consider whether possible changes or reform initiatives may cause various disorders in the running of the organization.

## Final Remarks

Like walking on thin ice, this chapter has aimed to understand autophagy as a complex process and phenomenon, and its potential role in social organizations. In particular, it is very difficult to provide a robust definition of the social autophagy process and to suggest a leadership model for organizations in today's chaotic environment.

However, this chapter has described a conceptual framework, based on the autophagy that occurs in medicine or biology, which might be helpful in developing leadership in social organizations. First, from the perspective of a social researcher, this chapter has attempted to explain how autophagy works in a biological organism by a simple framework, and it has then focused on understanding the potential role of autophagy in social organizations through use of metaphors, models, and theoretical explanations.

As a result, in this conceptual framework, eight suggestions to lead autophagy in social organizations are as follows: end needless practices, procedures, rules etc.; prioritize the internal resources of the organization during challenging periods; make the organization healthy and vigorous; make accurate diagnoses and give appropriate treatment; permit some kinds of organizational starvation; balance the benefits and risks of autophagy in the organization; protect autophagic mechanisms; and do not harm the running of the organization if it is functioning well.

Last, but not least, the explanations and the autophagic leadership model suggested in this chapter can be further developed and discussed. The author looks forward to seeing future studies further explain and contribute to the understanding of autophagy in social organizations, since each study to develop understanding of such mechanisms will have considerable potential to make social organizations more healthy and thus make the world a better place.

## References

BBC. (2018). *Can the science of autophagy boost your health?* By Laurel Iveshealth reporter. https://www.bbc.com/news/health-44005092. Accessed 10 Oct 2018.

China Daily. (2016). *Japanese scientist wins Nobel for cell 'self-eating' studies.* http://www.chinadaily.com.cn/world/2016-10/04/content_26967474.htm. Accessed 10 Oct 2018.

Mizushima, N., Levine, B., Cuervo, A. M., & Klionsky, D. J. (2008). Autophagy fights disease through cellular self-digestion. *Nature, 451*(7182), 1069–1075.

Mizushima, N., Yoshimori, T., & Levine, B. (2010). Methods in mammalian autophagy research. *Cell, 140*(3), 313–326.

Oxford Dictionary. (2018). *Autophagy.* https://en.oxforddictionaries.com/definition/autophagy. Accessed 5 Sep 2018.

# A Qualitative Research Analysis of Chaotic Circumstances Affecting the Happiness of Teachers

**Deniz Görgülü**

**Abstract** In research that is carried out to analyze the effects of chaotic circumstances influenced by the human factor in educational organizations on the happiness of teachers, a situational study pattern, which is a qualitative research method, is used. In the study described in this chapter, that approach was used to study the views of 24 teachers working in different cities in Turkey in the 2018–2019 educational year. Semistructured interview forms were used to collect the data. Of the participants, 71% were women and 29% were men, and 17% of the teachers had 1–5 years of teaching experience, 17% had 6–10 years of experience, 25% had 11–15 years of experience, 12% had 16–20 years of experience, and 29% had at least 21 years of experience. Among the teachers who participated in this research, the proportion who felt happy was found to be quite high (71%). It was found that chaotic circumstances that teachers came across during their first professional appointment mostly arose from environmental–physical, occupational, and sociocultural factors. During that period, teachers mostly resorted to other teachers, school administrators, and their own efforts to help them overcome those chaotic circumstances. As a result, the source of happiness of teachers during their first professional appointment was mostly associated with the student factor. The study also found that chaotic circumstances that teachers experienced during their recent working period centered around subjects related to parents (of their students), social factors, the educational system, students, and economic factors. The teachers mostly resorted to their own efforts to overcome chaotic circumstances they experienced during that period. Another finding was that professional success and student factors significantly influenced the happiness of the teachers.

**Keywords** Chaotic circumstances · Happiness · Teacher

D. Görgülü (✉)
Hacettepe University, Ankara, Turkey



Ş. Ş. ERÇETİN, Ş. N. AÇIKALIN (eds.), *Chaos, Complexity and Leadership 2018*, Springer Proceedings in Complexity, https://doi.org/10.1007/978-3-030-27672-0_20

## Introduction

The twenty-first century has brought a wave of change unlike anything seen before. With this wave of change, mass production and standardization of products have become outdated. Instead, a new paradigm in which diversity, flexibility, and personalization rule has arisen. This new paradigm is fostered by dynamic systems that are not linear but more in line with chaos and complexity theories (Alexander 2011).

The chaos concept, which is seen in systems and is defined as unpredictable movement, is a special form of complexity theory (Ertürk 2012). According to complexity theory, relationships in settings where complex systems (such as organizations) are dominant do not move in a linear direction, and unpredictable constructions that cause the emergence of unexpected results exist (Erçetin et al. 2013b). In other words, chaos and complexity theories have an understanding that emphasize the obscurity of organizations and see each organization as unique. This understanding conflicts with the positivist paradigm, which sees organizations as being no different from each other. According to the new paradigm, all organizations are unique in construction and the rules and principles developed for any organization may not have the same influence in other organizations.

It is possible to see the effects of chaos and complexity theories in the organizations that create the educational system. According to Cahill (2010), chaos has become a huge part of the educational system in various ways. A chaotic system has three basic characteristics. Kara (2008) listed them as follows:

1. The system does not work in a linear direction.
2. The system goes ahead in a loop. The outputs of this loop are used as inputs in the process.
3. Small changes in the starting conditions create great differences.

School administrators and all partners involved in educational services should know that the educational system today does not proceed in a linear direction. Whereas there is a simple cause and effect relationship in linear systems, that is not the situation in chaotic systems. Cause and effect relationships do not have a linear quality in chaotic systems (Erçetin et al. 2013a). Because of this, each incident in a chaotic system has its own specific characteristics and has to be analyzed in depth. It is also possible to come across situations in which a tiny incident creates a huge impact. This situation, which is called the butterfly effect, draws attention to the importance of overlooked details (Akdeniz and Anastasopoulos 2014).

As in all domains of life, chaotic circumstances are continuously encountered in educational organizations. The structure of educational organizations, which are open to unforeseen incidents, creates settings for the development of chaotic circumstances. The sources of chaos in educational organizations are shown in Fig. 1 (Bülbül 2007).

Nurettin Topçu (1997), who said, "In a country where the teachers are appreciated and respected, people are fortunate and virtuous" in his work *Education Lawsuit*, put forward a correlation between teaching proficiency and the degree of

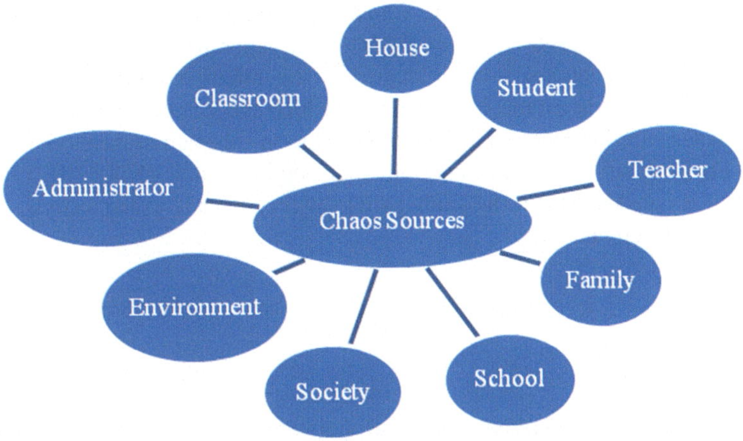


**Fig. 1** Sources of chaos in educational organizations

happiness of the individuals in a society. On the basis of Topçu's statement, it can be said that there is a connection between the human factor and the happiness of teachers. In this context, it is crucial to research the effects of chaotic circumstances influenced by the human factor in educational organizations on the happiness of teachers.

## Purpose of the Research

This research was conducted to analyze the effects of chaotic circumstances influenced by the human factor in educational organizations on the happiness of teachers. With this aim, answers were sought for the following subproblems:

1. What were the chaotic circumstances that teachers came across during the period when they were first appointed to their profession, and what were the resources they resorted to in order to overcome them?
2. What were the teachers' sources of happiness during the period when they were first appointed to their profession?
3. What were the chaotic circumstances that teachers came across during their recent working period, and what were the resources they resorted to in order to overcome them?
4. What were the teachers' sources of happiness during their recent working period?

## Methods

In this research, which aimed to analyze the effects of chaotic circumstances influenced by the human factor in educational organizations on the happiness of teachers, a situational study pattern, which is a qualitative research method, was used. Qualitative research is one of the ways of generating information so as to be able to discover the social order shaped by the common struggle of humanity. A situational study is a type of research in which the components influencing a circumstance, the influence of the circumstance on other circumstances, and the factors bringing out the circumstance are analyzed as a whole (Köse 2017). In this respect, it is considered that the study pattern will contribute to the analysis of the problem condition in detail.

## *Study Participants*

One of the purposive sampling techniques, homogeneous sampling, was preferred in determining the participants in this research. The aim in purposive sampling techniques is profound investigation of the conditions of the analyzed group rather than generalization (Ekiz 2017). In this research, which aimed to analyze the effects of chaotic circumstances influenced by the human factor in educational organizations on the happiness of teachers, the purpose of the research was explained to teachers working in different cities of Turkey by accessing them through social media. The characteristics of the study participants, who took part in the research on the basis of the voluntariness principle and expressed their opinions, are listed in Table 1.

When Table 1 is analyzed, it can be seen that 71% of the participants were women and 29% were men, and that 17% had 1–5 years of teaching experience, 17% had 6–10 years of experience, 25% had 11–15 years of experience, 12% had 16–20 years of experience, and 29% had at least 21 years of experience. Moreover, among the

**Table 1** Characteristics of the study participants

| Characteristic | | Frequency | Percentage |
|---|---|---|---|
| Gender | Male | 7 | 29 |
| | Female | 17 | 71 |
| Professional experience | 1–5 years | 4 | 17 |
| | 6–10 years | 4 | 17 |
| | 11–15 years | 6 | 25 |
| | 16–20 years | 3 | 12 |
| | ≥21 years | 7 | 29 |
| Degree of professional happiness of the teachers | Not happy at all | 1 | 4 |
| | Not happy | 2 | 8 |
| | Indecisive | 4 | 17 |
| | Happy | 15 | 63 |
| | Very happy | 2 | 8 |
| Total | | 24 | 100 |

teachers who participated in the research, the proportion who felt happy turned out to be quite high (71%).

## Data Collection

This research benefited from a semistructured interview technique with the aim of analyzing, in detail, the effects of chaotic circumstances influenced by the human factor in educational organizations on the happiness of teachers. Although the interview questions were prepared by the researcher in advance for use in the semistructured interview technique, the interview questions could be rearranged during the interview to better elicit the views of the participants (Ekiz 2017). In this way, the interview did not progress under the control of the researcher, and flexibility could be provided in the research. The semistructured interview technique was preferred because of this flexibility that it provides.

In determining the interview questions initially by analyzing the literature, the subject, the purpose, the problem condition, and the subproblems in the research were determined. After this step, the items that were considered to take part in the research were listed. Assessment of the items was done by two researchers with expertise in the field of measurement and evaluation. During preparation of the interview questions, the following principles, determined by Bogdan and Biklen (1992), were taken into consideration:

1. The questions needed to be understood easily.
2. The questions were not to be multidimensional.
3. The questions were not to be responding or directive.

After the expert consultation, a preliminary interview was done with ten teachers accessed through social media, and it was observed that all of the questions were easy to understand and the participants could easily answer them. Next, the process of identifying the teachers whose opinions would be sought was started. For this purpose, teachers were informed about the purpose and scope of the study, using a teacher consent form for participation in the study, which all participants signed. Thereupon, an interview was done with 24 teachers accessed through social media. The interviews were done via the internet in the case of participants who could not be observed directly, and this provided some advantages for the researcher (Creswell 2013), as there would otherwise have been only a small chance of accessing teachers working in different cities of Turkey directly.

## Process

In this research, conducted with the aim of analyzing the effects of chaotic circumstances influenced by the human factor in educational organizations on the happiness of teachers, at the beginning of the semistructured interview the teachers were informed about chaos and chaotic circumstances. Then the semistructured interview was carried out. The data obtained with the semistructured interview technique was analyzed in six stages (Creswell 2017):

1. Preparation and arrangement of the data for analysis
2. Reading and analysis of all of the data
3. Coding of the data
4. Creation of themes to describe the current situation
5. Description and presentation of the themes
6. Findings and interpretation of the findings

At the first stage, the interviews that were done were transferred onto a computer. At the second stage, through reading of all of the interviews, several notes were taken. At the third stage, the views of the participants were coded. At the fourth stage, themes were created to be able to describe the current situation. Finally, at the fifth and sixth stages, the findings were listed and interpreted.

In the process of separating the views of the teachers into themes, the reliability of the research was determined by expert consultation. In this determination process, the formula (Reliability = Consensus/(Consensus + Dissidence), defined by Miles and Huberman (1994), was used. As a result of the expert consultation, it was decided to change three themes. These were the second theme in the first subproblem (having a profession), the third theme in the third subproblem (environmental–physical factors, related to the problems teachers had during their first professional appointment), and the sixth theme in the fourth subproblem (information technology, related to the resources the teachers used to overcome the problems they came across during their recent working period). In this case, the reliability ratio was found to be 0.91. Since rates over 0.70 are considered reliable in the formula defined by Miles and Huberman (1994), it can be said that the results of this research were reliable.

In the following sections, all stages of the study are explained in detail so as to ensure the validity of the research. Moreover, the views of the participants are conveyed without any extra interpretation. For this purpose, a code number has been assigned to each participant. These code numbers include information about the gender and the professional experience of the teachers.

## Findings

This section describes the findings of the study.

## *Findings Related to Chaotic Circumstances That Teachers Came Across During Their First Professional Appointment and Resources They Used to Overcome Them*

In the first subproblem of the research, the question "What were the chaotic circumstances that teachers came across during the period when they were first appointed to their profession, and what resources did they resort to in order to overcome them?" was investigated. Information related to chaotic circumstances the teachers came across during their first professional appointment is given in Table 2.

When Table 2 is analyzed, it can be seen that the chaotic circumstances the teachers came across during their first professional appointment were mostly due to environmental–physical, occupational, and sociocultural factors. The resources that the teachers used to overcome those chaotic circumstances are given in Table 3.

According to Table 3, the resources that the teachers used to overcome chaotic circumstances they came across during their first professional appointment were mostly other teachers, followed by school administrators, individual efforts, the state/provincial National Education Directorate and institutional governors, and parents and acquaintances. Examples of the views of the teachers related to the chaotic circumstances they came across during their first professional appointment and the resources they used to overcome them are as follows:

**Table 2** Chaotic circumstances that teachers came across during their first professional appointment

| Line | Chaotic circumstances | Frequency |
|---|---|---|
| 1 | Occupational factors (T1, T6, T7, T8, T11, T13, T14, T17, T18, T21) | 10 |
| 2 | Sociocultural factors (T4, T7, T9, T12, T20, T22, T23, T24) | 8 |
| 3 | Environmental–physical factors (T2, T3, T4, T10, T12, T14, T18, T19, T20, T22, T24) | 11 |
| 4 | Social factors (T15, T16) | 2 |
| 5 | Other factors (T7) | 1 |

**Table 3** Resources that teachers used to overcome chaotic circumstances they came across during their first professional appointment

| Line | Resources used | Frequency |
|---|---|---|
| 1 | Teachers and school administrators (T1, T4, T8, T12, T13, T15, T16, T17, T21) | 9 |
| 2 | State/provincial National Education Directorate and institutional governors (T2, T9, T14, T15, T22) | 5 |
| 3 | Individual efforts (T6, T11, T13, T17, T18, T19, T24) | 6 |
| 4 | Parents and acquaintances (T6, T9, T14, T22, T23) | 5 |
| 5 | Family members (T18) | 1 |
| 6 | Students (T7) | 1 |
| 7 | Other resources (T3, T5, T10, T20) | 4 |

*The physical condition of the school I was first appointed to was very bad. To improve the physical condition of my school, I received support from the provincial National Education Directorate, the provincial Gendarmerie headquarters, and school principals in the main county town.* [T2/PE/11–15 years]

*Working in the county town and being far away from my hometown was a problem for me. I waited for the right time and wanted to be appointed at the first opportunity.* [T3/PE/16–20 years]

*The institution I was appointed to was not even in use. I was at a loss. They said to me, "You are a candidate teacher; you will devastate your teaching profession; do not say anything." I received treatment for depression.* [T5/PE/16–20 years]

*I had a language problem with the students. We created a language dictionary with the students who knew Turkish in order to solve this problem.* [T7/PE/11–15 years]

*The interference of the school administration with the work that was done was a problem from my point of view. Against this situation, I continually tried to tell people the right thing. However, in some cases, I kept to my own course.* [T11/PE/1–5 years]

*I had been regarding myself as incompetent professionally, so to overcome this problem I read many books about teaching proficiency and researched this subject. Moreover, I got help from my colleagues.* [T17/R/≥21 years]

*My classroom and school were in a physically inadequate condition. Besides, I had difficulty in preparing daily plans according to the levels of the children and communicating with the parents. I wanted to receive support from the provincial National Education Directorate but I could not get a positive response. My grandmother helped me financially; my spouse helped with improving the physical condition of my school. I got help with the daily plans from my friends who had been appointed to a school already. For the parents I sent notes all the time and gave information with messages.* [T18/PE/1–5 years]

*Being far away from my family and adapting to a totally unfamiliar culture were very big problems for me. I tried to get away from loneliness by making friends. I created a nice bond with the people of the region I was appointed to.* [T23/R/16–20 years]

## *Findings Related to Teachers' Sources of Happiness During Their First Professional Appointment*

In the second subproblem of the research, an answer to the question "What were the teachers' sources of happiness during the period when they were first appointed to their profession?" was sought. In this respect, the teachers' sources of happiness during their first professional appointment are given in Table 4.

**Table 4** Teachers' sources of happiness during their first professional appointment

| Line | Sources of happiness | Frequency |
|---|---|---|
| 1 | Students (T1, T2, T4, T6, T7, T8, T9, T11, T14, T15, T17, T18) | 12 |
| 2 | Having a profession (T5, T8, T9, T12, T16, T21, T23) | 7 |
| 3 | Environment (T3, T10, T24) | 3 |
| 4 | Being a teacher (T13, T19, T20, T22, T23, T24) | 6 |

When Table 4 is analyzed, it can be seen that the teachers' sources of happiness during their first professional appointment were mostly related to the student factor. In addition it stood out that the factors of having a profession and being a teacher influenced the happiness of the teachers. Another situation that stood out was that components with an environmental origin had the least influence on the happiness of the teachers. Examples of the views of the teachers about their sources of happiness during their first professional appointment are as follows:

*Knowing that little hearts would be shaped in my hands, learning new things with them every day, by standing on my own, seeing that thousands of kilometers away from my family, I held on to life with their love made me happy during the period when I was first appointed to my profession.* [T9/PE/6–10 years]

*Since I was thinking more idealistically even though there was no transportation or communication, the sincere behaviors of people were enough to make me happy during the period when I was first appointed to my profession.* [T10/R/≥21 years]

*Just to be appointed made me happy during the period when I was first appointed; other than that, the region I was appointed to and the living standards were too far away from my lifestyle.* [T12/PE/1–5 years]

*Despite the impossibilities in the region I went to, the light in the eyes of the children made me happy during the period when I was first appointed to my profession.* [T14/PE/11–15 years]

*To be able to attain my dream job and to gain financial independence made me happy during the period when I was first appointed to my profession.* [T23/R/16–20 years]

## Findings Related to Chaotic Circumstances That Teachers Came Across During Their Recent Working Period and Resources They Used to Overcome Them

In the third subproblem of the research, answers to the question "What were the chaotic circumstances that teachers came across during their recent working period, and what were the resources they resorted to in order to overcome them?" were sought. Information related to the chaotic circumstances that the teachers came across during their recent working period is given in Table 5.

**Table 5** Chaotic circumstances that teachers came across during their recent working period

| Line | Chaotic circumstances | Frequency |
|---|---|---|
| 1 | Economic factors (T1, T4, T7, T16, T23) | 5 |
| 2 | Students (T2, T3, T10, T17, T18, T19) | 6 |
| 3 | Parents (T1, T2, T3, T10, T11, T12, T15, T17, T18, T19) | 10 |
| 4 | Social factors (T4, T7, T9, T13, T15, T20, T23, T24) | 8 |
| 5 | Educational system (T1, T6, T7, T8, T14, T20, T21, T22) | 8 |
| 6 | Teachers and school administrators (T1, T17, T18, T22) | 4 |
| 7 | Physical conditions of schools (T5, T6) | 2 |

**Table 6** Resources that teachers used to overcome chaotic circumstances they came across during their recent working period

| Line | Resources used | Frequency |
|---|---|---|
| 1 | Teachers and school administrators (T6, T8, T12, T13, T15, T17, T21, T22) | 8 |
| 2 | State/provincial National Education Directorate (T15) | 1 |
| 3 | Individual efforts (T1, T3, T4, T5, T7, T9, T11, T18, T19, T23) | 10 |
| 4 | Parents and acquaintances (T2, T6, T15, T24) | 4 |
| 5 | Nongovernmental organizations (T16, T24) | 2 |
| 6 | Information technology (T2, T8, T21) | 3 |
| 7 | Occupational publications (T10, T17, T24) | 3 |
| 8 | Other resources (T20) | 1 |

When Table 5 is analyzed, it stands out that the chaotic circumstances the teachers came across during their recent working period varied and resulted from parents, social factors, the educational system, and economic factors. The resources that the teachers used to overcome the chaotic circumstances they came across during their recent working period are given in Table 6.

According to Table 6, the teachers generally tried to overcome chaotic circumstances they came across during their recent working period through their own individual efforts. Furthermore, other teachers and administrators were among the most frequently used resources in this regard. Examples of the views of the teachers related to chaotic circumstances they came across during their recent working period and the resources they used to overcome them are as follows:

*The negative perceptions of the parents and students concerning the teaching profession disturb me. By showing the right examples at every opportunity, I try to improve our occupational prestige.* [T3/PE/16–20 years]

*The classrooms being too crowded and preschool teachers not having break times are the problems I have. As my class is too crowded, I want the parents to take care of their children more. In addition, I get support from the school administration for the problems I have.* [T6/PE/6–10 years]

*The biggest problem for me is that the curriculum changes all the time and there is inconsistency in the educational policies. I try to reduce the effects of these problems to the minimum level by benefiting from my colleagues and information technology.* [T9/PE/6–10 years]

*Some behaviors of the parents and the students are the biggest problem from my point of view. To overcome this problem I mostly get help from occupational publications.* [T10/R/≥21 years]

*I argue that the quality of the education has decreased. With empty curricula, texts that are not interesting, and senseless questions, we raise children incapable of thinking. For the sake of increasing the quality of education, I cooperate with school administrators, teachers, and parents.* [T14/PE/11–15 years]

*Our schools and the teaching profession not being valued sufficiently and the indifference of the parents are the most important problems for me. I get help from my acquaintances, colleagues, and the provincial National Education Directorate for solutions to these problems.* [T15/R/1–5 years]

*I have financial difficulties. For the sake of solving these financial problems that teachers have, I serve actively in nongovernmental organizations.* [T16/R/6–10 years]

*I see parents who suppose they know everything, professionally incompetent teachers, and students who think they learn everything without attending classes as the problems. To deal with these problems, I resort to the views of my colleagues and experts.* [T17/R/≥21 years]

*The educational system that changes all the time and teachers losing caste are the two big problems I see. I try to get help from everyone [which is] essential for the problems I have recently.* [T20/R/≥21 years]

## Findings Related to Teachers' Sources of Happiness During Their Recent Working Period

In the fourth subproblem of the research, answers to the question "What were the teachers' sources of happiness during their recent working period?" were sought. The teachers' sources of happiness during their recent working period are given in Table 7.

From Table 7, it can be seen that professional success and student factors had the greatest effects on the happiness of the teachers during their recent working period. Examples of the views of the teachers related to their sources of happiness during their recent working period are as follows:

*Thanks to the smart board and the internet, the subjects have gotten easier to understand. Because of this situation, I am happy.* [T8/PE/≥21 years]

*Despite the unfair behaviors aimed at our profession, working in a place where my profession is respected makes me happy.* [T9/PE/6–10 years]

**Table 7** Teachers' sources of happiness during their recent working period

| Line | Sources of happiness | Frequency |
|---|---|---|
| 1 | Students (T3, T4, T5, T6, T14, T15, T17, T23, T24) | 9 |
| 2 | Professional development (T1, T2, T8, T9) | 4 |
| 3 | Professional success (T4, T7, T11, T12, T13, T16, T18, T19, T20, T21) | 10 |
| 4 | Environment (T4, T9, T23) | 3 |
| 5 | Being a teacher (T22) | 1 |
| 6 | Other resources (T10) | 2 |

*I could not come across anything to be happy about in recent years.* [T10/R/≥21 years]

*To be able to get the reward for the effort made makes me happy.* [T11/PE/1–5 years]

*The powerful bond of affection between me and my children [students] makes me happy.* [T14/PE/11–15 years]

*To close the door of the classroom and spend educational and entertaining time with my children away from all negativity make me pretty happy.* [T22/PE/11–15 years]

## Discussion

The Newtonian model—which describes the natural world as linear, decisive, sequential, and steady—fails to explain human interaction patterns conceptually. Research shows that the earth is much more than the particles that create it. This situation is valid in terms of human interaction and education as well. Education is a versatile and dynamic process. To be able to understand this dynamic process completely, components such as the curriculum, tuition, and student performance have to be evaluated from a holistic perspective (Crowell 1989).

Chaos theory, about which most scientists studying subjects related to social sciences and behavioral sciences do not acquire knowledge visibly, is quite effective in making inferences about human behaviors and educational research (Cziko 1989). Chaos theory paves the way for elimination of deficiencies in systems by providing exposition of conditions that cannot be observed. Thus, past phenomena are seen from a wider angle with a new perspective (Ural 2005)

Although the word "chaos" is perceived as if it means "disorder," actually an order formed by disorder lies beneath this phenomenon (Öge 2005). From this perspective, according to chaos theory, the ideal teacher is one who can carry on educational activities by minimizing the effects of chaotic circumstances. Among the results of this research, carried out with the aim of analyzing the effects of chaotic circumstances influenced by the human factor in educational organizations on the happiness of teachers, the high rate (71%) of professional happiness of the teachers stands out.

It can be seen that the chaotic circumstances that the teachers came across during their first professional appointment mostly originated from environmental–physical, occupational, and sociocultural factors. During that period, to overcome the chaotic circumstances they came across, the teachers preferred to use resources such as other teachers and school administrators, their own individual efforts, state/provincial National Education Directorate and institutional governors, parents, and acquaintances. When we look at the teachers' sources of happiness during their first professional appointment, it stands out that the happiness of the teachers was mostly due to their students.

In Turkey, most teachers who are appointed to their profession for the first time work in socioeconomically underdeveloped regions. Components such as the mother tongue of the people living in these regions being different, the inadequacy of the opportunities of educational administrations, and being professionally inexperienced are seen as chaotic circumstances from the perspective of the teachers. From this aspect, to minimize the effects of chaotic circumstances that newly appointed teachers experience, the physical structures of the educational administrations should be strengthened initially. It is an obligation for the teachers to have orientation training to be able to comply with the cultural structure of the region they are appointed to. Furthermore, institutions that educate teachers need to focus on practice for new appointed teachers so they do not feel professionally deficient.

As a result of this research, it can be seen that the teachers mostly tried to overcome chaotic circumstances with the help of their colleagues during their first professional appointment. However, the places that teachers are first appointed to are generally socioeconomically underdeveloped regions. Since most teachers working there are not experienced enough, experienced teachers who can be mentors in these regions are required. Thus, the Ministry of National Education should create policies for teacher appointments that take this situation into consideration.

Students, having a profession, and being a teacher were the factors that had the greatest influence on the happiness of the teachers during their first professional appointment. To increase the happiness of teachers, the number and the effects of chaotic circumstances should be decreased. Thereby, the circumstances that make teachers happy could be increased. For instance, working in more convenient conditions during their first professional appointment could have a positive effect on the happiness of teachers.

As a result of this research, it can be seen that the chaotic circumstances that the teachers came across during their recent working period were different from those they came across during their first professional appointment. While the teachers mostly saw the characteristics of the places where they worked as a source of chaos during their first professional appointment, they saw parents, social factors, the educational system, students, and economic factors as sources of chaos during their recent working period. As a consequence of this difference, it can be said that the teachers who participated in this research were appointed to more convenient regions as a result of replacement. However, the teachers with more convenient living conditions came across chaotic circumstances arising from parents and negative perceptions related to the prestige of the teaching profession in these places. Teachers who came across chaotic circumstances in this period tried to be happy with their professional achievements during this period.

It was seen that teachers resorted to their own individual efforts to overcome chaotic circumstances they came across during their recent working period, possibly because they felt alone. Teachers need more support to overcome chaotic circumstances they come across during their work and to increase their happiness. This support should be provided by all partners involved in educational services.

In conclusion, as a result of this research, which was carried out to analyze the effects of chaotic circumstances influenced by the human factor in educational organizations on the happiness of teachers, it was found that chaotic circumstances influenced teacher happiness. To be able to assess the degree of this effect more reliably, research studies with mixed patterns and broader participation are required.

## References

Akdeniz, K. G., & Anastasopoulos, N. (2014). Chaotic awareness and simulacra in the recent emergence of the self-organized multitude. In Ş. Ş. Erçetin (Ed.), *Chaos, complexity and leadership 2014* (pp. 23–28). Cham: Springer.

Alexander, G. (2011). Dealing with the complexities of integration in cultural diverse rural school communities in South Africa. *Online Journal of New Horizons in Education, 1*(1), 1–12.

Bogdan, R., & Biklen, S. K. (1992). *Qualitative research for education: An introduction to theory and methods*. Boston: Allyn and Bacon.

Bülbül, M. Ş. (2007). *Kaos ve eğitim*. Ankara: Beyaz Kalem Yayıncılık.

Cahill, K. (2010). *Even in chaos: Education in times of emergency*. New York: Fordham University Press.

Creswell, J. W. (2013). *Qualitative ınquiry & research design: Choosing among five approaches* (3rd ed.). Thousand Oaks: Sage.

Creswell, J. W. (2017). *Araştırma deseni* (Selçuk Beşir Demir, Çev.). Ankara: Eğiten Kitap Yayıncılık.

Crowell, S. (1989). A new way of thinking: The challenge of the future. *Educational Leadership, 18*(3), 60–63.

Cziko, G. A. (1989). Unpredictability and indeterminism in human behavior: Arguments and implications for educational research. *Educational Researcher, 15*(3), 17–25.

Ekiz, D. (2017). *Bilimsel araştırma yöntemleri*. Ankara: Anı Yayıncılık.

Erçetin, Ş. Ş., Bisaso, S. M., & Saeed, F. (2013a). Understanding chaos and complexity in education systems through conceptualization of fractal properties. In Ş. Ş. Erçetin & S. Banerjee (Eds.), *Chaos, complexity and leadership 2013* (pp. 147–161). Cham: Springer.

Erçetin, Ş. Ş., Potas, N., Kısa, N., & Açıkalın, Ş. N. (2013b). To be on the edge of chaos with organizational intelligence and health. In S. Banerjee (Ed.), *Chaos and complexity theory for management: Nonlinear dynamics* (pp. 184–203). Hershey, PA: IGI Global.

Ertürk, A. (2012). Kaos kuramı: Yönetim ve eğitimdeki yansımaları. *Kastamonu Eğitim Dergisi, 20*(3), 849–868.

Kara, A. (2008). *Understanding today's schools with chaos theory*. http://www.fountainmagazine.com/Issue/detail/Understanding-Todays-Schools-with-Chaos-Theory, Accessed 28 Dec 2018.

Köse, E. (2017). Bilimsel araştırma modelleri. In R. Y. Kıncal (Ed.), *Bilimsel araştırma yöntemleri* (pp. 99–122). Ankara: Nobel Yayın Dağıtım.

Miles, M. B., & Huberman, A. M. (1994). *Qualitative data analysis*. Thousand Oaks: Sage.

Öge, S. (2005). Düzen mi düzensizlik (kaos) mi? Örgütsel varlığın sürdürülebilirliği açısından bir değerlendirme. *Selçuk Üniversitesi Sosyal Bilimler Enstitüsü Dergisi, 13*, 285–303. Erişim: http://dergisosyalbil.selcuk.edu.tr, 27 Kasım 2018.

Topçu, N. (1997). *Türkiye'nin maarif davası*. E. Erverdi ve İ.Kara (Yay. Haz.). İstanbul: Dergâh Yayınları.

Ural, Ş. (2005). *Kozmozdan kaosa. Kaos, Mantık, Matematik ve Felsefe II*. Ulusal Sempozyumu, 21–24 Eylül 2004. Assos Çanakkale, Yay. Haz. Ural, Ş., Yüksel, Y., Koç, A., Şen, A., Hacıbekiroğlu, G., Özer, M., İstanbul, İKÜ Yayınları, pp. 353–363.

# Analysis of Organizational Memory in the Context of School Administrators

**Abdullah Balıkçı**

**Abstract** Organizations need various tools to achieve their goals. One of these tools is organizational memory. Organizational memory can be used effectively in the organization through administrators. Schools and school administrators also need to act in line with the approach mentioned in this context. This research study aimed to reveal the views of school administrators on organizational memory on the basis of the thoughts mentioned in the research. A qualitative research method and a phenomenological design were implemented in the research. Twelve school administrators participated in the study. A snowball and maximum sampling model with convenience sampling in a purposeful sampling method were used. The data were collected through interviews, observations, and investigations of the obtained documents. Triangulation, participant confirmation, expert review, and substantial identification strategies were carried out in the reliability–validity phase of the study. The findings and conclusions, discussions, and recommendations depending on these findings are provided in this research report.

**Keywords** Organization · School · School administrator · Qualitative research · Phenomenology

## Introduction

Organizations need to use various practices to maintain their existence in accordance with their objectives. One of the practices that organizations benefit from is organizational memory, which is accrued over time and is expected to serve as a bridge between the past and the future. Organizational memory has an important place in the organization in terms of revealing things that were done in the past and shedding light on what is to be done in the future. Today, organizations are

---

A. Balıkçı (✉)
İstanbul University-Cerrahpaşa, İstanbul, Turkey
e-mail: abdullah.balikci@istanbul.edu.tr



Ş. Ş. ERÇETİN, Ş. N. AÇIKALIN (eds.), *Chaos, Complexity and Leadership 2018*, Springer Proceedings in Complexity, https://doi.org/10.1007/978-3-030-27672-0_21

paying more and more attention to organizational memory and taking precautions to preserve this phenomenon. That organizational memory is becoming an effective practice in achieving the aims of the organization can be shown as the main reason for the precautions that are taken.

## Organizational Memory

Ackerman and Halverson (2004) interpreted organizational memory as a phenomenon including personnel–organizational studies as well as an element and process that should be available in the organization. According to Walsh and Ungson (1991), organizational memory generally refers to information that is stored in past experiences through which the organization has advantages in decision making. The stored information is affected by decisions taken by employees of the organization. Three points should be taken into consideration in this regard. These are the place of organizational memory in the organization, the results acquired through the process of obtaining and using it, and the presence of performance-based use. Stein (1995) defined organizational memory as acquisition, preservation, retention, and accessibility of information at the desired time. Similarly, Kruse (2003) described organizational memory as acquisition, protection, and accessibility of data and information forming the basis of organizational memory in ways that support each other. Wexler (2002) mentioned that organizational memory is a concept based on design, control, and management of memory and is related to putting forward an adaptation of the economy of the managed system. According to Şen (2014), knowledge and experience in organizations constitute the basis for organizational memory. Organizational memory has the characteristics of acquiring and preserving information in the organization, reviewing it according to the conditions, ensuring its security, and accessing it readily when needed. Another element of organizational memory that needs to be comprehended is its contribution to rational decision making through systematic storage of information. According to Moorman and Miner (1998), organizational memory is important for an organization, and different organizational memory sources should be utilized by the organization. However, humans are the most important factor for effective use of organizational memory. According to Dunham and Burt (2011), senior employees have an important place in organizational memory. The interaction between senior employees and the organization has benefits. While organizational memory refers to respect for and recognition of senior employees in the organization, and their psychological power, its meaning for the organization is that senior employees use their current influence on their colleagues in favor of the organization, since they have done the current work before. Atak (2011) and Ilkka Tuomi (2000) paid attention to the relationship between organizational memory and information management. Atak (2011) expressed organizational memory in terms of information management. He also considered information management as a process involving acquisition, use, and transmission of information. Ilkka Tuomi (2000) argued that organizational memory is a social process that should be considered together with information management. Most efforts in successful systems take technical and social aspects of information management and organizational memory into account.

## Benefits of Organizational Memory

Lehner and Maier (2000) listed the contributions of organizational memory to the organization as follows: Organizational memory facilitates organizational learning. It provides the organization with flexibility, effectiveness, and innovation. It also ensures management of change in the organization. It contributes to reconciliation of organizational culture and technology in the organizational environment in ensuring organizational effectiveness. According to Walsh and Ungson (1991), organizational memory plays three roles in the organization. Firstly, it includes effective and efficient information within it. As its second role, it functions as a control mechanism for new decision making. Thirdly, it has a political role because of the power associated with having the information. Perez and Ramos (2013) mentioned gains derived from use of organizational memory as facilitating the roles of the affected employees. In particular, administrators are in a position to make decisions on the basis of information and data held by the organization. Organizational memory can affect the future of the organization because of its importance in decision making. Thus, there is a link between the effectiveness and performance of the organization and its organizational memory. According to Kaya (2005), organizations need to use quality documentation and archival systems to sustain gains and to create organizational policies. The archives should be traceable, testable, and measurable in order to be valuable. Arvas (2015) urged that acquisition, storage, and points of access to the information should be clarified to obtain the desired efficiency from organizational memory. The inability to make adequate use of organizational memory can adversely affect employees' learning of work and effective execution of work. To avoid negative situations, measures should be taken to manage information and to preserve organizational memory. According to Walsh and Ungson (1991), these measures can be effective provided that the factors that affect the place of memory in the organization are known. These factors are retention and preservation of the information, the status of employees, the organizational culture, changes in the organization, the organizational structure, the environment, nonorganizational archival sources, the characteristics of the stored information, and the accessibility of the information. Shadow (1999) evaluated the importance of organizational memory in terms of the risks that can be experienced by organizations. One of the most dangerous situations for organizations is loss of organizational memory and the possibility that an administrator with a lot of data can make a wrong decision. According to Kızıldağ et al. (2011), loss of organizational memory can adversely affect the functioning of the organization and can be difficult or even impossible to compensate for. For this reason, the required measures should be taken within the organization to prevent organizational memory loss. Systematic structuring within the organization and appropriate management structures and strategies should be generated as a precaution. The research studies mentioned above clearly suggest the importance of the contribution of organizational memory to the organization. Similarly, Stein and Zwass (1995) emphasized that preservation of organizational memory is a fundamental issue for organizations. In addition, they defined organizational memory as transfer of knowledge from the past to the present and indicated that the organizational memory information system they proposed would allow the organization to obtain current benefit from previously obtained information. They emphasized that in this way the effectiveness of the organization can reach the desired level.

## Organizational Memory Instruments

It is possible to benefit from various instruments for formation of organizational memory. Baransel and Tabak (2006) stated that documents are used to perform the functions of organizations and that nowadays these documents are prepared on computers and on the internet. Computer-aided preparation of documents positively affects work speed, efficiency, and organizational memory. Aydın (2010) mentioned that institutions prefer the electronic environment on account of the requirements of the new age. Speed, access, and cost can be seen as the main reasons for this. The purpose here is to prevent problems with the integrity and validity of the documents. The terms "archive" and "archiving" are also used in the context of organizational memory. Archiving can be considered as "preserving all of the original data and information contained in the document" (p. 163). Arısoy and Özdemirci (2017) and Özbal (2013) emphasized the importance of regulations on legal use of documents and archiving of works that can be considered as one of the important factors constituting the basis for organizational memory. These arrangements are important both for security and for access when needed. The main point that should be considered is that processing of documents must be carried out in accordance with the purpose even if an electronic document (e-document) system is in use. Rowlinson et al. (2010) drew attention to the human factor in the organization in terms of creation, use, and protection of organizational memory. Similarly, Jimerson (2003) considered archives that can be evaluated in terms of organizational memory as reliable sources that can help employees in the organization. Thus, he emphasized the importance of organizational memory in human and archive dimensions. In the literature, some model suggestions for organizational memory are also available. One of these suggestions comes from Anand et al. (1998). The model they proposed in their study on organizational memory includes an approach to surviving and competing in a dynamic and complex organizational environment. The features of this model can be listed as including information management based on information technologies, providing guidance in understanding organizational processes and techniques that will improve use of information and data at important stages for researchers, and, finally, being appropriate for use by all parties to the organizational memory.

## The Organizational Memory–Administrator Interaction

The most important task in the organization in terms of organizational memory belongs to the administrator. Stein (1995) listed the contribution of organizational memory to the organization in general and to the administrator in particular as follows:

- It assists administrators to make strategic decisions in time.
- It protects the organization from situations that cause new and old problems over time.
- It helps employees in the organization to generate a new working style.
- It facilitates organizational learning.
- It strengthens organizational identity.
- It provides guidance for new employees on what previous employees have done.

Seçkin-Halaç and Seçkin-Çelik (2016) listed administrative requirements in terms of organizational memory as follows:

- The administrator should give importance to organizational memory and make the organizational culture flexible and open to innovation by taking organizational memory into account.
- It is necessary to develop the infrastructure so as to bring about innovations in the context of organizational memory.
- Assistance should be obtained from experts in terms of organizational memory.

As in every organization, organizational memory is used in educational organizations. In the Turkish education system, there are also bases and practices based on them with regard to organizational memory. Table 1 lists the legal regulations that are considered relevant to organizational memory in the Turkish education system.

As the scope, number, and content of the relevant legal texts are thought to be greater than is mentioned in Table 1, use of the regulations was preferred for the purposes of this research. It is understood that there are 121 regulations when the related web page is examined. In accordance with the concept of the "archive" that is the basis of organizational memory, the regulations were searched. In addition, it can be seen that the mentioned regulations are based on the "Regulation on the Archive of the State Personnel Presidency" and that the regulations are generated in accordance with this regulation. The "Ministry of National Education Regulation on Repealing the Regulation on the Archive Services" may be shown as a basis for the mentioned issue. When Table 1 is examined, it can be comprehended that organizational memory is included in 17 legal texts and various items (numbers 1–9) in the national education legislation. This indicates that organizational memory is important in legal texts. Another point that is noteworthy in these legal texts is that

**Table 1** Legal regulations related to organizational memory in the Turkish education system

| Item number | Legal basis | Related article number |
|---|---|---|
| 1 | Ministry of National Education Regulation on Preschool Education and Primary Education Institutions | 75 |
| 2 | Ministry of National Education Regulation on Secondary School Institutions | 76, 219, 221 |
| 3 | Ministry of National Education Regulation on Special Transportation Services, Vocational Training, and Development Courses | 13, 14, 29 |
| 4 | Ministry of National Education Regulation on Lifelong Learning Institutions | 3, 5, 19, 21, 28, 30, 31, 52, 84 |
| 5 | Ministry of National Education Board of Education and Discipline | 13 |
| 6 | Ministry of National Education Regulation on School Parents' Associations | 16, 20, 23 |
| 7 | Ministry of National Education Regulation on Repealing Regulations on Archive Services | 1 |
| 8 | Ministry of National Education Regulation on Legal Consultancy | 5 |
| 9 | Ministry of National Education Regulation on Teacher Assignment and Relocation | 14 |
| 10 | Ministry of National Education Regulation on Guidance Services | 21, 34 |
| 11 | Ministry of National Education Regulation on Opening, Closing, and Naming of Institutions | 5 |
| 12 | Ministry of National Education Regulation on Personnel to Be Assigned Permanently to Overseas Organizations | 12 |
| 13 | Ministry of National Education Regulation on Primary School Teachers' Health and Social Assistance Funds | 23 |
| 14 | Ministry of National Education Regulation on Committees of Inspection | 29 |
| 15 | Ministry of National Education Regulation on Private Motor Vehicle Drivers' Course Regulation | 18, 42 |
| 16 | Ministry of National Education Regulation on Provincial and District National Education Directorates | 20, 21 |
| 17 | Ministry of National Education Regulation | 8, 18, 69, 83 |

Source: http://www.meb.gov.tr/mevzuat/liste.php?ara=6 (Accessed 20 Oct 2018)

the work and arrangements are essentially expected to be done by administrators. This situation shows that the administrator has an important function in the management of organizational memory.

When the related literature is reviewed, it can be observed how important organizational memory is for an organization and what it provides. Furthermore, it can be understood from Table 1 that practices related to organizational memory exist in the Turkish education system and that there are some expectations that essentially should be met by administrators. However, in the related literature, especially when the Council of Higher Education (CHE) thesis bank was scanned, it was determined that there was no research examining the interaction between organizational memory and school administrators in the Turkish education system. The stated reasons make it necessary to reveal the place of organizational memory within the education system from the perspective of the school administrator.

## Aim

The aim of this research was to reveal the views of school administrators on organizational memory. Answers were sought for the following questions:

1. How is the relationship between the school administrator and the school perceived by school administrators?
2. How does the school administrator perceive the organization and organizational memory?
3. How does organizational memory reflect on the organization according to the school administrator?

## Methods

The methods used in this research are presented in this section. In this context, the following issues are explained: the method and design of the research, the participants in the research, the data collection and analysis, the reliability–validity strategies used in the research, the ethics of the research, and the role of the researcher in the research.

### *Research Methods and Design*

School administrators representing the target group in the research and academicians who had studied this field and used a qualitative research method in their research were interviewed for the purpose of determining the methods used in the study. Qualitative research was preferred since experts studying this field had mentioned that it might be appropriate to reflect the experiences of school administrators. Patton (2014a) stated that it is appropriate to answer some questions in determining methods in research. Some of them are related to the purpose of the research, the target group, and the data sources. In the present study, since the relationship between organizational memory and school administrators was to be investigated on the basis of the experiences of school administrators, qualitative research was preferred because the school administrators participating in the research considered interviews appropriate, and interviews, observations, and documents related to the research could be used. According to Creswell (2016) and Merriam (2015a), the purpose of qualitative research is to reveal how people make sense of their lives through their experiences. A phenomenological design was preferred as the research design since the aim was to reveal findings based on experiences in this study. According to Ersoy (2016), the most important point in phenomenology that should be paid attention to is to try to reveal and understand the phenomenon (the perceived situation) as a whole.

## *Participants in the Research*

The participants in the research were 12 school administrators. In determining the size of the sample, the literature on qualitative research was taken into consideration. In the relevant literature, there was no recommendation of a specific sample size. The questions of the researcher, the data obtained, and the resources used were considered to be determinants of the number of sampling (Merriam 2015b; Patton 2014b).

The following principles were followed in selecting the participants in the research:

- Participation in the research would be voluntary.
- The researcher would conduct the research through formal and informal meetings and interviews.
- The data obtained from interviews conducted by the researcher would contribute to the research (through understanding of the questions and allowing the answers to provide substantial descriptions).

The participants were selected through convenience sampling using purposeful sampling methods, since the participants were expected to contribute voluntarily and effectively to the research on the basis of their observations in formal and informal interviews. Yıldırım Şimşek (2011) stated that convenience sampling is appropriate in terms of economics and facilitating the progress of the research. In addition to convenience sampling, since the participants were able to contribute to the research, and as recommended by experienced school administrators, snowball sampling was also utilized. According to Miles and Huberman (2015, p. 28), the purpose of the snowball strategy is "to define the situation of those who know who is familiar with what information is substantial in knowledge." Finally, since the views of school administrators working in different types of schools were utilized, maximum diversity sampling was conducted in determining the participants. According to Miles and Huberman (2015, p. 28), the purpose of maximum diversity sampling is "to document the different varieties and to identify important de facto designs." While the views of the participants were included, coding was used instead of real names. The school administrators were coded as SA 1, SA 2, SA 3, and so on. All of the participants were selected from those working in public schools.

## Data Collection

A semistructured interview form (applied to 12 school administrators) generating the data for the study, observations on each theme, and document analysis based on the 121 regulations listed in Table 1 were used in this research. The questions were prepared for the semistructured interview form, and the opinions of two Turkish teachers and two school administrators—in terms of the language and expressions used on the interview form—were sought from a technical point of view. To prevent any loss of data obtained during the interviews, the interviews were recorded with a voice recorder and shared with the participants, and their sensitivities and approval were taken into consideration.

## Data Analysis

Upon documentation of the audio records, the data analysis was started. In this study, inductive analysis was implemented. According to Patton (2014c), as patterns in phenomena are understood through inductive analysis, the categories or dimensions of the analysis emerge. As a result of the analysis carried out in the current research, three themes emerged. In the reporting of the interview analysis, the aim is that the reader will better perceive and understand the research through the use of direct quotations. The quotations are presented directly so as to achieve the stated purpose in this qualitative research.

## Reliability–Validity Strategies Used in the Research

Four of the strategies indicated by Merriam (2015c) were used for reliability and validity in this study. The reflections of these strategies in the research are as follows:

1. *Triangulation*: In the study, 12 participants were interviewed, the observations recorded on each theme were reviewed, and document analysis was carried out on the basis of the 121 regulations listed in Table 1. Thus, interviews, observation, and document analysis were implemented in the research.
2. *Participant confirmation*: After examining the content of the interviews and identifying which content could be used in the research, it was sent to the

relevant participants to obtain their comments and approval for the content of the interviews.
3. *Expert review*: A report of the study was sent to two academicians with research experience in this field, and the research was reviewed on the basis of the feedback received from them.
4. *Substantial description*: The views of the participants were reflected in the research through use of their direct quotations, and the content of the research was enriched by comments on the quotations.

### Research Ethics

After the participants were informed about the purpose of the interview, the required consent was received and appointments related to the research were made. The consent received from the participants was confirmed with the participants by including the question "Do you allow this interview to be used and published in this research?" in each interview.

### The Role of the Researcher

The fact that the researcher had experience of both teaching and school administration, as well as being involved in academic studies, was thought to have a positive effect on the perception and interpretation of the data and the conduct of the research process appropriate to its purpose.

## Findings and Comments

The findings and comments based on these findings of the research are presented in this section. According to the study, six categories and 21 subcategories of three themes appeared (see Table 2).

### Theme 1: Findings and Comments on the School Administrator–School Relationship

Two categories and seven subcategories were obtained from the data on this theme. The categories that were acquired from the data were the school administrator–school relationship and the organizational perception of the school administrators.

**Table 2** Themes and categories emerging in research

| Theme | Category | Subcategory |
|---|---|---|
| Theme 1: School administrator–school relationship | School administrator | Leadership |
| | | Guiding, directing |
| | | Organizational memory |
| | Organization | Specific and common purposes |
| | | Human factor |
| | | Positive perception |
| | | Participation and contributions of stakeholders |
| Theme 2: School administrator–organizational memory relationship | Organizational memory perception | Experience |
| | | Knowledge |
| | | Achievement |
| | | Recording |
| | Reasons for creating organizational memory | Evaluation |
| | | Criteria |
| | | Source |
| | | Instrument |
| Theme 3: Use of organizational memory | Actions to be taken to create organizational memory | Participating in administration |
| | | Specialization |
| | | Communication |
| | | Written and digital instruments |
| | Instruments to be used to create organizational memory | Written instruments |
| | | Technology-based instruments |

## The School Administrator–School Relationship

At this point, it was seen that the participants made efforts to identify themselves with the school and that the school administrators reflected their individual and professional characteristics in their perceptions. In addition, the participants intensively reflect both their individual and professional characteristics to the school. The participants also agreed that the school administrator is a vital element in the school. The different and undeniable functions of the school administrator were pointed out. It was determined that three points came to the fore in the opinions. The first of them was that the school administrator is the leader of the school. Secondly, the school administrator is the one who directs and manages the school. The third point was that the school administrator is the brain and the memory of the school. The main point in the opinions was the functions fulfilled by school is influenced by the administrator's perspective and attitude. This situation was expressed as "school influenced by the administrator's perspective and attitude, and the administrator is the tram driver of the school." The participants also pointed to the importance of the ability of the administrator to mobilize their stakeholders, knowledge, and skills to achieve the responsibilities of the school. In the opinions that were put forward, it was understood that the school administrator always moves the school forward and that they move from the idea of development-oriented work. The assessments related to the subject were as follows:

SA 1: "The school administrator should adopt the school more than the school staff, and should have total information about the school. Our school is our second family."

SA 2: "The school administrator must lead and be an example for the people around him."

SA 4: "The school administrator is the person who manages and coordinates the school. He/she coordinates and manages the school for certain purposes."

SA 5: "In my opinion, what a school administrator is, is the school."

SA 7: "The school administrator is the most important organizational memory of the school in which the past achievements, physical condition, problems, and solutions of the school are available."

SA 8: "The school administrator is not the locomotive of the school, but the tram driver himself/herself."

## Organizational Perception

The participants think of the organization as a concept that includes positive meanings based on their experiences in the school. They also preferred to express these positive meanings through values that could be expressed individually and professionally: cooperation, being unified, development, and so on. Thus, they demonstrated that they adopted the concept of "organization." For the organization, the participants prominently emphasized the following as fundamental requirements of the organization:

- To struggle for the same and specific purposes
- To bring the human element to the forefront
- To see benefits in the school environment
- To include those who are closely or relatively related to the work done

The evaluations of the participants on the subject were as follows:

SA 3: "We can call it a community of people coming together for a particular purpose."

SA 6: "It is a stacking community that acts in response to common decisions with the same benefits created together with the stakeholders."

SA 9: "The organization in the school environment refers to forming a whole together with the school administration, teachers, school staff, students, and parents."

SA 10: "It is the whole of the structure that shares the same goal, takes the same responsibility, and creates the vision and mission together."

SA 11: "They are the people running toward the same goal together in a positive direction."

SA 12: "I think it is a community that comes together for a particular vision or purpose."

According to the observations of the researcher, the participants appeared to be sincere and explicit, and they responded to the questions by basing their answers on their experiences. In addition, the different analogies and examples they used in explaining their opinions also indicated that the questions were correctly perceived. The attitudes of the participants could be considered as important points in terms of achieving the aim of the research. In addition, when all of the statements are taken into consideration, the school administrator influences the school through what he/she says and does not say, and what he/she does and do not do, and the idea that he/she can shape the school emerges. This can make the responsibilities and duties of the school administrator more intensive and heavy, and can cause expectations to increase. The fact that students and teachers came with questions and expectations at different points during the interviews provided the researcher with an opinion in terms of the workload of the administrator. In addition, it was observed that the participants postponed dealing with those questions and expectations to focus on the interview with the researcher. This situation showed that the participants cared about the research and led to a positive impression of the purpose of the research.

## *Theme 2: Findings and Comments on the School Administrator–Organizational Memory Relationship*

Two categories and eight subcategories on this theme emerged. These categories were the causes of organizational memory perception and organizational memory.

### Organizational Memory Perception

The participants highlighted different points in explaining their views on organizational memory in line with their views related to the organization. According to the participants, the basic elements in organizational memory are recordings of past experiences created in different ways—electronic or written—in order not to forget them, to preserve the organizational memory in a way that is based on information and experience, to keep the organizational memory archived in case it is needed later, and to use the experience in the creation of organizational memory. It was also understood that the participants presented organizational memory as a form of communication and tried to create an organizational culture. In other words, it is necessary to perform studies, to record and archive the work done, to benefit from this experience and knowledge when necessary, and, finally, to create an organizational culture. It was thought that the participants tried to keep the organizational memory alive and functional by basing it on the mentioned perceptions. The evaluations of the participants on the subject were as follows:

SA 1: "Organizational memory is having prior knowledge about the school and managing the school in the light of previous experiences."

SA 2: "Whereas previously recorded statistical data, filing, file systems, and previous studies create the archive, all kinds of administrative staff, students, parents, and school-related studies constitute the organizational memory."

SA 3: "When we talk about organizational memory in our schools, all of the work done in schools before can be called organizational memory."

SA 4: "[It] is the record of everything that has been done in the institution since the institution was opened."

SA 5: "I can specify this as the work and operations of the organization in the context of implementation, adaptation, and transportation of all kinds of elements and the past experiences of the organization in any visible or invisible system. The organizational memory fulfills the information needs of the organization."

## Reasons for Creating Organizational Memory

All of the participants considered it necessary to create organizational memory. They described the reasons for the necessity of organizational memory as follows:

- Organizational memory allows evaluations within the organization; thus, it provides the opportunity to change and develop.
- Organizational memory is a starting point, since it shows what has been done before. It also eliminates uncertainty about what has been done or what has not been done in the organization before. It can be seen as a criterion.
- Organizational memory refers to a source for understanding the current situation and reviewing the work done previously before starting new work.
- Organizational memory is an instrument and an intermediary for accessing organizational knowledge and experience.

It was also pointed out that organizational memory contributes to saving time and effort in the organization. In this way, it can be said that organizational memory enables organizational work to be done in a systematic way and that management processes to make decisions in the organization can be implemented effectively. The evaluations of the participants related to the subject were as follows:

SA 7: "It should definitely be created, because when we think about what we need to pay attention to in social activities at the school and how social activities are done, it takes a lot of organization and attention if this is being done for the first time. If we take this work into organizational memory, it doesn't take much effort when we do the same or similar work. At the same time, we can correct the mistakes made in past years and make better programs."

SA 10: "Of course, when the school administrator has left the school, the success of the school, the statistical data, and the happiness of the teachers, students, and parents will be the clearest evidence of the occurrence of organizational memory."

SA 12: "It should be created. As a result, we always deal with people as the school. Therefore, it is necessary to have organizational memory if we think it may be required from the beginning and graduation of the person to the death of the person."

According to the observations of the researcher, the fact that participants explained what they had done for their organizational memory and the practices they followed by showing some examples—an electronic information management system (EIMS), and so on—indicated that their actions were consistent with their answers. It was also observed that there was an improvement in the context of better and faster access to different information sources and that both individual and organizational contemporary approaches were exhibited by using technology and internet-based instruments together, as well as traditional instruments (paperwork, blue-colored files, and yellow-colored personnel files).

## *Theme 3: Use of Organizational Memory by the School Administrator*

Two categories and six subcategories on this theme emerged. These categories were the instruments that can be used to create organizational memory and the activities that the administrator does so as to create organizational memory.

### Actions to Be Taken to Create Organizational Memory

The fact that the participants considered it necessary to create organizational memory in the school environment demonstrated the actions to be taken to create organizational memory as being participation of stakeholders in the administration, specialization of staff by getting education from experts in this field, establishment of a healthy communication setting, and use of written and digital instruments. In particular, the participants cared about giving information to stakeholders in various settings, receiving information from those who had previously served at the school, benefiting from their experiences and files, and giving in-service training to support or use organizational memory. It was said that implementation of the stated understanding contributes to participation of stakeholders in management, ensuring the desired level of communication and effective use of organizational memory. The opinions of the participants on the subject were as follows:

SA 2: "They can use the staff, the parents, the environment, the statistical data, and the technical facilities. They can also use facilities such as computers or written documents. They can use the archive and the files."
SA 3: "They may increase the participation of stakeholders in administration."
SA 4: "Specifically, the data should be created and stored in a digital environment."
SA 5: "To have a belief in creating organizational memory, the school administrator can get support from experts in the field."
SA 6: "First of all, the actions to be done can be listed as identifying the characteristics of the school, analyzing the event, selecting the staff that recognize the individual and the environment, and exchanging information with other schools and observing them."

## Instruments to Be Used to Create Organizational Memory

The participants in this research basically divided the instruments that can be used to create organizational memory into two groups: written instruments and technology-based instruments. Whereas written instruments generally refer to a copy of correspondence and their storage in filing systems, technical instruments can be created by using internet-based instruments (computers, web pages, etc.) or non-internet-based ones (cameras, videos, flash memory, compact disks, etc.). It was understood that the participants use the instruments in both groups together under the current conditions. The participants could be said to act situationally at this point. The use of both printed and electronic tools, as well as the use of previous data and the environment from time to time on behalf of the organization, show that different sources of organizational memory are utilized. The main purpose of using the specified instruments of different types and functions is to support the memory with data obtained from different instruments, control and backup, and, most importantly, prevention of any problems such as malfunction, loss, and virus transmission that may occur in any of the instruments.

SA 1: "Correspondence, meetings, school websites . . . "
SA 6: "The internet, visual media, TV, writers, historical sources, books describing past events . . . "
SA 8: "The organizational staff is the most important instrument in creating organizational memory. Records of work and transactions carried out in the organization also support organizational memory. Watching video recordings of previous work in the organization at ceremonies and meetings . . . "
SA 11: "Computers, SSD [solid-state drive] resources for computer storage, social–digital media, document presentation programs . . . "

According to the observations of the researcher, the participants cared about all of the instruments they use and develop various measures for their use, protection, and control. The purpose of doing this is to facilitate their work, to be prepared for developments that may occur at unexpected moments, and not to face any unwanted situations. At this point, the participants showed the researcher examples of the practices and systems they used, such as showing the library to the researcher, sharing images of activities, and sharing their impressions about the activity with the researcher. The expression of the opinions of the participants in the manner indicated above suggested that they tried to integrate what they had said and what they had done.

According to Table 2, three themes, six categories, and 21 subcategories appeared in the research. It can be seen that the views on the second theme were more intense than those on the other two themes. This also indicates that the themes and categories emerging in Table 2 can be accepted as a remarkable link between the school administrator and organizational memory.

## Conclusion, Discussion, and Recommendations

The following results were obtained from this research conducted using a qualitative research method and a phenomenological design with 12 participants: The school administrator is in an effective position that directs the school to reach its goals. The effectiveness of the mentioned position is directly proportional to the participation of stakeholders in administration. Drucker (2011), Küçükali (2011), and Beyer and Ruhl-Smith (1998) previously revealed that school administrators can be effective and perform their functions together with stakeholders in the school. It can be said that this previous research overlaps with the findings of this present research in revealing the importance of school administrators in schools. The results with regard to organizations and organizational memory obtained from the research can be listed as follows: The organizational approach takes into consideration the unity of the human factor and the achievement of its aims. Organizational memory is based on the experience created by recording and bringing experiences together. Organizational memory is an instrument used in achieving the goals of the organization. This instrument also constitutes the basis for showing the starting point and the point to be reached from what has been done. All stakeholders in the school should consider themselves as people who are valued for their views related to the school for formation of organizational memory. Since organizational memory requires technical knowledge and skills, there is a need for experts in the field. Different types of instruments are used in a coordinated manner in creating organizational memory. The understanding mentioned above is necessary to reduce or eliminate the impact of risky situations. It is understood from the stated results regarding organizational memory that the human factor contributes to the organization, expertise, experience, and tools, and that organizational memory contributes to the organization. When the related literature is reviewed, in studies regarding organizational memory conducted by Dunham and Burt (2011), Harvey (2012), and Karabal (2015), it can be seen that those experienced workers who have worked in the organization before are important for organizational memory. The contribution of this experience to the organization is to introduce the organization to new employees and to transfer information about the organization in a fast and effective way. Abel (2008) stated that the contribution of organizational memory to the organization is to manage the information in the organization and to enable employees to learn information about the organization. The concept of organizational memory contributes to the organization of knowledge management. Organizations need to benefit from memory as a learning organization memory, contributing to acquisition and management of knowledge, and explaining the competencies that people will gain. According to Walsh and Ungson (1991), there is an undeniable contribution of employees to obtaining, storing, and using information to create organizational memory. In their research, Fiedler and Welpe (2010) revealed the relationship between organizational memory and specialization. This relationship is based on knowledge and experience acquired by employees in the organization over time. In other research, Scalzo (2006) drew attention to the instruments used in the formation of organizational memory. These instruments are the people working in the organization, social networks, and computer-based and other information-providing instruments. In a similar study, Johnson and Paper (1998) revealed that organizational memory includes the memory of the individuals in the organization, routine work procedures, behavior regarding roles, and environmental factors. Organizational memory must be structured, and this memory consists of information stored by means of specific instruments for preservation and design of the organizational memory. These instruments are electronic practices, records, and archives related to operations. Furthermore, Akgün et al. (2005) revealed the contributions of organizational memory to the organization. These contributions are basically acquisition of information by assuming it can be reused and storing it, and providing an opportunity to use it when necessary. Research by Kılıç (2007) showed how important organizational memory is for the organization. Stein and Zwass (1995) explained the contribution of organizational memory to the organization in terms of its functions. These functions can be listed as acquisition, retention, and preservation of information, allowing its use for research and its accessibility when needed. Stein and Zwass (1995) defined another contribution of organizational memory to the organization as its contribution both to decision-making processes of employees working in the organization and to continuing and showing what is learned in the organizational environment. According to the research, an organization must have organizational memory to perform its function effectively and efficiently. With its organizational memory, the organization can look at the future with confidence and achieve its goals effectively and efficiently. Research related to this issue in the literature supports the results achieved in this present research. However, it is thought that it is not sufficient to adequately explain the relationships among education, school administrators, and organizational memory.

On the basis of the findings of this research, the following recommendations can be made for practitioners: By reviewing the training that contributes to the role of the school administrator in terms of organizational memory, the required measures may be taken to increase the knowledge and skills that will contribute to specialization in this field. The communication of the administrator at the point of sharing information with people to contribute to the formation of organizational memory can be made systematic. "Archiving" training can be given for proper functioning of organizational memory. This research can be developed with stakeholders other than the school principal who can contribute to organizational memory.

On the basis of this research, the following recommendations can be made for researchers: The themes and categories created can be used in scale preparation studies to be carried out regarding organizational memory. More research examining the relationships among organizational memory, education, and school administration should be conducted.

## References

Abel, M. H. (2008). Competencies management and learning organizational memory. *Journal of Knowledge Management, 12*(6), 15–30. https://doi.org/10.1108/13673270810913595.

Ackerman, M. S., & Halverson, C. (2004). Organizational memory as objects, processes, and trajectories: An examination of organizational memory in use. *Computer Supported Cooperative Work, 13*(2), 155–189.

Akgün, A. E., Keskin, H., & Günsel, A. (2005). Örgütsel hafıza ile enformasyon teknolojileri arasındaki ilişkiler. *Celal Bayar Üniversitesi İ.İ.B.F. Yönetim ve Ekonomisi Dergisi, 2*(12), 1–16.

Anand, V., Manz, C. C., and Glick, W. H. (1998). An organizational memory approach to information management. The Academy of Management Review, 23(4), 796–809.

Arısoy, Y. E., & Özdemirci, F. (2017). Kurum ve kuruluşların bakış açısından e-belgelerin arşivlenmesi. *DTCF Dergisi, 57*(2), 1250–1270.

Arvas, M. Ü. (2015). *Bilgi yönetimi ile örgütsel hafıza arasındaki ilişki: bir kamu kurumu uygulaması* (Yayımlanmamış Yüksek Lisans Tezi). Okan Üniversitesi, Sosyal Bilimler Enstitüsü, İstanbul.

Atak, M. (2011). Organizational knowledge management and investigation in learning organization literatüre. *İşGüç Industrial Relations and Human Resources Journal, 13*(2). ISSN: 1303-2860.). https://doi.org/10.4026/1303-2860.2010.178.x.

Aydın, C. (2010). *Elektronik belgelerin arşivlenmesi ve erişim* (Yayımlanmamış Doktora Tezi). Ankara Üniversitesi Sosyal Bilimler Enstitüsü, Ankara.

Baransel, A. E., & Tabak, İ. İ. (2006). Document management and work flow systems on the way to etransformation. *Görüşler/Opinion Papers Türk Kütüphaneciliği, 20*(1), 89–100.

Beyer, B. M., & Ruhl-Smith, C. (1998). The principal's role in school restructuring and reform: An examination of self-perceived leadership styles. *The Journal of Leadership Styles, 5*(2), 116–128.

Creswell, J. W. (2016). Nitel araştırma tasarımı. A. Budak ve İ. Budak, (Çev.), M. Bütün ve S. B. Demir, (Çev. Ed.), *Nitel araştırma yöntemleri* (s. 42–68) içinde. Ankara: Siyasal.

Drucker, P. F. (2011). *Etkin yöneticiyi etkin yapan nedir?* (M. İnan, Çev.), Liderlik (s. 35–53). İstanbul: Optimist.

Dunham, A. H., & Burt, C. D. B. (2011). Organizational memory and empowerment. *Journal of Knowledge Management, 15*(5), 851–868. https://doi.org/10.1108/13673271111174366.

Ersoy, A. F. (2016). Fenomenoloji. A. Saban ve A. Ersoy (Ed.), *Eğitimde nitel araştırma desenleri* (s. 51–110). Ankara: Anı.

Fiedler, M., & Welpe, I. (2010). How do organizations remember? The influence of organizational structure on organizational memory. *Organization Studies, 31*(04), 381–407. ISSN 0170–8406.

Harvey, J. F. (2012). Managing organizational memory with intergenerational knowledge transfer. *Journal of Knowledge Management, 16*(3), 400–417. https://doi.org/10.1108/13673271211238733.

Ilkka Tuomi, I. (2000). Data is more than knowledge: Implications of the reversed knowledge hierarchy for knowledge management and organizational memory. *Journal of Management Information Systems, 16*(3), 103–117.

Jimerson, R. C. (2003). Archives and memory. *OCLC Systems & Services: International Digital Library Perspectives, 19*(3), 89–95. https://doi.org/10.1108/10650750310490289.

Johnson, J. J., & Paper, D. J. (1998). An exploration of empowerment and organizational memory. *Journal of Managerial Issues, 10*(4), 503–519.

Karabal, C. (2015). *Örgütsel hafızanın örgütsel değişime direnç üzerindeki etkisi: İstanbul'da inşaat sektöründe örnek bir uygulama* (Yayımlanmamış Doktora Tezi). Beykent Üniversitesi, Sosyal Bilimler Enstitüsü, İstanbul.

Kaya, A. (2005). *Kurumlarda evrak ve arşiv yönetimi ile ilgili proseslerin ISO – 9000 serisi standartlarına uyarlanması* (Yayımlanmamış Yüksek Lisans Tezi). Marmara Üniversitesi, Türkiyat Araştırmaları Enstitüsü, İstanbul.

Kılıç, T. (2007). *Örgüt kültürü, örgütsel hafıza arasındaki ilişkiler ve firma performansı üzerine etkileri* (Yayımlanmamış Yüksek Lisans Tezi). Kadir Has Üniversitesi, Sosyal Bilimler, İstanbul.

Kızıldağ, D., Ağca, V., & Uğurlu, Ö. A. (2011). Örgütlerde hafıza kaybı ve etkileri: tekstil sektöründen bir yansıma. *Atatürk Üniversitesi İktisadi ve İdari Bilimler Dergisi, 25*(1), 193–206.

Kruse, S. D. (2003). Remembering as organizational memory. *Journal of Educational Administration, 41*(4), 332–347. https://doi.org/10.1108/09578230310481612.

Küçükali, R. (2011). *Yönetim felsefesi.* Ankara: Nobel.

Lehner, F., & Maier, R. K. (2000). How can organizational memory theories contribute to organizational memory systems? *Information Systems Frontiers, 2*(3/4), 277–298.

Merriam, S. B. (2015a). Nitel araştırma nedir? S. Turan, (Çev.), S. Turan, (Çev. Ed.), *Nitel araştırma yöntemleri* (s. 3–19) içinde. Ankara: Nobel.

Merriam, S. B. (2015b). Araştırmanın desenlenmesi ve örneklem seçimi. S. Turan ve D. Yılmaz, (Çev.), S. Turan, (Çev. Ed.), *Nitel araştırma yöntemleri* (s. 55–82) içinde. Ankara: Nobel.

Merriam, S. B. (2015c). Nitel araştırmalarda geçerlik, güvenirlik ve etik. E. Dinç, (Çev.), S. Turan, (Çev. Ed.), *Nitel araştırma yöntemleri* (s. 199–228) içinde. Ankara: Nobel.

Miles, M. B., & Huberman A. M. (2015). Veri toplamaya odaklanmak ve veri toplamayı sınırlamak: anlamlı bir başlangıç. D. Örücü, (Çev.), S. Akbaba Altun ve A. Ersoy (Çev. Ed.) *Nitel veri analizi* (s. 16–39) içinde. Ankara: PegemA.

Moorman, C., & Miner, A. S. (1998). Organizational improvisation and organizational memory. *The Academy of Management Review, 23*(4), 698–723.

Özbal, G. (2013). *Kurumsal belge ve arşiv yönetimi uygulamaları: Kalkınma Bakanlığı örneği* (Yayımlanmamış Yüksek Lisans Tezi). Hacettepe Üniversitesi, Sosyal Bilimleri Enstitüsü, Ankara.

Patton, M. Q. (2014a). Nitel araştırmanın doğası. M. Bütün, S. B. Demir (Çev.), M. Bütün, S. B. Demir (Çev. Ed.), *Nitel araştırma ve değerlendirme yöntemleri* (s. 1–36) içinde. Ankara: PegemA.

Patton, M. Q. (2014b). Nitel araştırmada stratejik temalar Y. Dede ve F. I. Bilican (Çev.), M. Bütün ve S. B. Demir (Çev. Ed.), *Nitel araştırma ve değerlendirme yöntemleri* (s. 37–74) içinde. Ankara: PegemA.

Patton, M. Q. (2014c). Nitel araştırma tasarımı. B. Tarman, M. F. Yiğit (Çev.), M. Bütün, S. B. Demir (Çev. Ed.), *Nitel araştırma ve değerlendirme yöntemleri* (s. 209–258) içinde. Ankara: PegemA.

Perez, G., & Ramos, I. (2013). Understanding organizational memory from the integrated management systems (ERP). *Journal of Information Systems and Technology Management, 10*(3), 541–560. ISSN online: 1807-1775. https://doi.org/10.4301/S1807-17752013000300005.

Rowlinson, M., Booth, C., Clark, P., Delahaye, A., & Procter, S. (2010). Social remembering and organizational memory. *Organization Studies, 31*(01), 69–87. ISSN 0170–8406.

Scalzo, N. J. (2006). *Radical organizational change and organizational memory systems: A qualitative case study of tacit and explicit knowledge* (Doctoral dissertation). George Washington University. Retrieved from Pro Quest Dissertations and Theses (UMI No. 3199653).

Seçkin-Halaç, D., & Seçkin-Çelik, T. (2016). Örgütsel hafıza ve yenilikçi iş davranışı ilişkisinde örgütsel hikâye anlatımının aracı rolü. *Galatasaray Üniversitesi İletişim Dergisi, 25*, 261–285. https://doi.org/10.16878/gsuilet.283074.

Şen, Y. (2014). *Örgütsel hafıza sistemi ve bir model önerisi: karar verme üzerindeki rolü ve bir uygulama* (Yayımlanmamış Doktora Tezi). İstanbul Üniversitesi, Sosyal Bilimler Enstitüsü, İstanbul.

Shadow, P. R. (1999). Strategy and cognition: Understanding the role of management knowledge structures organizational memory and information overload. *Creativity and Innovative Management, 8*(2), 140–148.

Stein, E. W. (1995). Organizational memory: Review of concepts and recommendations for management. *International Journal of Information Management, 15*(2), 17–32.

Stein, E. W., & Zwass, V. (1995). Actualizing organizational memory with information systems. *Information Systems Research, 6*(2), 85–117.

Walsh, J. P., & Ungson, G. R. (1991). Organizational memory. *The Academy of Management Review, 16*(1), 57–91.

Wexler, M. N. (2002). Organizational memory and intellectual capital. *Journal of Intellectual Capital, 3*(4), 393–414. https://doi.org/10.1108/14691930210448314.

Yıldırım, A., & Şimşek, H. (2011). *Nitel araştırma yöntemleri*. Ankara: Seçkin.

# A Way for Organizations to Cope with Uncertainty: Mimetic Isomorphism

**İnci Öztürk**

**Abstract** Uncertainty, which means absence of sufficient clarity in obtained information and lack of full or correct information, is experienced by managers and employees because of dynamic changes within the organizational environment. Mimetic isomorphism, which is due to the attempts of organizations to minimize uncertainty, occurs when organizations mimic or take well-established and successful organizations as a model to legitimize themselves, especially where the objectives are uncertain. Competition, which accompanies globalization and massification, allows higher education institutions to react to changing social demands, with restructuring of existing programs and elimination of older ones. In this chapter, the ways in which higher education institutions mimic other such institutions when they encounter uncertainty are elaborated with examples. It is concluded that higher education institutions may adopt faulty applications as well as positive applications as a model for themselves.

**Keywords** Mimetic isomorphism · Uncertainty · Coping with uncertainty · Higher education institutions · Bologna Process

## Definition of Uncertainty

The concept of uncertainty is used to describe both a situation in the organizational environment and the perceptions of individuals who do not have the necessary information regarding the environment. While the first definition indicates that it is possible to objectively analyze how much the organizational environment is uncertain, the second definition implies that environmental uncertainty may be perceived from the perspective of an observer. These two perspectives have led to the emergence of different definitions of uncertainty (Milliken 1987). Garner [(1962), cited in Duncan (1972)] defined uncertainty as a logarithm of possible outcomes

İ. Öztürk (✉)
Ankara University, Ankara, Turkey



Ş. Ş. ERÇETİN, Ş. N. AÇIKALIN (eds.), *Chaos, Complexity and Leadership 2018*, Springer Proceedings in Complexity, https://doi.org/10.1007/978-3-030-27672-0_22

of an event. Lorsch [(1967), cited in Duncan (1972)] described three characteristics of uncertainty: absence of sufficient openness in obtained information, existence of long intervals between feedbacks, and existence of general uncertainty about the relationships between causes and effects. Downey and Slocum (1975) developed a psychological definition of uncertainty, according to which uncertainty points to a situation in which the individual intentionally displays behavior based on information regarding the possible outcomes of the relation between the relationship of the individual with his/her environment, the individual's behaviors in relation to environmental variables, and the individual with his/her existing environment. According to this definition, psychological uncertainty is not only the individual's loss of control of the situation that the individual experiences, but also the individual's loss of his/her own control.

When the matter is taken into consideration from the perspective of management, uncertainty is the lack of full or correct information (Refsgaard et al. 2007). Bordia et al. (2004) suggest a three-factor conceptualization for description of uncertainty in the process of organizational change: strategic, structural, and job-related uncertainties. *Strategic uncertainty* refers to uncertainties regarding issues at the organizational level, such as the reasons for change, planning, the business environment, future trends, and the sustainability of the organization. *Structural uncertainty* stems from internal functioning of the organization, such as the functions of different units. *Job-related uncertainty* involves changes in job security, career opportunities, and job descriptions. Uncertainty is not just experienced by members who have recently joined the organization; it is also experienced continuously by members faced with changes within the organizational environment. Similarly, decision makers encounter uncertainty coming from the environment and stakeholders (Kramer 2015). In almost every area, organizational leaders are expected, in chaotic environments created by risk and uncertainty, to act more consciously and consistently than other members and to manage the process with confidence (Neyişci and Potas 2014). It may be said that within the organizational framework, all managers and workers in an organization routinely experience uncertainty. In a rapidly changing environment, there is a transformation from classical organizational values to neoclassical, chaotic, and uncertain organizational values (Erçetin and Kayman 2014).

## Coping with Uncertainty

There are various methods that organizations use to cope with uncertainty. One of them is the use of simultaneous structures. Work units may be shown as an example of use of simultaneous structures in an organization. Work units are expected to carry out a series of different functions in terms of complexity and predictability. Additionally, work units have to cope with external environmental factors such as government regulations, suppliers, and technological changes, besides other work units within the organization. There may be a stable and certain environment in other work units in the organization; however, government regulations that affect a particular work unit may lead to formation of unstable and uncertain environments. Therefore, a work unit may be exposed to different situations, each of which is more or less uncertain at the same time. It is up to the manager to use the structural dimensions in various combinations such as which decisions are centralized and to what degree, cooperation, and execution of rules and procedures. In the case of uncertainty, increasing centralization is observed. Indeed, this situation, which is not routine, allows the manager to think about which action could be taken and how to proceed. Lack of rules, processes, and guidelines—in other words, a need for formalization—lead subordinates to collect and process more information. It is then possible for the manager who trusts the expertise and knowledge of the subordinates to make the final decision (McDonough and Leifer 1983).

Strategic management is one of the methods that organizations use in coping with uncertainty because the strategy developed by organizations assists them in successfully coping with competitive difficulties (Parnell et al. 2000). According to Yoo and Digman (1987), strategic management is necessary to cope with uncertainty in decision making and to manage it. Strategic management eases future problems and opportunities, determining open objectives for the future of the organization, providing timely information for decision makers, providing cost savings, and allowing workers to perform better, making organizations that embrace strategic management more effective [see Hegerty (1976) and Wood and LaForge (1979), both cited in Yoo and Digman (1987)].

In addition to uncertainty at the organizational level, as explained above, we may also consider uncertainty at the sectoral level. Organizations that are located in the same sector and face the same market uncertainty have similar motivations. Organizations in the same sector establish relations with their environment in accordance with the same reasoning as the organization that they take as an example, for the purpose of increasing their credibility (Beckman et al. 2004). Organizations that are located in the same sector are exposed to similar environmental pressures and gain similar structural and operational characteristics parallel to the expectations and constraints of the environment. Thus, isomorphism emerges. In other words, organizations in the same field of activity are similar to each other in terms of their structural and operational characteristics (Koçel 2011). Isomorphism is a limiting process that forces a unit in a population to become similar to other units encountering the same environmental conditions (DiMaggio and Powell 1991).

Besides their structural isomorphism, organizations become isomorphic with the environment against environmental pressures. The organizations try to provide certainty regarding how various activities will be realized and try to prevent practices that change from person to person against expectations as to how certain work is to be done (Koçel 2011). An organization's establishment of a system and operation of that system by getting rid of personal dependency—in other words, institutionalization (Öztürk Fidan 2018)—is the emergence of patterns that allow the organization to gain a particular character and to reach a certain capability or the organizations adopt practices that prevent them from demonstrating their competence to perform (Manning 2013).

There are three types of isomorphism mechanisms—coercive, normative, and mimetic isomorphism—that take place alongside organizational isomorphic change. *Coercive isomorphism* originates from the formal and informal pressures applied to the organization by other organizations where the organizations are dependent upon cultural expectations in the societies where they operate (DiMaggio and Powell 1991). As the hegemony of rational states and other massive rational organizations in the social life increases, organizational structures gradually start reflecting institutionalized rules, which are legitimized by/in the state (Meyer and Rowan 1977). *Normative isomorphism* originates from a situation where organizations make the dominant norms a base for their socialization and feel that the members of a profession have the ability to form their own policies (Levy 2006). Professional and commercial associations are a tool for defining and disseminating normative rules regarding organizational and professional behavior (DiMaggio and Powell 1991). Not every isomorphic tendency may originate from coercive or normative reasons. Uncertainty, which encourages mimicking, is also a reason for isomorphism as powerful as the mandatory reasons (Fidan 2017). *Mimetic isomorphism*, which is one of the means used by organizations to cope with uncertainty, is elaborated in detail in the following sections.

## Mimetic Isomorphism

Among organizations that have the same environmental conditions, those that encounter uncertainty may mimic others to get rid of uncertainty. According to Haveman (1993), organizations reduce their costs when they encounter uncertainty (Cyert and March 1963), and they mimic the effectiveness of other organizations by replacing technical rules with institutional rules (Meyer et al. 1983). Mimetic isomorphism, which is realized by organizations' attempts to minimize uncertainty, is realized through their imitation of well-established and successful organizations for the purpose of legitimizing themselves, especially where their objectives are uncertain. Mimetic isomorphism happens where the actors are not clear regarding what they need to do and thus copy successful organizations (Levy 2006). In other words, when organizational technologies are not sufficiently understood, the objectives are uncertain, or the environment creates symbolic uncertainty, organizations may model themselves on other organizations. When an organization encounters uncertain effects or uncertain solutions, its search for a solution to problems may provide it with an economically applicable solution (DiMaggio and Powell 1991).

Modeling is a response to uncertainty. The modeled organization may be unaware of the modeling, or that organization may not wish to be copied. The modeled organization serves only as a suitable source of applications that the modeling organization may use. Models may be tacitly spread by organizations (such as consultancy firms or industrial–commercial associations), through workers transferring or quitting, unintentionally, or indirectly. Even innovation may be explained by organizational modeling (DiMaggio and Powell 1991). Organizations have a tendency to take similar organizations in their field as a model when they have a perception that those organizations are more legitimate or successful. The existence of certain structural arrangement types everywhere is caused by universality of mimetic processes rather than concrete evidence that the embraced models increase productivity (DiMaggio and Powell 1991). If the relationship between the organizational structures, processes, and outputs is not clear, the modeling actions become very important because the modeling organization achieves certainty without incurring any costs and displaying any effort to understand the connection between the inputs and outputs (Kondra and Hurst 2009). A strategy to avoid such a risk provides certainty of organizational outputs. This strategy may be intentionally applied or it may be taken for granted (Kondra and Hinings 1998).

Alongside the organizational structures that are put in place when an organization is established, the role of the person who founds the organization is immense in its ongoing survival, because the founder plays a pivotal role in the history of the organization in terms of its survival, strategy, and performance. Social actors, which are perceived as an important mechanism in the establishment of an organization and in the emergence of new organizational forms, structure the action and contribute to the persistence of organizational elements through their actions and interactions (Johnson 2007). Successful institutionalization of social patterns is dependent upon purposeful, competent, and motivated actors that choose the correct and necessary action (Meyer 2010). When social actors carry out certain things in a certain way, certain courses of action become taken for granted or institutionalized. Thereafter, other social actors undertake that course of action without thinking too much about it. If a social actor embraces a certain course of action, similar social actors will also mimic that social actor (March 1981). For instance, if small firms in certain sectors incline toward a new market segment, other small firms will also enter that market segment (Haveman 1993). The social constructive perspective—which describes obedience to rules and mandatory action developed against willfulness, which is one of the assumptions of the classical choice theories—also explain mimetic isomorphism taking place in organizational forms (March 1981; Haveman 1993).

Description of regular behavior perceives an action as a match of the behavior and the circumstances. The rules help us to understand both how organizations maintain their stability and the ways in which organizations change. When the existence and continuity of rules merge with idiosyncratic independence of the members of an organization, it enables the organization to function reasonably in a dependable and rational way. At the same time, change in organizations are closely related to rules. Most organizations easily and regularly respond to the environment by relying upon processes that are connected to environmental impetus. Organizational adaptation is based on observing the organization and the environment for the purpose of receiving messages regarding the circumstances of the world and performing actions that are appropriate for the rules. Rules are effective tools that allow complex organizations to respond to variations in circumstances. Organizations adapt to their environment by changing the rules. However, this type of change process is slower than adaptation to changing circumstances within the rules. One of the ways of understanding which rules change in what ways is observation of the contagion effect. The basic idea is that rules may change relatively through the processes of slow learning and choice; however, rapid changes in the rules emerge through imitation. Organizations copy each other. Observing the contagion effect is much easier than observing the effect of learning and choosing (March 1981).

An example may be given for the contagion effect: If it is desired to produce an explanation regarding the acceptance of accounting conventions, the ways in which standard accounting processes spread between organizations are investigated. It is seen that individual organizations accept best-practice examples approved by professional associations and applied by leaders (March 1981). Another example is a study that Tolbert and Zucker (1983) described, regarding how public service reform spreads in cities where reforms are mandatory and in those where they are not. When public service reform is rendered mandatory by the state, it spreads from the state to each of the cities in a fast and direct way. When the public service processes are not made mandatory at the state level, they spread among cities gradually through social impact. If a gradually increasing number of organizations embrace a program or policy, that program or policy either becomes institutionalized or starts becoming a component of a rationalized organizational structure. Additionally, when cities embrace public service reform earlier, this is related to organizational necessities, whereas when cities embrace it later, this is related to institutional definitions of legitimate structural forms. Briefly, an organization's embrace of a policy or a program is determined by that policy's or program's legitimation through the law or in a gradual way—in other words, by its level of institutionalization.

It may be argued that coercive pressures such as legal arrangements shape organizations fast and directly, and that an organization mimics another organization that it perceives as being successful and achieving its objective in cases where the former experiences uncertainty or is newly established. Thus, an organization experiencing uncertainty in its structure and operation is motivated to achieve the same legitimacy as the organization that it takes as a model through modeling.

## Mimetic Isomorphism in Higher Education

Mimetic isomorphism generally emerges as an outcome of competitive market forces (Bess and Dee 2008). Global competition has become a strong structure shaping the economy of a country for reasons such as the growth of multinational companies and organization of international production. Governments have created an internal market of universities by mimicking newly emerging implementations of global commerce that organize the internal market (Dominelli and Hoogvelt 1996). The forces of competition that come with globalization and massification require higher education institutions to respond to fast-changing labor markets and the benefit of student programs. Higher education institutions increasingly seek autonomy and flexibility. Thus, these institutions react to fast-changing social demands, restructuring existing programs and eliminating older programs (Dill 2010). In the global world, the latest developments bring changes and variety in doctoral programs to the forefront (Balcı 2011).

In settings lacking central authority, organizational structures may display similarities due to competitive and mimetic processes (Scott 1991). In the event that an organizational field has a fragmented decision-making structure with many and uncoordinated authority and impact sources, it is possible for heterogeneity to emerge in organizational elements (Meyer et al. 1987). In such an organizational field, competitive arrangements, norms, and cognitive models may develop for these organizations because there is no central authority to coordinate activities or resolve disagreements (Hoffman 1999). Organizations that have similar products and services and are located close to each other will encounter relatively strong competition as they feed from the same limited consumer pool, because competition may increase the possibility of these organizations underperforming. Thus, these organizations will require organizational change in order to differentiate themselves from their competitors and to avoid losses (D'Aunno et al. 2000). In this case, in the higher education sector, shaped by market conditions, it may be argued that one of the aspects in which higher education institutions will be in competition is in attracting good students, and they will follow or model the best and the most successful among themselves regarding innovation (Öztürk Fidan 2018).

As an example of a university's modeling of another university, when a university sets up high-speed internet access at its student dorms, it then becomes possible for its rival universities to increase their internet capacity. Similarly, universities spend significant resources to develop their interuniversity athletics programs. The university leaders believe that many students and alumni expect their universities to be successful in sports matches. If a rival university achieves success, other universities will also attempt to catch the same advantage (Bess and Dee 2008). With regard to academic departments, Riesman [(1956), cited in Bess and Dee (2008)] suggested that less prestigious institutions are forced to follow those that are considered admirable. The common goal among university leaders is to improve their university's Carnegie classification or transform their institutional identity into one of a more prestigious university (Morphew 2002). The higher education institutions strive to develop graduate programs, find research grants, and increase acceptance conditions to attain a more prestigious position in the academic hierarchy (Bess and Dee 2008).

An example of mimetic isomorphism is the functioning of the Bologna Process (Dokuz Eylül Üniversitesi, 2011a) may be given whose foundations are laid by the Sorbonne Declaration in 1998. The Bologna Process, in which each country voluntarily participates, aims to make the higher education systems of the participating countries uniform in order to make them comparable to and compatible with each other by protecting their individual differences. Thus, it is planned to ease the transition from one country or one higher education system to the next, to provide learning and lecturing mobility, and to increase employment (Dokuz Eylül Üniversitesi, 2011c). The objectives of the Bologna Process may be listed as creating comparable higher education degrees, applying the European Credit

Transfer System (ECTS), providing mobility of students and faculty members, creating quality assurance systems in higher education, encouraging lifelong learning, making the European Higher Education Area (EHEA) more attractive, and creating synergy between the European Research Area (ERA) and the EHEA.

According to the governance model envisaged by the Open Method of Coordination (OMC), which is the ontology of the Bologna Process, the member countries of the European Union became actively involved in the Bologna Process, in which they are competitive and ambitious, because the member countries of the European Union have protected the momentum of the Bologna Process and created standards for prospective comparisons. For instance, the multicolored scorecards of the countries submitted in stock-taking reports are one of the most widespread monitoring techniques in the Bologna Process. The scorecard system shapes institutions and people. The visuals used in the stock-taking reports display big data, make it easier to have access in different geographies, provide information regarding nation states, and support the demand for establishing the EHEA (Dokuz Eylül Üniversitesi, 2011b). Each participant country attempts to perform well in the active areas of the Bologna Process because other participant countries also attempt to perform well. The participant countries mimic the desire of other countries to survive. Thus, it may be said that mimetic desire has competitive characteristics and this is a strong emotional factor that accelerates the application process (Thrift 2007). Countries display their reputation and success on public webpages, with follow-up mechanisms such as stock-taking reports and scorecards. The follow-up mechanisms create change and progress by triggering a mimetic desire that reflects the tension between the risk of shame and the chance of fame. In other words, the countries have embraced a governance model in which they feel the pressure of other countries through follow-up and exposure (Brøgger 2016).

Within the scope of the Bologna Process, the class of "Community Service Practices" (YÖK 2006), which was included in university programs in Turkey by the Higher Education Council in the 2006–2007 educational year, constitutes an example of mimetic isomorphism. Within the scope of the Bologna Process, the need for restructuring of educational programs has emerged in universities, and it was decided to support university students to conduct studies regarding solutions to social problems with reference to the principle that universities need to be integrated with society and social services need to be made a function of universities (Anadolu Üniversitesi 2006).

Mimetic isomorphism in higher education is also seen between the focus institution and the spawning institutions that got separated from the focus institution. There is research indicating that the Faculty of Education at Ahi Evran University and the Faculty of Education at Kastamonu University, which separated from Gazi University in 2006, mimicked the Faculty of Education at Gazi University, which they found to be "successful" in operation and "they previously knew well." For instance, when the personnel in the Faculty of Education at Kastamonu University had problems when entering student grades into the computer for the first time, they stated that they consulted Gazi University and took Gazi University as a model. Similarly, members of the Faculty of Education at Ahi Evran University were of the opinion that their university took the functioning of Gazi University as a model because the faculty members had previously worked at Gazi University and/or had obtained graduate education at Gazi University. Researchers have ascertained that those higher education institutions that separated from Gazi University consulted Gazi University when they initially opened their graduate programs. The spawning institution took the focus institution as a model in matters such as preparation of course content (Öztürk Fidan 2018).

It may be argued that institutions take negative applications from each other besides their sharing of positive applications. An application that an institution has modeled from another institution may be unethical and/or illegal. It is possible for unethical behaviors to become institutionalized in organizations. According to Hosmer (1987), inappropriate management behaviors regarding business ethics can be encountered in big companies. For instance, General Electric cheated the US government at different times in the sales of rockets with nuclear warheads and satellites, and E.F. Hutton & Company, which provided brokering services, got involved in check fraud in which 400 banks incurred losses. Defense industry companies have been involved in many unethical actions, and those companies have cheated the US government in military procurement. McDonnell Douglas made faulty decisions that created high costs; Rockwell International kept their labor fees high; General Dynamics overcharged the government; Boeing provided substandard replacement products; and United Technologies gave gifts as inducements and practiced faulty pricing. According to Brief et al. (2001), the repetitive cycles of those types of routinized activities has been defined as institutionalization of corruption.

Routinized applications that are against regulations at universities in Turkey are discovered through auditing by the Turkish Court of Accounts and internal auditing done by the Board of Internal Auditing at the universities. Failure to establish data recording systems providing correct and trustworthy reporting and failure to establish internal control systems in accordance with regulations and the preliminary financial control unit are discovered through external auditing done by the Turkish Court of Accounts. According to the *External Auditing General Evaluation Report* of 2017, issued by the Office of the Turkish Court of Accounts in September 2018 (Sayıştay 2018) as a consequence of auditing done at the universities, it was determined that some universities had carried out their operations in breach of the relevant regulations. Some of the faulty applications detected in the external auditing were failure to carry out the rental of real estate properties in accordance with the tendering method stipulated in the relevant laws, failure to follow up on financial fixed assets in accounting records, failure to include the circulating capital enterprises of universities in the internal control systems of the universities, failure to use accounts that were supposed to be used for following up the allowances given within the scope of scientific research projects and expenses that were incurred, transfer of revenues that pertained to the acceptance examination of foreign students to the circulating capital enterprises instead of the special budget and distribution of those revenues as supplementary payments, employment of faculty and administrative personnel at educational units that had no students at the time, and several problems regarding the circulating capital enterprises of the university research and training hospitals. As a consequence of the internal auditing that the universities carried out, it was also determined that there were various faulty applications regarding additional course accruals, circulating capital contribution payments, and salary accruals (Fırat Üniversitesi 2011). It is possible that whether those faulty practices were carried out intentionally or unintentionally, they were modeled by a newly established university as a result of creation of an agenda by a professor who had been appointed as the president of that university while serving as a professor at another university, on the basis of his/her experience gained from his/her original university, and that new structures and strategies were created by the same professor making a comparison between the two universities regarding university organization. These examples show the impact of modeling of an administrator (Öztürk Fidan 2018). It may be argued that the inexperience of the administrator in administrative and financial matters, and the administrator's failure to encourage the internal stakeholders of the university to make appropriate decisions, caused faulty practices in the universities and resulted in them becoming institutionalized.

## Discussion and Conclusion

In this chapter, an attempt has been made to discover how higher education institutions mimic other higher education institutions when they encounter uncertainty. One of the ways in which higher education institutions cope with uncertainty is mimetic isomorphism. Mimetic isomorphism generally emerges as a result of competitive market forces. Social, environmental, technological, and economic changes—such as an increase in labor variety, increased use of subcontractors, the spread of flexible employment, and increased automation—have continuously shaped global labor markets (DESA/DPAD 2017), and intensified competition (Madden et al. 2017). The competition that comes with globalization has also influenced higher education institutions. As higher education institutions diversify in terms of their size, scope, and complexity, the higher education sector has become a more competitive ecosystem (Baruch 2013). Because different actors have different agendas and interests, academia displays characteristics of a complex ecosystem where change is almost unpredictable (Şimşek and Louis 1994). Under these conditions of uncertainty and unpredictability, higher education institutions resort to continuously restructuring their existing programs according to market conditions. Thus, these institutions attempt to mimic other higher education institutions that they perceive as being successful and well regarded in the sector and as having reached a certain level of stability.

On the other hand, universities may also mimic the faulty practices of the universities that they take as a model. In particular, such faulty practices, whether they are carried out intentionally or unintentionally, may become institutionalized over time. By adhering to relevant regulations, university administrations should not allow faulty practices to occur. Using internal auditing, university administrations should detect any faulty practices that breach regulations and rectify them.

Under conditions where coercive processes such as legal regulations and normative processes such as professional and ethical values are not sufficiently influential, mimetic processes fill this gap (Baruch and Fidan 2019). Universities experiencing the same environmental conditions start becoming similar to each other in terms of their structural and functional characteristics. Investigating this isomorphism allows us to understand how organizational structures and functions are formed in higher education; in other words, it allows us to understand which mechanisms are institutionalization mechanisms in higher education.

**Acknowledgements** I would like to thank Dr. Tuncer Fidan for his valuable opinions regarding the auditing.

## References

Anadolu Üniversitesi. (2006). *Topluma hizmet*. http://iibf.anadolu.edu.tr/?page_id=440. Accessed 10 Feb 2015.

Balcı, A. (2011). Eğitim yönetiminin değişen bağlamı ve eğitim yönetimi programlarına etkisi. *Eğitim ve Bilim, 36*(162), 196–208.

Baruch, Y. (2013). Careers in academe: The academic labour market as an eco-system. *Career Development International, 18*(2), 196–210.

Baruch, Y., & Fidan, T. (2019). The Turkish academic labor market as an ecosystem. In T. Fidan (Ed.), *Vocational identity and career construction in education* (pp. 37–57). Hershey, PA: IGI Global.

Beckman, C. M., Haunschild, P. R., & Philips, D. J. (2004). Friends or strangers? Firm-specific uncertainty, market uncertainty, and network partner selection. *Organization Science, 15*(3), 259–275.

Bess, J. L., & Dee, J. R. (2008). *Understanding college and university organization: Theories for effective policy and practice. Volume I: The state of the system*. Sterling: Stylus.

Bordia, P., Hunt, E., Paulsen, N., Tourish, D., & DiFonzo, N. (2004). Uncertainty during organizational change: Is it all about control? *European Journal of Work and Organizational Psychology, 13*(3), 345–365.

Brief, A. P., Buttram, R. T., & Dukerich, J. M. (2001). Collective corruption in the corporate world: Toward a process model. In M. E. Turner (Ed.), *Groups at work: Theory and research* (pp. 471–499). Mahwah: Lawrence Erlbaum.

Brøgger, K. (2016). The rule of mimetic desire in higher education: Governing through naming, shaming and faming. *British Journal of Sociology of Education, 37*(1), 72–91.

Cyert, R., & March, J. (1963). *Behavioral theory of the firm*. Englewood Cliffs, NJ: Prentice Hall.

D'Aunno, T., Succi, M., & Alexander, J. A. (2000). The role of institutional and market forces in divergent organizational change. *Administrative Science Quarterly, 45*(4), 679–703.

DESA/DPAD [Department of Economic and Social Affairs/Development Policy & Analysis Division]. (2017). *Frontier issues: The impact of the technological revolution on labor markets and income distribution*. United Nations. https://www.un.org/development/desa/dpad/publication/frontier-issues-artificial-intelligence-and-other-technologies-will-define-the-future-of-jobs-and-incomes/.

Dill, D. D. (2010). Quality assurance in higher education: Practices and issues. In P. P. Peterson, E. Baker, & B. McGaw (Eds.), *International encyclopedia of education* (3rd ed., pp. 377–383). Oxford: Elsevier.

DiMaggio, P. J., & Powell, W. W. (1991). The iron cage revisited: Institutional isomorphism and collective rationality in organizational fields. In W. W. Powell & P. J. DiMaggio (Eds.), *The new institutionalism in organizational analysis* (pp. 63–82). Chicago: University of Chicago Press.

Dokuz Eylül Üniversitesi. (2011a). *Dokuz Eylül University Bologna Process* (In Turkish). http://webb.deu.edu.tr/bologna/index.php/bologna-suereci-nedir. Accessed 6 Mar 2019.

Dokuz Eylül Üniversitesi. (2011b). *Bologna Process member states* (In Turkish). http://webb.deu.edu.tr/bologna/index.php/bologna-suerecine-ueye-uelkeler. Accessed 6 Mar 2019.

Dokuz Eylül Üniversitesi. (2011c). *10 action titles of the Bologna Process* (In Turkish). http://webb.deu.edu.tr/bologna/index.php/bologna-suerecinin-10-eylem-bal. Accessed 6 Mar 2019.

Dominelli, L., & Hoogvelt, A. (1996). Globalization, contract government and the Taylorization of intelluctual labour in academia. *Studies in Political Economy, 49*, 71–100.

Downey, H. K., & Slocum, J. W. (1975). Uncertainty: Measures, research and sources of variation. *Administrative Science Quarterly, 18*, 562–577.

Duncan, R. B. (1972). Characteristics of organizational environments and perceived environmental uncertainty. *Administrative Science Quarterly, 17*(3), 313–327.

Erçetin, Ş. Ş., & Kayman, E. A. (2014). How to be a quantum leader in an intelligent organization? In S. Banerjee & Ş. Erçetin (Eds.), *Chaos, complexity and leadership 2012* (pp. 247–252). Dordrecht: Springer.

Fidan, T. (2017). Kurumsalcılık yaklaşımları ve yeni kurumsalcılık perspektifinden eğitim örgütleri. *Medeniyet Eğitim Araştırmaları Dergisi, 1*(1), 1–16.

Fırat Üniversitesi. (2011). *Fırat Üniversitesi internal audit unit annual report (year 2011)* (In Turkish). http://web.firat.edu.tr/icdenetim/FR/F%C3%9C%20%C4%B0DB%20Faaliyet%20Raporu%202011.pdf. Accessed 13 Mar 2019.

Haveman, H. A. (1993). Follow the leader: Mimetic isomorphism and entry into new markets. *Administrative Science Quarterly, 38*(4), 593–627.

Hoffman, A. J. (1999). Institutional evolution and change: Environmentalism and the US chemical industry. *Academy of Management Journal, 42*, 351–371.

Hosmer, L. T. (1987). The institutionalization of unethical behavior. *Journal of Business Ethics, 6*(6), 439–447.

Johnson, V. (2007). What is organizational imprinting? Cultural entrepreneurship in the founding of the Paris Opera. *American Journal of Sociology, 113*(1), 97–127.

Koçel, T. (2011). *İşletme yöneticiliği* (13. Baskı). İstanbul: Beta Basım A.Ş.

Kondra, A. Z., & Hinings, C. R. (1998). Organizational diversity and change in institutional theory. *Organization Studies, 19*(5), 743–767.

Kondra, A. Z., & Hurst, D. C. (2009). Institutional processes of organizational culture. *Culture and Organization, 15*(1), 39–58.

Kramer, M. W. (2015). Uncertainty and communication in organizations. *The International Encyclopedia of Interpersonal Communication*, 1787–1866. https://onlinelibrary.wiley.com/doi/10.1002/9781118540190.wbeic001.

Levy, D. C. (2006). How private higher education's growth challenges the new institutionalism. In H. D. Meyer & B. Rowan (Eds.), *The new institutionalism in education* (pp. 143–161). Albany, NY: State University of New York Press.

Madden, L., Kidder, D., Eddleston, K., Litzky, B., & Kellermanns, F. (2017). A conservation of resources study of standard and contingent employees. *Personnel Review, 46*(3), 644–661.

Manning, K. (2013). *Organizational theory in higher education*. New York: Routledge.

March, J. G. (1981). Footnotes to organizational change. *Administrative Science Quarterly, 26*(4), 563–577.

McDonough, E. F., III, & Leifer, R. (1983). Using simultaneous structures to cope with uncertainty. *The Academy of Management Journal, 26*(4), 727–735.

Meyer, J. W. (2010). World society, institutional theories, and the actor. *Annual Review of Sociology, 36*, 1–20.

Meyer, J. W., & Rowan, B. (1977). Institutionalized organizations: Formal structure as myth and ceremony. *American Journal of Sociology, 83*(2), 340–363.

Meyer, J. W., Scott, W. R., & Deal, T. E. (1983). Institutional and technical sources of organizational structure: Explaining the structure of educational organizations. In J. W. Meyer, & W. R. Scott (Eds.). *Organizational environments*. (p. 45–70). Beverly Hills, CA: Sage.
Meyer, J. W., Scott, W. R., & Strang, D. (1987). Centralization, fragmentation, and school district complexity. *Administrative Science Quarterly, 32*, 186–201.
Milliken, F. J. (1987). Three types of perceived uncertainty about the environment: State, effect, and respeonse uncertainty. *The Academy of Management Review, 12*(1), 133–143.
Morphew, C. (2002). "A rose by any other name": Which colleges became universities. *Review of Higher Education, 25*(2), 207–223.
Neyişci, N., & Potas, N. (2014). Avoidance behaviors of school managers in uncertain and chaotic environments. In S. Banerjee & Ş. Erçetin (Eds.), *Chaos, complexity and leadership 2012* (pp. 219–227). Dordrecht: Springer.
Öztürk Fidan, İ. (2018). *Üniversitelerde iz bırakma ve kurumsallaşma* (Yayımlanmamış doktora tezi). Ankara Üniversitesi Eğitim Bilimleri Enstitüsü, Ankara.
Parnell, J. A., Lester, D. L., & Menefee, M. L. (2000). Strategy as a response to organizational uncertainty: An alternative perspective on the strategy-performance relationship. *Management Decision, 38*(8), 520–530.
Refsgaard, J. C., van der Sluijs, J. P., Højberg, A. L., & Vanrolleghem, P. A. (2007). Uncertainty in the environmental modelling process—A framework and guidance. *Environmental Modelling & Software., 22*, 1543–1556.
Sayıştay. (2018). *2017 yılı dış denetim genel değerlendirme raporu: Eylül 2018*. Ankara: Sayıştay Başkanlığı.
Scott, W. R. (1991). Unpacking institutional arguments. In P. J. DiMaggio, & W. W. Powell (Eds.). *The new institutionalism in organizational analysis*. (p. 164–182). Chicago, IL: University of Chicago Press.
Şimşek, H., & Louis, K. S. (1994). Organizational change as paradigm shift: Analysis of the change process in a large, public university. *The Journal of Higher Education, 65*(6), 670–695.
Thrift, N. (2007). *Non representational theory: Space, politics, affect*. New York: Routledge.
Tolbert, P. S., & Zucker, L. G. (1983). Institutional sources of change in the formal structure of organizations: The diffusion of civil service reform, 1880–1935. *Administrative Science Quarterly, 28*, 22–39.
Yoo, S., & Digman, L. A. (1987). Decision support system: A new tool for strategic management. *Long Range Planning, 20*(2), 114–124.
YöK. (2006). *Dersler ve içerikleri*. Ankara: Yükseköğretim Kurulu Başkanliği.
Yükseköğretim Kurulu. (2006). *Eğitim fakültelerinde uygulanacak yeni programlar hakkında açıklama*. http://www.yok.gov.tr/documents/10279/49665/aciklama_programlar/aa7bd091-9328-4df7-aafa-2b99edb6872f. Accessed 17 Feb 2017.

# Index

**A**
Achilles' heel, 123–124
Adenauer's leadership style, 111
Akaike information criterion (AIC), 69
Alija Izetbegovic
  Bosnians, 29–32
  family member, 28
  Muslims, 29–32
  Quantum Mechanics, 27
Allee effect, 97, 98, 102, 106
Angela Merkel
  CDU/CSU–FDP coalition, 114–116
  first grand coalition, 112–114
  second grand coalition, 116–118
AR models, 141
Artificial intelligence (AI)
  chaos theory, 204, 208
  decision-making technologies, 208
  description, 203
  futurism, 204–206
  golden rules, 204
  strange attractor, 206–207
  world of possibilities, 206
Association of Chartered Certified Accountants (ACCA), 219
"ATM card switch on/off" facility, 167
Attractor, 42
Autophagic leadership
  accurate diagnosis, 230
  autophagic mechanisms protection, 231
  benefits and risks balancing, 231
  conceptual framework, 229
  internal resources prioritization, 229
  organization health, 229, 231
  practices/procedures/rules, 229
  starvation, 230
Autophagy
  autophagic leadership (*see* Autophagic leadership)
  cellular autophagic process, 228
  definition, 227
  Dysfunction, 228
  human organization, 228
  "self-eating mechanism", 228
  social organizations, 228

**B**
Banking chaos and complexity
  consumer satisfaction, 157
  digital banking services, 158–168
  financial institution services, 157–158
Belarus, 80–83
Berlin Declaration, 113–114
Bifurcation diagram, 105
Bologna Process, 277, 278
Brexit process, 92
Bundestag election, 111

**C**
Carbon emissions trading, 149
Causality, 68, 69, 75
Center for Strategic Continuing Education and Training (CRESCENT), 216
Central Limit Theorem, 34


Ş. Ş. ERÇETİN, Ş. N. AÇIKALIN (eds.), *Chaos, Complexity and Leadership 2018*, Springer Proceedings in Complexity, https://doi.org/10.1007/978-3-030-27672-0

Chancellor of unity
elements, 108
historical background, 109
party leadership, 110–111
personal engagement, 111
personal prestige, 110
political communication, 112
principle, 109–110
Chaos theory, 21, 78
incalculability/unpredictability, 79
in international relations, 88–89
nonlinearity, 78
pro-Western leadership, 94
relevance, 128
Russian factor, 94
self-similarity, 79
sensitive dependence, 78
Chaotic circumstances
Chaos theory, 244
educational organizations, 234, 235
human factor influence, 245
professional appointment, 245
research findings, 238–244
research methods
data collection, 236–237
process, 237
study participants, 236
teachers happiness, 245
teachers perspective, 245
Chaotic dynamic systems, 138
Chaotic system
characteristics, 234
linear quality, 234
unpredictable movement, 234
Chaotic time series, 140
Chartered Institute of Management Accountants (CIMA), 220
Civic society, 17
Classical physics theory, 77
Coercive isomorphism, 274
Cointegration, 68, 69, 72, 75
Cold War, 87
Colonization, 195
Color revolutions, 78
Belarus, 81
civil revolutions, 82–83
media, 82
NGOs, 81
organizational scheme, 80
pro-Western governments, 83–84
youth organizations, 81, 83
Comparative analysis
assurance, 164, 165
empathy, 163
reliability, 163
responsiveness, 164
SBI/ICICI, 165, 166
securities and services, 164
Conceptual framework, readiness for change
attitudes and behaviors, 183
cognitive dimension, 182
Education, 182
operating order, 182
readiness for change, 182
stakeholders relationship, 183
teacher participation, 183
Conservatives and Liberal Democrats, 2
Coping
complexity and predictability, 272
mimetic isomorphism, 274
organization's establishment, 273
rganizational level, 273
strategic management, 273
structural dimensions, 273
structural isomorphism, 273
Cost–benefit analysis, 213
Council of Higher Education (CHE) thesis bank, 255
Cronbach alpha reliability, 185
Customer satisfaction
internet banking, 157
null hypothesis, 161, 162
quantitative feedbacks, 160
service quality, 166

**D**
Data analysis
customer expectation, 160
customer perceptions, 160
interpretation, 161
online banking customer responses, 161
Digital banking
customer satisfaction, 157
definition, 156
measures and analysis (*see* Data analysis)
objectives, 157
recommendations, 166–167
research findings, 166
research methodology, 157–160
Dynamical systems, 47–53

**E**
Economic Commission for Africa (ECA), 200
Economic data, 137
Educational systems

chaos, 234
chaotic circumstances, 242
school administrators, 234
Egyptian President Hosni Mubarak, 8
Einstein Field Equations, 36–37
Electroencephalogram (EEG) signals, 24–25
Electronic information management system (EIMS), 263
Error-prone variables
Eurasianism, 94
European Atomic Energy Community, 146
European Coal and Steel Community (ECSC), 146
European Credit Transfer System (ECTS), 278
European energy policies
competitiveness, 151
ECSC, 147
energy resource concerns, 146
European Commission, 147, 148
implications for Turkey, 152–153
security of supply, 150–151
security politics, 146
sustainability, 148–150
European Higher Education Area (EHEA), 278
European project, 95
European Research Area (ERA), 278
Executive Board, 31

**F**
Federal government funding
capacity development, 218
education loans, 218
IIUM, 216
international rankings, 217
land grants, 216
operating costs, 216
quality assurance measures, 217
research activities, 217
research grants, 216
scholarships, 217
Federal Land Development Authority (FELDA), 219
Final prediction error (FPE), 69
Fixit, 93
Foreign and Commonwealth Office (FCO), 2, 6
Former USSR countries, 78
Fourier analysis, 21
Fractal dimension, 43
Fractals, 39–42
Fundamental Research Grant Scheme (FRGS), 216
Futurism
bifurcation, 205
golden rules, 204

**G**
Generalized variance decomposition, 69
German Association for Foreign Affairs, 30
Grexit
economic crisis, 90
eurozone, 91–92
financial support, 91
Greek pensions and wages, 90
UK and EU, 92

**H**
Hausdorff-Besicovitch dimension, 43–45
Higher education (HE)
funding problems, 213–214
government funding, 213
Malaysia (*see* Malaysia, HE)
poor facilities and programs, 214–215
population growth, 212
student enrollments, 212
Higher education institutions (HEIs)
budget funding, 221
funding, 221
Gazi University, 279
globalization and massification, 276
sectors, 277
Ugandan National Council of Higher Education, 221
Human immortality, 207

**I**
IIUM Center for Continuing Education (ICCE), 216
Illegal migration
clandestine, 194
colonization, 196
Europe, 194
measurement
brain drain, 200
economic opportunity, 199
globalization, 200–201
human rights, 199
labor shortages, 200
unemployment, 199
policy measures, 199–201
Immortality
atmospheric events, 207
instinctual immortality plan, 207
technology-supported people, 204
Impulse response (IR) method, 69, 73

Information and communications technologies (ICT), 214, 216, 223
Insightful leadership
  attitudes, 177–178
  characteristics, 178–179
  qualities, 177
Instinctual immortality plan, 207
Integer complexity, 38–39
International Energy Agency, 151
International Islamic University Malaysia (IIUM), 216
International Standards Organization (ISO), 217
International System of Units, 34
Internet of things (IOT), 156
Islamic Conference of 1997, 31
İstanbul stock exchange
  emerging markets, 68
  G8 countries, 68, 74–75
  generalized impulse responses, 74
  Granger causality test, 69
  implementation, 69
  multivariate cointegration test results, 72
  price movements, 67
  psychological effects, 67–68
  Turkey, 68
  unit root test results, 70, 71
Iterated Function System (IFS), 40–41

J
Jabatan Perkhidmatan Awam/Public Service Department (JPA), 217
Job-related uncertainty, 272

K
Keretapi Tanah Melayu Berhad (KTMB), 219
Key performance indicators (KPIs), 217
Kuwait National Assembly, 9–10
Kyoto Protocol, 114, 149, 152

L
Leadership, 31, 32, 129
  processes, 131–132
  situational, 129
  strategic, practices, 130
  sustainable competitive advantage, 131
  three-dimensional, 129
Linearity, 37
Lisbon Treaty, 148
Low-level measurement error, 143
Lyapunov exponents, 58–59
  dynamic system, 140

M
Machine learning (ML), 156
Majlis Amanah Rakyat/People's Trust Council (MARA), 217
Malaysia, HE
  federal government funding, 215–218
  government-linked companies roles, 219
  private sectors roles, 219–220
  state governments, 218–219
Malaysian Agricultural Research and Development Institute (MARDI), 217
Malaysian Qualifications Agency (MQA), 217
Malaysia Research Assessment Instrument (MyRA), 217
Mass spectroscopy, 34
Mathematics
  and Chaos
    bifurcation diagram, logistic map, 56–58
    cobweb diagram, 55
    dense periodic points, 55
    dynamical systems, 47–53
    logistic growth, 54, 55
    Lyapunov exponent, 58–59
    phase space, 60
  and complexity
    cognitive structures and natural expectations, 46
    fractal dimension, 43
    fractals, 39–42
    fundamental notions, 46
    Hausdorff-Besicovitch dimension, 43–45
    integer, 38–39
  macrosimplified model, 34
  mass spectroscopy, 34
  and modeling
    axioms, 35
    Borel sigma-algebra, 35–36
    Einstein Field Equations, 36–37
    Euclidian geometry, 35
    geometry, 35
    linearity, 37
    natural sciences, 36
    Newtonian mechanics, 36
    Planck length, 36
    quantum mechanics, 37

Newtonian mechanics, 34
probability, 33–34
qualitative/quantitative characterization, 33
quantum mechanics, 34
and randomness
Brownian motion, 63–64
dart games and archery, 60
entropy, 63
Kolmogorov's Axioms, 60
probability distributions, 63
thermodynamics, 60, 63
real-world problem, 34
sensors and measuring devices, 34
Matlab statements, 40
Measurement error, 138, 139
Meter-kilogram-second (MKS), 34
Middle East and North African (MENA), 68
Migration
clandestine/illegal, 194
colonization, 196–197
definition, 193
globalization, 198
post-colonial era, 197–198
sub-Saharan Africa (*see* Sub-Saharan Africa migration)
Mimetic isomorphism
competitive market forces, 280
HEIs (*see* Mimetic isomorphism, HEIs)
modeled organization, 274
social actors, 275
social constructive perspective, 275
types, 274
uncertain objectives, 274
Mimetic isomorphism, HEIs
autonomy and flexibility, 276
Bologna Process, 277, 278
competitive ecosystem, 280
competitive market forces, 276
EHEA, 278
follow-up mechanisms, 278
fragmented decision-making structure, 277
Gazi University, 279
negative applications, 279
OMC, 278
organizational change, 277
routinized applications, 279
university's Carnegie classification, 277
university's modeling, 277
Ministry of Education (MOE), 216
Ministry of Science, Technology and Innovation (MOSTI), 216
Monte Carlo simulations, 140
Multimedia University (MMU), 219

**N**
Natural gas, 151
Networked world
Arab Uprisings, 3, 4, 6–7
Bahrain, 7–12
Blair Government, 2–3
British foreign policy, 4
Conservative-led coalition, 3
economic/societal networks, 4
informal networks, 2, 5
International Relations, 4
Liberal Conservatism, 6–7
liberalism, 3
Libya, 13–17
Manuel Castells stresses, 5
mission statement, 2
nodes, 4
policy-making, 4
political–and human–cost, 3
qualitative elements, 3
quantitative analysis, 3
realism, 3
type of, 5
Newtonian mechanics, 36
Newtonian model, 244
Normative isomorphism, 274
North Atlantic Treaty Organization (NATO), 84, 116, 146
Numerical simulation, 102–105

**O**
Open Method of Coor dination (OMC), 278
Organizational insight
characteristics, 176
communication, 176
definition, 175
dynamism, 176
insightful leadership (*see* Insightful leadership)
pivotal element, 176
quantum entanglement, 176
Organizational memory
administrator interaction (*see* School administrator)
"Archiving" training, 267
benefits, 251–252
categories, 264
characteristics, 250
definition, 250
instruments, 252–253, 265, 266
loss of, 251
organization needs, 249

Organizational memory (*cont.*)
- organizations functions, 266
- phenomenon, 250
- roles, 251
- school administrator, 266
- senior employees, 250
- social process, 251

Organization for European Economic Cooperation (OEEC), 146

**P**

Party of Democratic Action, 29
Petroleum Exporting Countries (OPEC), 150
Petronas University of Technology (UTP), 219
Phase space, 60
Phenomenology, 256
Planck length, 36
Pneumocardiogram (PNCG) signals, 23
Principals
- attitudes and behaviors, 182
- dependent variable, 185
- evaluation system, 184
- readiness for change, 184
- role, 183
- stakeholders, 187
- trust relations, 183

Probability, 33–34

**Q**

Qualitative research, 255–256
Quantum mechanics, 37
Quantum theory, 77

**R**

Rating System for Malaysian Higher Education Institutions (SETARA), 217
Readiness for change
- conceptual framework, 182–183
- data analysis, 186
- data collection tools, 185–186
- incentives, 188
- limitations, 188
- relationship, 187
- research context, 183–184
- research model, 185
- reserch findings, 186–187
- sample, 185
- theoretical and researcher recommendations, 188
- trust-based relationships, 182
- trust in principals, 187, 188

Reliability–validity strategies, 257–258
Renewable energy, 150
Research University (RU), 217
Reserve Bank of India (RBI), 167
Rio Summit, 149
R Project program, 140
Rubber Research Institute (RRI), 219

**S**

SBI Quick app, 167
Scale index method
- fields, 23
- inner scalogram, 22
- linear models, 21
- parameter, 22–23
- practical application, 22
- seismic waves, 23
- social and political sciences, 22
- wavelet transform, 22
- weak periodic signal detection, 23–25

School administrator
- aim, 255
- CHE, 255
- contributions, 253
- fndings and comments
  - creating organizational memory, 262–265
  - organizational memory relationship, 261–262
  - organizational perception, 260–261
  - school administrator–school relationship, 258–260
- national education legislation, 253
- research methods and design
  - data analysis, 257
  - data collection, 257
  - ethics, 258
  - participants, 256
  - phenomenological design, 256
  - qualitative research method, 255
  - reliability–validity strategies, 257–258
  - rsearcher rle, 258
- Turkish education system, 253, 254

School of Professional and Continuing Education (SPACE), 216
Science, technology, and mathematics (STEM), 223
Seismic waves, 23
Semistructured interview technique, 238
SERVQUAL model, 159
Shadow Foreign Secretary, 6
Snowball strategy, 256
Sociology, 37

South China Growth Triangle, 68
Space of fractals, 40
Stability analysis, 99
  local, 99–102
Strange attractors, 206
Strategic management
  chaotic systems, 128
  definition, 126
  history, 127
  process, 130
Strategic uncertainty, 272
String theory, 179
Structural uncertainty, 272
Sub-Saharan Africa migration
  1884–85 Berlin Conference, 196
  colonization, 195
  economic, 195
  "emigrant producing countries", 195
  independent era, 196
Supranational bodies, 146

T
Teacher's happiness
  educational organizations, 236, 244
  human factors, 235
  sources of, 240–241, 243
Tenaga Nasional owns Tenaga Nasional University (UNITEN), 219
Thirring model, 23
Time delay model
  characteristic equation, 100
  delay-induced system, 100
  differentiating equations, 101
  imaginary solutions, 100
  local stability, 99–100
  uniform persistence, 102
Treaty of Paris, 150
Treaty of Rome, 150
Trust in principals
  definition, 183
  dimension, 185
  perceptions, 187, 188
  relational survey model, 185
Turkey
  adoption national program, 152
  energy capacities, 146
  energy efficiency law, 153
  energy resources, 152
  energy sector, 152
  energy trade, 146
  harmonization process, 152
  Kyoto Protocol, 152
Turkish education system, 253

U
Uganda, HE
  advise universities, 223
  education loans, 221, 222
  facilities, 221
  funding, 222
  government projects, 221
  incentives and facilitation, 221
  low-cost ways, 223
  public institutions and universities, 223
  strategic plan, 222
  tax waivers and fiscal incentives, 223
UM Center for Democracy and Elections (UMCEDEL), 216
Uncertainty
  centralization, 273
  characteristics, 272
  coping with uncertainty, 272–274
  description, 271
  Mimetic isomorphism (*see* Mimetic isomorphism)
  perspectives, 271
  psychological, 272
  three-factor conceptualization, 272
United Nations Development Program (UNDP), 194
Universal primary education (UPE), 212
Universal secondary education (USE), 212
Universiti Kebangsaan Malaysia/National University of Malaysia (UKM), 216, 217
Universiti Teknologi Malaysia/Malaysian University of Technology (UTM), 216, 217
Unpredictable movement, 234

V
Variance decomposition (VDC) method, 69, 73
Variance inflation factor (VIF), 186
Vector autoregression (VAR) model, 69

W
Wavelet transform, 21–22
Weak periodic signals (WPS) detection, 23–25
Wiener process, 63–64
World War I, 87

The manufacturer's authorised representative in the EU is Springer Nature Customer Service Centre GmbH, Europaplatz 3, 69115 Heidelberg, Germany. If you have any concerns regarding our products, please contact ProductSafety@springernature.com

Printed and bound by CPI Group (UK) Ltd, Croydon, CR0 4YY
22/09/2026
02221060-0011